Agriculture in Western Europe

Challenge and Response 1880-1980

Second Edition

Michael Tracy

GRANADA

on Toronto Sydney New York

Granada Publishing Limited – Technical Books Division
Frogmore, St Albans, Herts AL2 2NF
and
36 Golden Square, London W1R 4AH
866 United Nations Plaza, New York, NY 10017, USA
117 York Street, Sydney, NSW 2000, Australia
100 Skyway Avenue, Rexdale, Ontario, Canada M9W 3A6
61 Beach Road, Auckland, New Zealand

ISBN 0 246 11446 0

First published in Great Britain 1964 by Jonathan Cape
under the title *Agriculture in Western Europe – Crisis and Adaptation since 1880*
Second Edition 1982 by Granada Publishing

Printed in Great Britain by Mackays of Chatham

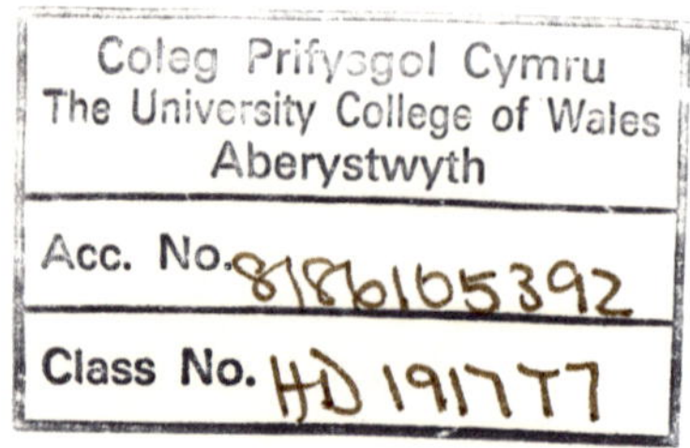

Contents

Introduction

The problems of agricultural policy have become increasingly acute throughout western Europe. The continued introduction of new technology, applied by millions of individual farmers, has brought about a growth of output exceeding that of demand. Farm product prices have come under pressure, while the costs of their inputs have risen. Farm incomes have remained on the whole unsatisfactory, with wide disparities within the farming sector, yet price support policies have become increasingly expensive. In the European Community, the common agricultural policy (CAP) takes up the greater part of the total budget, a situation which by the end of 1980 had become a major issue within the Community and had led to growing pressures for 'reform' of the CAP. While western Europe has become more self-sufficient in temperate foodstuffs, world agricultural markets have been unstable and generally depressed, in spite of the large potential needs of hungry people in the 'Third World'.

This book sets out to explain how these problems arose, as a result of the responses to successive challenges over the past hundred years.

The first challenge was the Great Depression which began in the 1880s, when for the first time competition from outside Europe, in particular from North America, made itself felt on a devastating scale. This occurred at a time when most European countries had nearly renounced tariff protection on agricultural as well as on industrial goods. The resulting slump in the prices first of grains, later to some extent of other agricultural products as well, gave rise to different reactions in the various European countries. Britain held determinedly to its policy of Free Trade and laissez-faire, and allowed much of its agriculture to undergo a long and drastic decline. Denmark too adhered to Free Trade, but carried out a thorough adaptation of its agriculture to the new conditions, as a result of which Danish agricultural exports became established on the markets of Britain and Germany. The Netherlands followed a similar course. Most other countries, in particular France and Germany, shielded their agriculture behind high tariff walls, and relatively little adaptation took place. These contrasting policies had an important influence on the subsequent development of agriculture in the various countries.

The second great convulsion occurred when European agriculture was drawn into the economic crisis of the 1930s: markets contracted, prices fell and many farmers went bankrupt. This time, laissez-faire was abandoned even in Britain, and all countries attempted to help their farmers – first by raising tariffs against outside competition, then, as these proved inadequate, by more direct restrictions on trade which led to extensive measures of intervention on domestic markets as well. These measures were maintained up to the Second World War, but in most cases they constituted merely a series of expedients. Few countries developed a coherent policy for agriculture; the major exception was Germany under the National Socialist regime.

These past developments are studied in this book with reference primarily to the north-western parts of Europe (Italy is covered, but not the other southern European countries). As it would be impracticable to examine in detail all the countries in this region, special attention has been given to Britain, France, Germany and Denmark. The choice of these four is a fairly obvious one: they are significant in the present European agricultural scene, and their past development – as has been seen above – presents marked contrasts. In dealing with these countries, an attempt has been made to give a reasonably complete explanation of why they adopted the policies they did. This explanation has to be sought not only in developments in agriculture itself but also in the general economic policy of the country, in its attitude to foreign trade, in overall political forces, in the activity of farmers' organisations, sometimes even in personalities.

The period since the Second World War has seen a third set of challenges, arising first from rapid economic growth which left farm incomes behind, subsequently from the still more difficult problems of agricultural adjustment in an economy suffering from 'stagflation' following the oil crisis of 1973. Policies to meet these challenges, moreover, had to be shaped in the context of the European Community, which created opportunities but also new problems for agriculture in the various member states, and which brought into existence a new framework of decision-making. Further, the international context had to be taken increasingly into account. While on the one hand agricultural exporting countries throughout the world were, as in the past, affected by policies pursued in western Europe, the tendency within the region for agricultural production potential to grow faster than requirements meant that prospects for western European agriculture depended increasingly on world market outlets. By the end of 1980, responses to these issues were still being worked out.

Note on the second edition

The first edition of this book (with the sub-title *Crisis and Adaptation since 1880*) was written in 1963. In this new edition, Part III has been almost completely rewritten to cover the developments just described, bringing the story up to 1980 – the last year of the Community of Nine, before the accession of Greece on 1 January 1981 added a new dimension. (The book takes account of events up to 31 December 1980; reference has been made to a few works published in early 1981.)

The work has also been extended backwards in time. A fuller – though still brief – account is given of ways in which agrarian structures were shaped in feudal times, tracing major developments through reform or revolution up to 1848, and taking account of the changing role of farming in society and the economy. This gives a better understanding of the context in which the challenge of the 1880s was faced, and helps to explain the different reactions of the various countries.

Throughout the book, revisions have been made to take account of works which have appeared since the first edition was prepared: these are referred to in the bibliographies at the end of each chapter. Such recent works have thrown more light on developments up to the nineteenth century and also, for example, on conditions in various parts of the United Kingdom during the Great Depression; on rural conditions in France in the nineteenth century and on the growth there of farm organisations; on the activities of the *Bund der Landwirte* and other farm organisations in Germany. As regards the 1930s, there has been comparatively little revision, except for a fuller treatment of National Socialist policy in Germany.

A 'schematic summary' has been added (see pages x-xi) as an aid in visualising the progression of events over this large field of space and time.

Acknowledgements

In a work such as this, one incurs pleasant debts to many people. Those who helped in the first edition need not be mentioned again, though I wish to recall in particular my debt to Paul Lamartine Yates: I like to think that this book is a continuation of his pioneering work in the comparative study of European agriculture.

For comments on draft chapters (or parts of them), I am grateful to the following: Professor Jerome Blum of Princeton University (chapter 1); John Higgs of the Arkleton Trust (chapter 2); Ian Farr of the University of East Anglia (chapter 4); Professor Erik Helmer Pedersen of Copenhagen University (chapter 5); J.E. Farquharson of Bradford University (chapter 9); Dr Hans-Broder Krohn, formerly Director-General in the Commission, and F. Rossi, formerly Director-General in the Council Secretariat (chapter 12); Michael Franklin, formerly Deputy Director-General in the Commission, now in the British Cabinet Office (chapter 13); Graham Avery, Head of Division in the Commission (chapter 14); Gérard Viatte, Deputy Director in OECD, and Marion Bywater, formerly of European Report, now in the Hong Kong Government service (chapter 15). Several of my colleagues in the Council Secretariat have also helped to check my drafts: D. Vignes (parts of chapters 12 and 14), R. Bandilla (parts of chapter 15) and J. Keller-Noellet, who has provided constructive suggestions on both chapter 3 and parts of chapter 14. Rosalind Falvey, formerly of the British Ministry of Agriculture, now in the British Council in Brussels, has made many valuable suggestions concerning both the historical chapters and recent events in agricultural policy.

Several officials in the Commission, in the European Communities' Statistical Office and in FAO have helped with data.

A special tribute is due to L. Goebel, in charge of the Council library. It would have been impossible for me to carry out this work in the limited time available without his efficiency in identifying and obtaining the material I needed, and without the patient assistance of all his staff. Linda Bull, librarian of the British Council in Brussels, has also helped whenever required.

W. Bader of the European Communities' Publications Office, was responsible for drawing the diagrams, and skilfully resolved technical problems in reproducing the maps.

I have benefited from efficient and willing secretarial help from Christine Dickert, Elly Blom and Marie-France Goeminne.

I am myself responsible for translations of texts quoted from foreign languages.

Finally, I want to record a debt to the students who in recent years have followed my course on agricultural policy at the College of Europe in Bruges. Their active participation and questioning has helped to shape this new edition. I hope that future students at the College and elsewhere will find some of their questions answered here.

The opinions expressed in this work are entirely my own responsibility and do not necessarily correspond with the views of the Council of the European Communities or its Secretariat by which I am employed. This book does not aim to reflect the policy of any Community institution, though proposals, memoranda, etc. of the Commission are given their due weight.

January 1981

In memoriam

Finn Olav Gundelach, Vice-President of the Commission, died of a heart attack on 13 January 1981, after ceaseless endeavours for progress in both agricultural and fisheries policy. His role with regard to the CAP is mentioned in chapter 14 of this book. A more personal memory concerns his contribution to a symposium in 1979 at the College of Europe in Bruges, which it was my privilege to direct, on prospects for the CAP in the world context. On that occasion, departing from his prepared text, he responded to the themes of the conference in a speech which reflected deep personal convictions (some passages from this speech are quoted in chapter 15). His loss is severe: his mastery of his subject and his dedication to internationalism were needed for the difficult adjustments still to come.

Brussels, 2 February 1981

Michael Tracy

SCHEMATIC SUMMARY

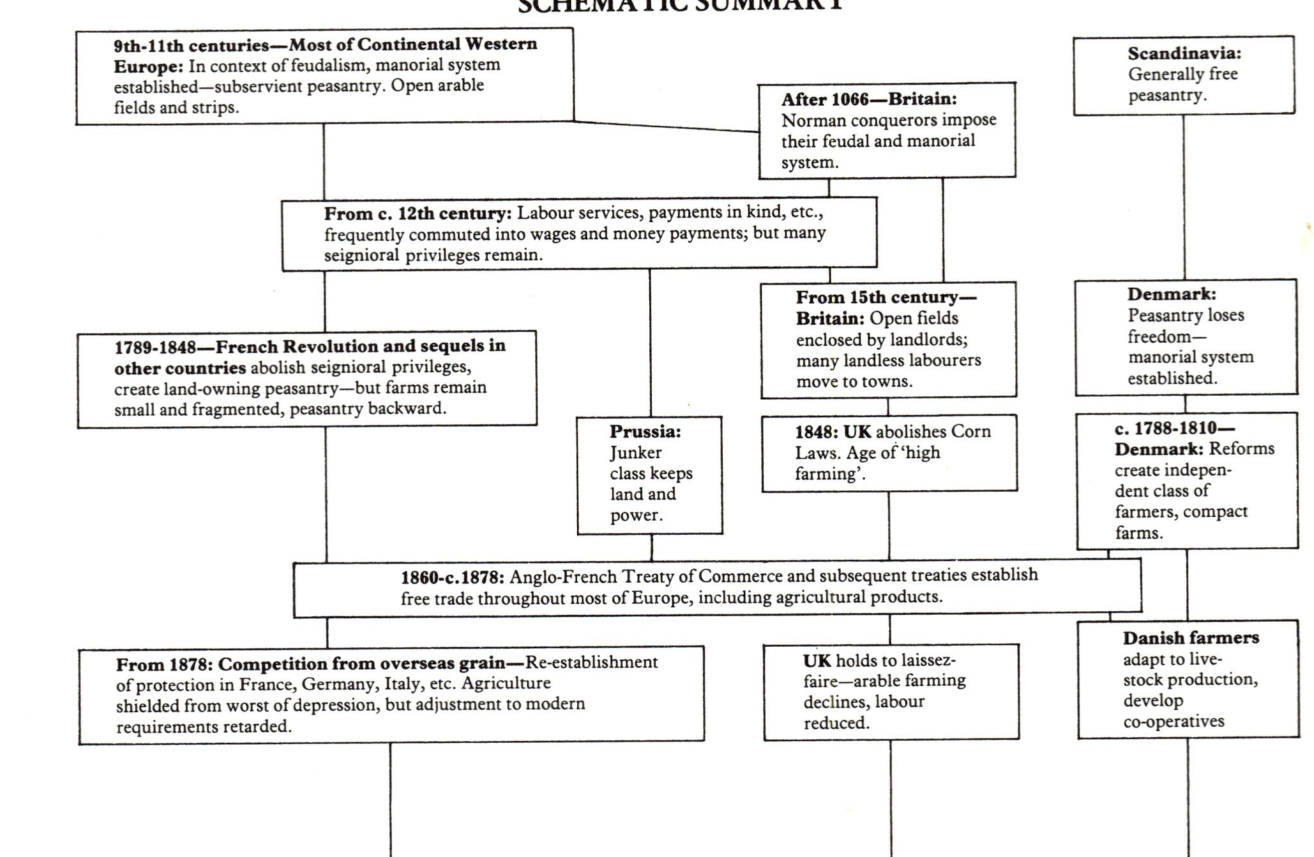

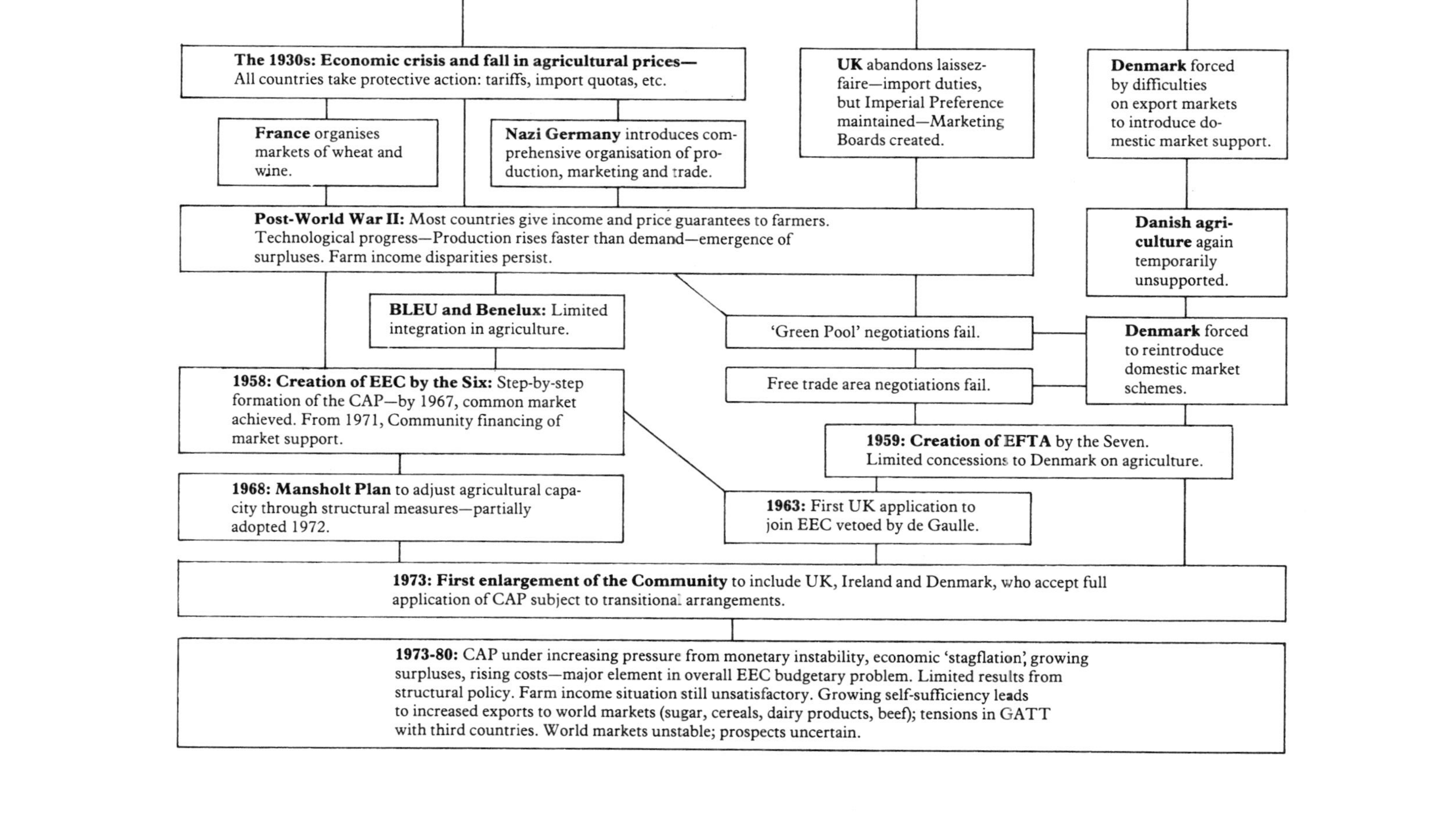

The 1930s: Economic crisis and fall in agricultural prices—
All countries take protective action: tariffs, import quotas, etc.
UK abandons laissez-faire—import duties, but Imperial Preference maintained—Marketing Boards created.
Denmark forced by difficulties on export markets to introduce domestic market support.
France organises markets of wheat and wine.
Nazi Germany introduces comprehensive organisation of production, marketing and trade.
Post-World War II: Most countries give income and price guarantees to farmers. Technological progress—Production rises faster than demand—emergence of surpluses. Farm income disparities persist.
Danish agriculture again temporarily unsupported.
BLEU and Benelux: Limited integration in agriculture.
'Green Pool' negotiations fail.
Denmark forced to reintroduce domestic market schemes.
Free trade area negotiations fail.
1958: Creation of EEC by the Six: Step-by-step formation of the CAP—by 1967, common market achieved. From 1971, Community financing of market support.
1959: Creation of EFTA by the Seven. Limited concessions to Denmark on agriculture.
1968: Mansholt Plan to adjust agricultural capacity through structural measures—partially adopted 1972.
1963: First UK application to join EEC vetoed by de Gaulle.
1973: First enlargement of the Community to include UK, Ireland and Denmark, who accept full application of CAP subject to transitional arrangements.
1973-80: CAP under increasing pressure from monetary instability, economic 'stagflation', growing surpluses, rising costs—major element in overall EEC budgetary problem. Limited results from structural policy. Farm income situation still unsatisfactory. Growing self-sufficiency leads to increased exports to world markets (sugar, cereals, dairy products, beef); tensions in GATT with third countries. World markets unstable; prospects uncertain.

Part I:

The Great Depression, 1880–1900 – The First Wave of Protectionism

Chapter 1

General

This book is primarily concerned with the challenges which have faced western European agriculture since the advent of cheap overseas grain around 1880, and with the responses of the farming sector and of governments to these successive challenges. To set the scene, however, and to understand why these responses differed from one country to another, it is necessary to look further back in time. By the mid-nineteenth century, farm structures – the number, size and shape of holdings – and forms of land tenure had been established which formed the basis of agriculture today. This was the outcome of the various ways in which agriculture had emerged from the typical 'manorial' pattern established in feudal times, through evolution, revolution and reform up to 1848. The general social and economic context in each country had also influenced the extent of technical progress in farming, as well as the relative importance of agriculture in the economy and the position of the farmer in society.

These early developments will be reviewed briefly below; fuller accounts are given in the four following chapters of Part I.

From feudalism to emancipation

Feudalism in medieval Europe implied in the strict sense a system of allegiances between lords and vassals, allegiances which could exist at various levels from the king downwards. The bonds thus created ensured mutual loyalty, necessary in troubled times: the vassal offered his services, the lord gave protection and a fee or benefit (*beneficium*), which frequently took the form of land.

The lords holding land as a fee from their king, besides reserving some for their own use, granted tenures of land to peasants. The latter paid for their use of this land by working on the lord's reserve and by dues in cash or, more usually, in kind. Property rights in a modern sense did not exist under such a system: a plot of land could in effect 'belong' simultaneously to a peasant, his lord, possibly other lords above this one, and ultimately to the king. Much land was held by the Church including the monasteries.

This was the basis of the medieval manorial system. Over most of the plain areas of Europe where arable cultivation predominated, manorial estates were associated with village settlements and with open-field farming: the arable land was arranged in two or three large fields, a simple crop rotation (involving fallow every second or third year) took place between these fields, and each peasant had land in each of the fields. These holdings were arranged in strips, to facilitate ploughing; the land in the lord's own use was often mingled with the peasants' strips. In each open field, the same crop had to be grown, and the operations of ploughing, sowing and reaping had to be carried out simultaneously. After the harvest, the animals were let on to the stubble. The need for such co-ordinated cultivation reinforced the cohesion of the village community, which also regulated the use of common pastures. The open-field system, however, was not a necessary characteristic of the manor: outside the arable areas, a variety of field arrangements could be found.

All peasants in a medieval manor paid dues to their lord, and in a basically

non-monetary economy these dues were mainly in kind. From the outset, tithes paid (in principle) to the Church were also required. Over time, numerous other seigniorial rights came to be established at the expense of the peasantry: monopolies in the form of compulsion to use, and pay for, the lord's mill, his oven for baking bread, his wine-press, his brewery; while hunting and fishing rights meant that peasants could not protect their crops from the depredations of game birds or animals nor even from the incursions of the hunters.

Some peasants, moreover, were not merely 'villeins' in the strict original sense of one living on a manor (*villa*), but also serfs (*servi*), tied to their master by hereditary personal bonds, forbidden to leave his service, unable without his authorisation to marry a partner who was not also in his service and unable to bequeath possessions to their children. Distinctions became blurred over time, because, on the one hand, some or all of these servitudes could be redeemed and, on the other, the freedom of ordinary villeins to leave the manor was often restricted.

Finally, the lord of the manor in medieval times exercised jurisdiction over his peasants, usually without appeal.

Such, in broad outline, was the system developed around the eighth century in the Frankish lands between the Seine and the Rhine and spread by the Carolingian empire over most of continental western Europe except for some of the maritime fringes. It was adopted by the Viking invaders of Normandy and subsequently carried by the Normans to England, Sicily and southern Italy and as far afield as Syria. It spread to Germany east of the Rhine and in later centuries a derived form of estate organisation was imposed by colonising knights east of the Elbe. It was adopted in northern Italy and in the Iberian peninsula it dominated in Catalonia and Aragon. (In southern Spain and Portugal, great *latifundia* were created at a later period after reconquest from the Moors.)

The manorial system, however, was not all-pervasive, even in France. It entered the Celtic lands of the British Isles (which had their own varieties of feudalism in their clans and small kingdoms) only when imposed by Anglo-Norman invaders; and it never reached Scandinavia. The Scandinavian countries traditionally had a free peasantry. The regions of Sweden and Norway where grassland and livestock farming predominated, were not characterised by village settlements. A pattern of isolated farmsteads and the combination of forestry or fishing with farming, helped to reinforce the independent character of the rural population. In Denmark, however, village settlement and open-field cultivation was normal. In the decades after 1660, the sale of land by the Danish Crown produced a new class of commercially-minded landlords and the peasants lost their freedom, being tied to their place of origin ostensibly for reasons of military service, but in fact as a source of labour for their landlords. In Norway (then part of the same Kingdom), the Crown land was acquired by the peasants, who thus strengthened their independent position. The evolution in Sweden was different. In the seventeenth century, successive wars reinforced the position of the nobility as military commanders: they gained two-thirds of the land in Sweden and Finland, and created manorial-style estates. Swedish peasants, however, never became serfs, indeed the peasantry remained the Fourth Estate in the Diet, and they were assisted by effective pressure from the king upon the nobles to reduce their possessions. While complying, the nobles were able to rearrange their land into contingent domains. Later, in 1827, legislation was introduced promoting enclosures and throughout the nineteenth century farmers consolidated their holdings, creating unified farms.

In the feudal countries, over the centuries, bonds of allegiance were loosened. One factor was the Black Death in the early fourteenth century, when many peasants died and the normal life of village communities was disrupted. Subsequently, lords had difficulty in regaining control of the peasantry. More generally, the growth of population before the Black Death, and especially its recovery from the mid-fifteenth century onwards, meant that compulsory labour services became less important to lords. Moreover, as money became more plentiful from about the thirteenth century, salaried work took the place of bound services, cash rents were substituted for rents in kind, various dues could be redeemed and serfs could buy their freedom. Reference to custom for the definition of rights and obligations came to be replaced in part by written contracts, particularly in France where some *seigneurs* drew up village charters listing rights and obligations. In France and western Germany, landlords often found it more profitable to have as many rent-paying tenants as possible rather than farm the land themselves and in leasing more land from their reserves to peasants, they also reduced the burden of compulsory labour service.

The feudal bond between lord and vassal existed only so long as they both lived. Consequently the lord could in principle take back land on the death of his vassal. Usually a new bond of allegiance was entered into with the vassal's son or sons and in practice leases became hereditary (though subject to dues at each succession). Landlords were generally opposed to sub-division among heirs. Inheritance by the eldest son was not always the rule and often the landlord himself chose who should take over the holding. But the sub-division of holdings could not be avoided and as generations passed strips were divided and re-divided, and interchanged as a result of marriage, until in many regions of continental western Europe a single peasant might be farming dozens of tiny plots, often widely separated.

England followed a different path. From the fifteenth century, landlords began to enclose sections of the big open fields with fences or hedges and to create relatively large, unified farms which they farmed themselves or let to a reduced number of tenants. A growing demand for wool meant that putting land down to grass for sheep was profitable, and much less labour was required. The process increased the number of landless labourers, many of whom were forced off the land and into the growing towns. The creation of substantial farms was indirectly reinforced by Henry VIII's dissolution of the monasteries, as much of the former monastic land was eventually sold or rented to rural gentry as well as to the noble landlords. (In the Protestant German states, Church lands expropriated at the Reformation appeared to be taken over mainly by state rulers or nobility). Further enclosures and improvements in agrarian structures were systematically carried out in the late eighteenth and early nineteenth centuries. The move from the land was stimulated by the Industrial Revolution and though the rural population continued to increase, it did so more slowly than the urban population, so that the population living from agriculture formed a smaller proportion of the whole than in any other country. Further, overall economic progress was reflected in agriculture. The structure of agriculture, with large farms and a class of wealthy landowners ready to invest in their estates, made British farming particularly receptive to technological progress and in the middle of the nineteenth century it was the most advanced in the world.

In southern regions of the Low Countries, the agrarian situation in general reproduced approximately the forms and characteristics found in western Germany and northern France. In the north, however, after the United Provinces had been formed in 1579, political and social changes in the wake of the Reformation and the struggle for independence led to the general suppression of seigniorial rights and to an extensive parcelling-out of land, much of which came into the hands of the rich bourgeoisie. This resulted in the introduction of modern and remunerative methods of farm management. At the same time, considerable capital was spent on reclaiming new land through the drainage of marshy areas and estuary lands and the creation of *polders*.

The eighteenth century was a dynamic period throughout Europe: population was growing and markets were expanding. Progressive farmers could take advantage of new methods, in particular new crops and more elaborate and productive rotations. By including temporary pasture in the rotation, by sowing soil-restoring fodder crops such as clover and lucerne and by introducing turnips and potatoes, the frequency of unproductive fallow years could be reduced, the number of livestock could be increased, more manure could be obtained and soil fertility could be improved. The total output of both crops and livestock products could be raised to supply a growing market. But on the open fields, with co-ordinated cultivation and time-honoured rotations, such improvements were difficult. The mass of the peasantry opposed consolidation and though there were a few enterprising landowners who tried to promote consolidation and the new methods, they remained a minority.

In France, however, many landlords at this time sought to profit from the expanding market by enlarging their reserve, taking over peasant land whenever an excuse presented itself, such as arrears of dues, and even enclosing former common land. They also stepped up their exactions of dues of various kinds from the peasantry. In Germany too, particularly in Prussia, many landlords practised *Bauernlegen* – the taking-over of peasant land.

Most of the monarchs of continental Europe, and the rulers of the smaller states, sought to restrain the greed of the landed nobility and to improve the lot of the peasants. Their motivations, in the Age of Enlightenment, may have been partly idealistic but they also had practical considerations in mind: a numerous and prosperous peasantry was essential for the strength and finances of the State. In the years before the French Revolution, the Empress Maria-Theresa and the Emperor Joseph II carried out significant reforms in the Austrian Monarchy;

in Denmark, Crown-Prince Frederik and a reform-minded government freed the peasantry from bondage and compulsory labour service; in Bavaria and some of the smaller German states, the rulers freed serfs at least on their own lands but without succeeding in having their example followed by the nobility or by the Church; the Duke of Savoy freed all serfs. In Prussia the Hohenzollerns tried to restrain *Bauernlegen* without much success and introduced reforms on their own estates. But as Blum (1978, page 218) observed, 'the rulers of France, alone among the major sovereigns of servile Europe, did not make even a limited or half-hearted sustained effort to persuade seigniors to agree to reforms in the status and obligations of their peasants'.

Sporadic peasant revolts occurred in various parts of Europe in the second half of the eighteenth century, usually without any preconceived aim of overthrowing the established order. In 1789 in France, peasant agitation combined with the urban rebellion. The outcome was the overthrow of all forms of seigniorial privilege and peasants became full owners of the land they farmed. They did not, however, necessarily acquire more land. Though most estates of the Church and of *émigré* nobles were sold, peasants managed to acquire only part of this. The Revolution instituted individual freedom in the cultivation of holdings in place of co-ordinated village management but the vast majority of peasants still had small holdings fragmented into many tiny strips. Further, the Napoleonic Civil Code consecrated the practice of division among heirs at each generation.

The principles of the French Revolution were extended to the vast areas which were overrun by Napoleon's armies or came under his control. In the Rhineland, the Low Countries, in Switzerland (where the peasantry had already gained substantial independence in previous centuries) and in Italy, all surviving feudal privilege was wiped out, the Civil Code was introduced and monasteries were dissolved. Among the sixteen theoretically independent states of the Confederation of the Rhine formed by Napoleon (which included most of Germany outside Prussia) the extent of reform varied. But as in France, these reforms did not directly put more land into the hands of the peasantry. Indeed, more complete reform by the rulers was probably retarded by the fears engendered by the French Revolution. In Austria, for example, the reform movement inspired by Joseph II was halted.

It was not until the further revolutionary wave in 1848 that the process of emancipation from feudal privilege was completed in the various west German states. Generally, however, the peasants had to redeem the land of which they obtained ownership, usually through repayments over a lengthy period of years to a government-instituted fund which compensated the previous owners. The peasantry had thus gained personal and economic freedom, and equality before the law; but the farm structure was only marginally changed.

In Prussia the need for reform was particularly great. In most of the provinces the landlords had created large estates which they managed themselves as commercial enterprises, producing grain for export to the west; peasant rights had been steadily eroded. Defeat by Napoleon and realisation of the need for national unity brought about a reform law in 1807, but its purpose was subsequently deflected to such an extent that the landlords actually managed to further increase their estates and reinforce their position. In southern Portugal, Spain and Italy, the *latifundia* remained intact.

In only one country was the agrarian structure totally reshaped. In Denmark – where, as has been seen, subjugation of the peasantry was comparatively recent – the reforms did not merely involve liberation from servitude; the strips were consolidated and enclosed, credit was provided to enable the liberated peasants to buy the land they farmed and, in many cases, to instal new farm-houses and buildings on the unified farmland. In a few decades the Danish landscape was transformed and viable farms, in the hands of an independent peasantry capable of meeting the impending challenges, had been formed.

Map 1 Vorderdenkental, Baden-Württemberg, 1696
From one of the earliest land surveys. Strips belonging to the three inhabitants of this hamlet were scattered over the whole area. On the original (coloured) map, the three open fields appear clearly. A survey of the same area in 1823 showed much the same pattern of strips.
Source: Fischer, A. (undated), *Die Vereinodung in Oberschwaben,* reproduced by *Beiträge zur Landeskunde.* (Staatsanzeiger für Baden-Württemberg), reproduced by permission of the Landesvermessungsamt, Stuttgart.

Map 2a Rempertshoven, Baden-Württemberg, 1803, before consolidation
The strip system of a larger village. Areas in black all belong to a single farmer. This village, with a total area of 252 ha, had thirteen farmers cultivating on average 20 ha each; there were 710 plots, averaging 0.35 ha each. The pattern is typical of conditions over a large part of continental western Europe.

Map 2b As 2a, after consolidation in 1803

A fairly rare case of successful early consolidation. The number of plots was reduced to thirty-six, averaging now 7 ha each; six farmsteads were moved out of the village. The black areas indicate the plots now held by the same farmer as in the previous map.

Rempertshoven was fortunate in that the farm structure did not seriously deteriorate again after this consolidation. The number of farms had increased to twenty-one by about 1965; in 1980 there were fourteen.

Source: Fischer, A. (undated), *Die Vereinödung in Oberschwaben,* reproduced by permission of the Landesamt für Flurbereinigung und Siedlung Baden-Württemberg.

Map 3a Langley, Essex, 1851 before enclosure

Map 3b Langley, Essex, 1851 after enclosure

A typical nineteenth-century English enclosure. Some enclosures had been carried out earlier, but the fields were still open and there were numerous strips. In 1851 a complete rearrangement was made.

Reproduced by permission of the Essex County Council Record Office.

Map 4 Ørsløv, Denmark, 1798, after enclosure

Before 1798, this village area consisted of open fields with a multitude of strips exceeding even that of Map 2a. The rearrangement of 1798 created a systematic and concentric field pattern, permitting farmsteads to be subsequently moved from the village to their land.

Reproduced by permission of the Matrikeldirectorat, København.

No example is given for France (the destruction of records during the Revolution makes it difficult to find early maps). The situation shown in Map 2a, however, would have been representative of much of France and little change in the arrangement of plots took place at the Revolution.

Map 1: Vorderdenkental, Baden-Württemberg, 1696

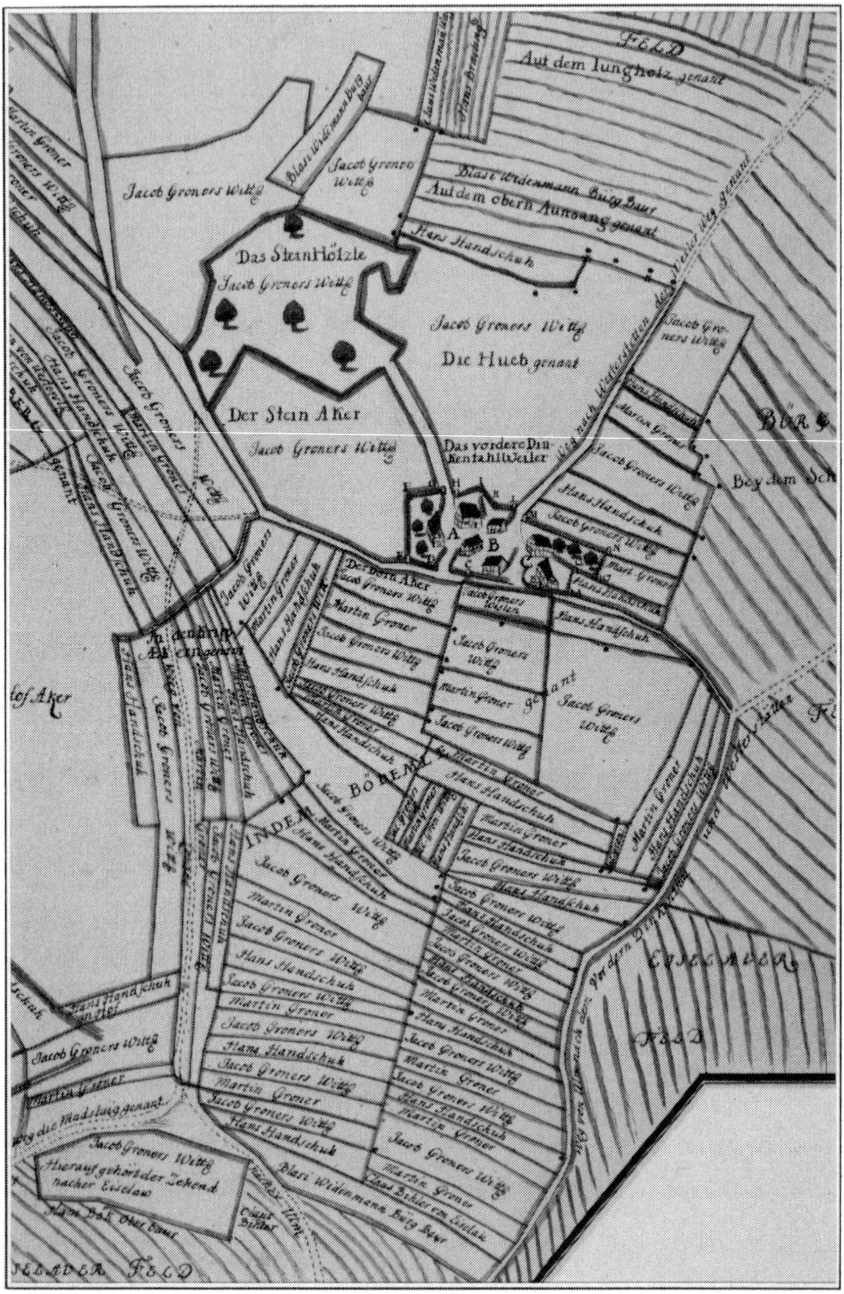

Map 2a: Rempertshoven, Baden-Württemberg, 1803
before consolidation

Map 2b: Rempertshoven, Baden-Württemberg, 1803 after consolidation

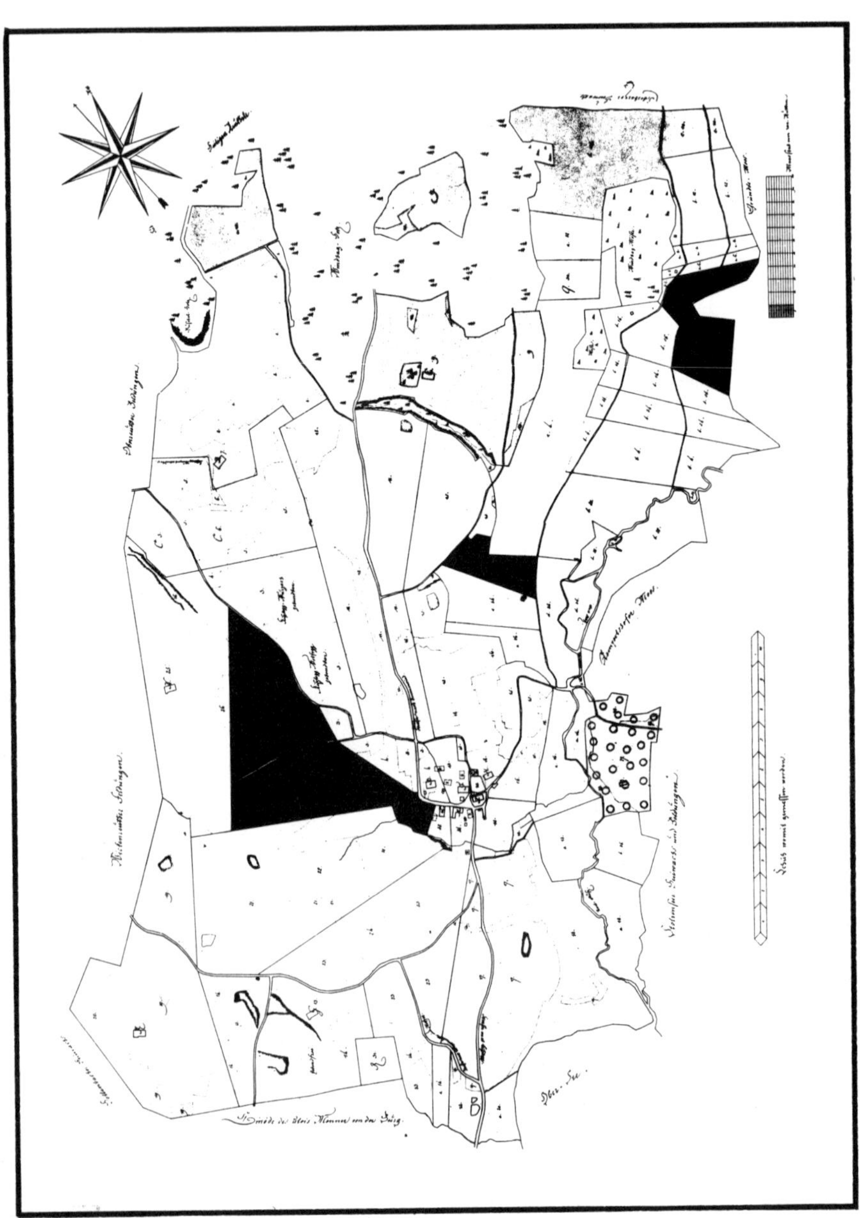

Map 3a: Langley, Essex, 1851 before enclosure

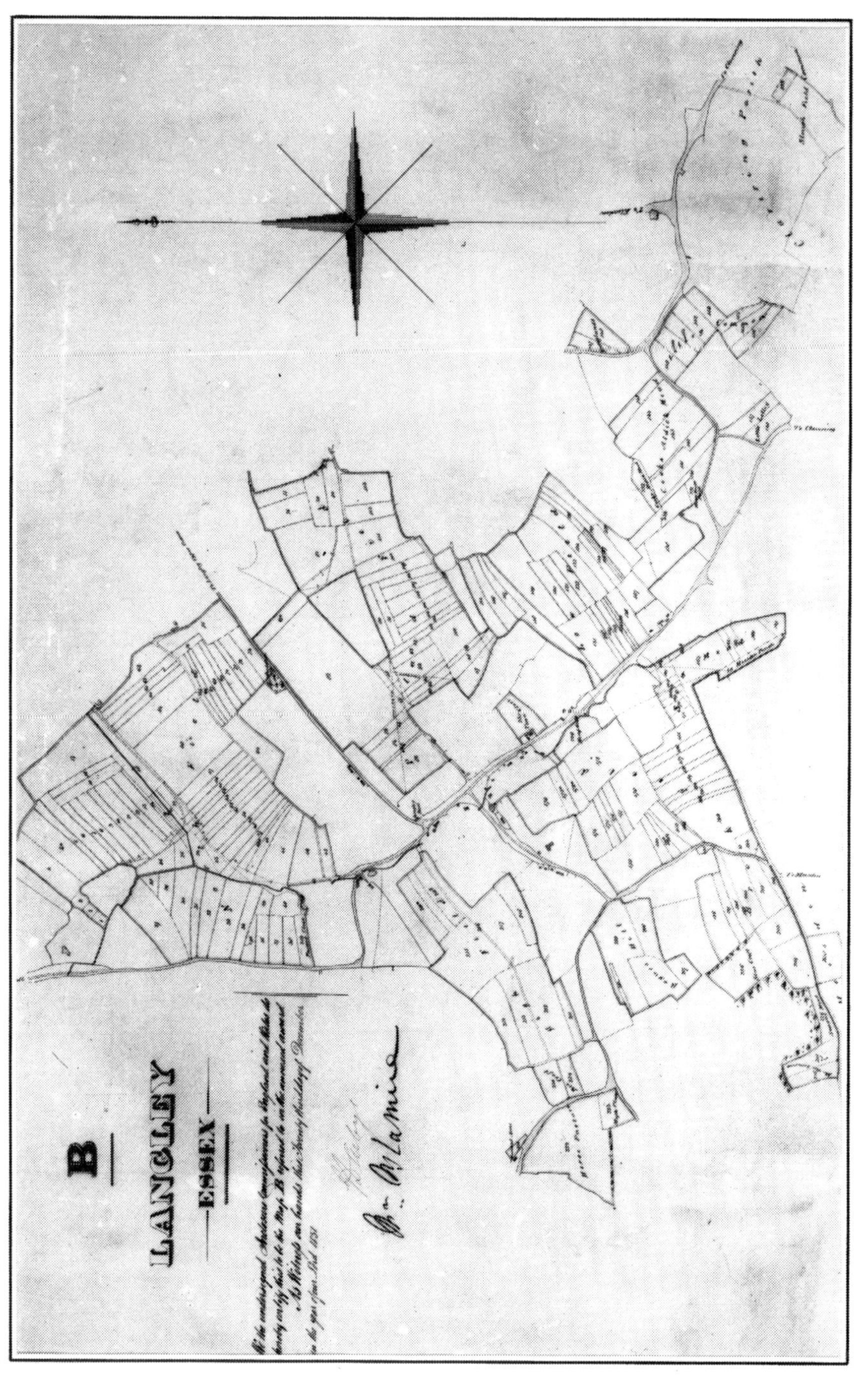

**Map 3b: Langley, Essex, 1851
after enclosure**

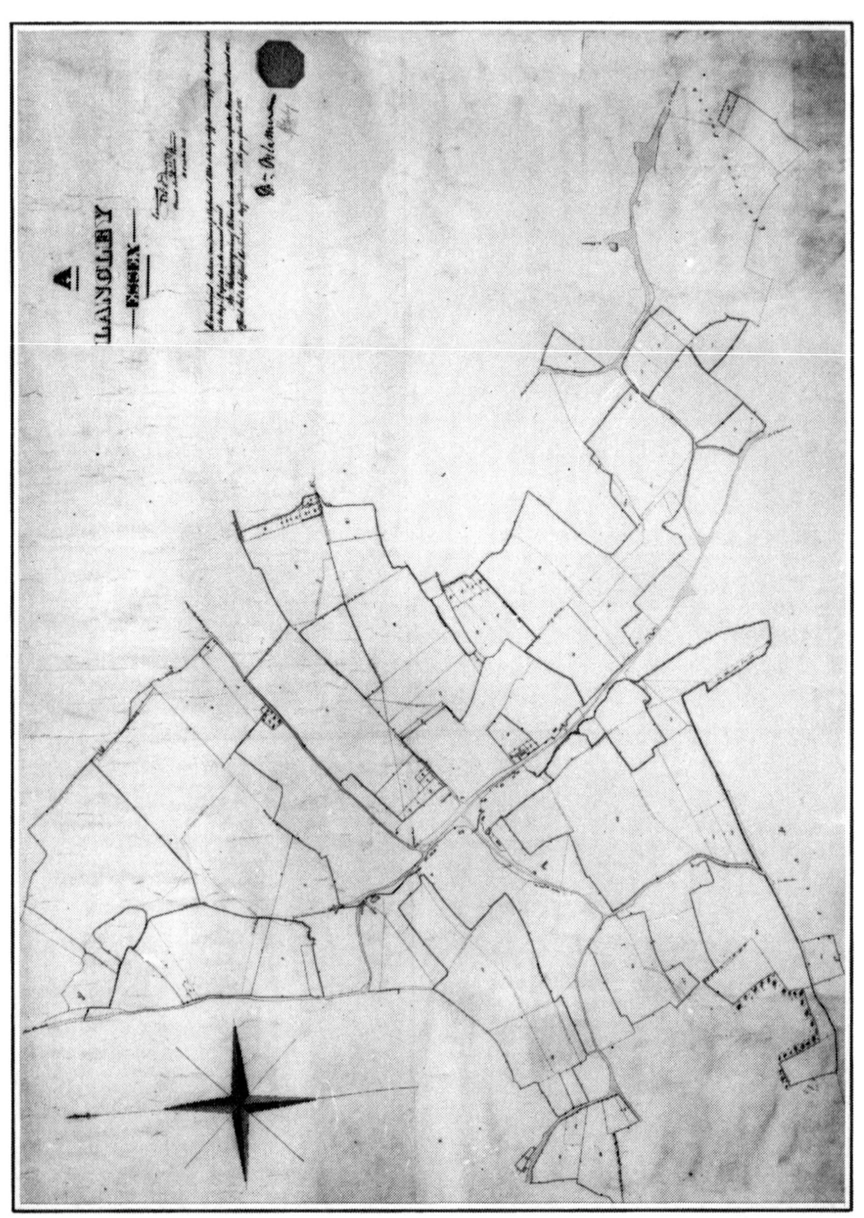

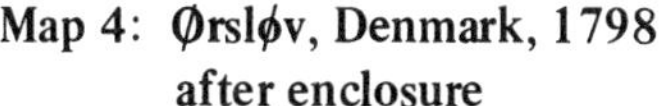

Map 4: Ørsløv, Denmark, 1798 after enclosure

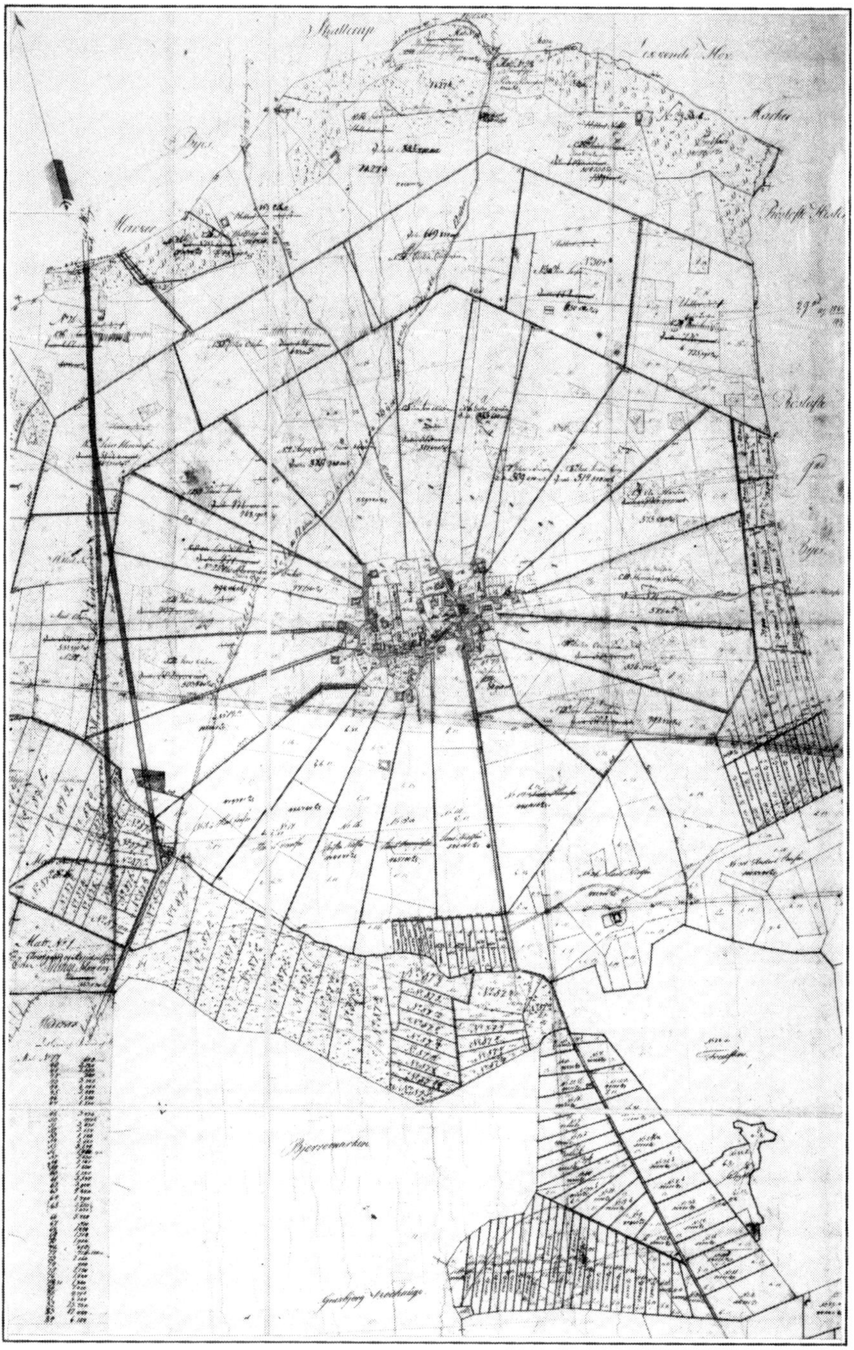

Agriculture and trade in the mid-nineteenth century

The nineteenth century brought immense possibilities for technical improvement in agriculture. In England the age of 'high farming' arrived: soil was improved by drainage and fertilisation, crop yields were increased, livestock breeds were improved, new machinery was introduced and steam power became available for threshing and even ploughing. Agricultural research progressed in Germany, particularly as regards soil chemistry and the use of artificial fertilisers. The Low Countries made progress especially in horticulture and fruit-growing. Wherever there were progressive landlords or educated farmers with sufficient resources, the new technology could be adopted.

But where the peasantry had barely emerged from centuries of oppression, where few of them were even literate, and where their holdings remained small and fragmented, they could not be expected to take advantage of these possibilities. Moreover, in most of the countries of the Continent, by the middle of the century, the Industrial Revolution had hardly begun to influence agriculture. Farming remained the most important occupation; dissatisfied peasants had not much opportunity to move into expanding industries. There was as yet hardly any stimulus for agriculture to improve its methods: transport and communications were deficient and there was little flow of new ideas into the countryside. Rural education was often rudimentary, and agriculture remained backward.

The contrast with the New World was underlined by de Tocqueville:

> The Americans do not use the word 'peasant'; they do not use the word, because they do not have the concept; the ignorance of early society, the simplicity of the countryside, the rustic nature of the villages have not been preserved in their country, and they cannot imagine either the virtues or the vices, the rude habits or the naive charms of an amerging civilisation. [De Tocqueville (1835) *De la Démocratie en Amérique,* Livre I]

The latent strength of American agriculture, however, had not yet made an impact on Europe. The advance of technology had not reached the stage when the agricultural produce of the New World could be transported quickly and cheaply by land and sea. Agricultural trade was still very largely an intra-European affair, and generally remained unimportant in relation to production. Britain, as a result of its expanding industrial population, had already become a large importer of food; but these imports, in spite of the repeal of the Corn Laws in 1846, hardly constituted a serious danger for British agriculture. France imported some grains, especially wheat, and exported others. Germany imported grain for the consuming and livestock-producing areas of the west, and exported wheat and rye from the great estates of the east. Denmark at this time also exported grain. Her export trade in livestock products had scarcely begun. The largest exporter of grain in the world was Russia.

The Free Trade interlude

For a short period before the Great Depression, trade between European countries became more nearly free from tariffs and other restrictions than at any other time in history. The impetus came from Britain, where the Free Trade movement had gathered strength throughout the early nineteenth century and achieved its most spectacular success in the repeal of the Corn Laws in 1846. This movement drew its inspiration from the teaching of Adam Smith and his disciples, but it corresponded also to the commercial interests of Britain at the time: with undisputed leadership in industry and commerce, Britain had everything to gain and nothing to lose from the greatest possible freedom of trade.

Having committed herself to free trade, Britain therefore sought to induce other countries to follow along the same path. The liberal doctrines of the Manchester School were not unknown in Continental countries, but the inferiority of their industry as compared with that of Britain made protection seem essential. Somewhat unexpectedly, Britain found an ally in France: Napoleon III, who in 1852 had established himself as emperor, with extensive powers, was favourably disposed towards free trade. He was moreover anxious, for political reasons, to establish close relations with Britain. In 1860, against the opposition of his Parliament, he signed the Anglo-French Treaty of Commerce.

This treaty was of significance not only for the reductions it brought about in French duties (and the comparatively minor concessions by Britain), but also for its consequences as the corner-stone of a series of subsequent treaties between France and practically all other European countries. Each of these was based on the 'most favoured nation' principle, so that the concessions granted to one country were automatically extended to the others. Thus, throughout most of Europe, trade was freed or subjected only to mild duties.

Trade in agricultural products became even more free than that in manufactured goods. By 1860 Britain had abolished duties not only on grain but on almost all other agricultural imports, leaving only a few revenue duties. France practically removed agricultural protection in the 1860 treaty, and in 1861 abolished her sliding scale of grain duties. Farmers in Germany were still interested mainly in exporting grain and therefore wanted free trade: the Zollverein's duties on grain had been abolished in 1853. In Italy the moderate Piedmontese duties formed the basis of the tariff for the unified kingdom, and after treaties with France and other countries, agriculture was protected only by low duties on grains. Belgium in 1871 decreed free entry for the main foodstuffs. The Netherlands dropped its grain duties in 1862. In most other countries agricultural trade was free or nearly so.

The challenge of overseas competition

The agricultural scene was transformed from the 1870s onwards by the immense increase in imports of cheap grain coming from North America and also to some extent from Russia. This was the result of the opening up of virgin lands and of revolutionary improvements in methods of transport. In North America especially the availability of vast land resources enabled extremely cheap production, and from the middle of the century steadily increasing use was made of farm machinery – reapers, binders and, later, efficient combine harvesters. Russia's advantage lay in having plenty of land and cheap labour.

In the 1850s American railways began to tap the Great Plains, and by 1884 the Rockies could be reached by seven different railway routes in the United States or Canada. In Russia too, new railways brought grain to ports of the Baltic and the Black Sea. At the same time there was rapid progress in shipping: in the second half of the century ships began to be built not with wood but with iron, later with steel, and this made possible a great increase in their size and carrying capacity. In addition, sail gave way to steam, and the compound engine, which greatly reduced fuel costs, came into use from the 1860s. The extent of the fall in transport costs is indicated in Table 1.1: during the 1870s and 1880s, the cost of transporting wheat both by rail from Chicago to New York and by

Table 1.1: Freight rates and prices for wheat

	1870–74	1875–9	1880–84	1885–9	1890–94	1895–9
			pence per quarter			
Freight rates:						
Chicago to New York, by rail	113	72	63	61	53	47
New York to Liverpool, by steamer	66	60	35	25	20	23
Price of US wheat from Atlantic ports, c.i.f. Liverpool	625*	568*	531	402	379	356

* Including wheat from Pacific ports from 1871 through 1875.

Source: Board of Trade (1903).

See page 401 for explanation of symbols etc.

steamer from New York to Liverpool was cut by about half. The price of American wheat arriving in Liverpool fell by even more than the amount of the reduction in transport costs.

The effect of these technical developments might have been felt even sooner but for the Crimean War of 1853–6, which closed the Baltic and the Black Sea to Russian exports, and the American Civil War of 1861–4, which curbed US exports to Europe. The peace which prevailed in America and Europe for forty years after the Franco-German War of 1870–71 provided the conditions necessary for a great expansion of trade.

From 1877 North America had the benefit of four consecutive seasons in which harvests were excellent. In Western Europe, on the other hand, these were bad years: the harvest of 1879 was disastrous. Previously, small crops had been accompanied by higher prices which helped to maintain returns to farmers. Now,

Table 1.2: Exports of wheat

	USA	Canada	Russia	India	Australasia
			million bushels		
1851–60	5	–	41	–	–
1861–70	22	–	75	–	–
1870–74	59	1	55	1	–
1875–79	107	3	71	6	–
1880–84	136	4	65	29	–
1885–89	110	3	95	36	–
1890–94*	170	9	104	30	8
1895–99*	184	16	107	15	3

* Data include wheat flour (in wheat equivalent) – the amounts involved, however, are relatively small.

Sources: Clough and Cole (1952); Crawford (1895); Farnsworth (1934).

however, exports from the United States rose sharply, doubling within a few years (Table 1.2). Prices in Europe fell instead of rising, and farmers suffered severe losses. From then on grain prices fell steadily. The whole period from about 1873 to 1896 was one of general economic depression and falling prices, which aggravated the problems of agriculture.

In the late 1880s, American grain exports for a time ceased to rise and there was a slight recovery in prices. But the 1890s brought a renewed crisis. In 1891 the United States had a record crop and exports once again rose substantially. At about the same time, Russia was increasing its exports, Canada was beginning to make an impact on the world market, and the Suez Canal was facilitating trade from India and Australia. Grain prices fell once more in 1892, and in 1895 they reached their lowest point: the world price of wheat was then little more than half the pre-depression level.

Reduced wheat crops in 1895 and 1896, partly due to some contraction in the wheat area in response to the lower prices, helped to relieve the pressure. At the same time, rising production costs in the United States made themselves felt: by about 1890 all the productive land had been occupied and the price of land began to rise. Also, domestic consumption in the United States was rising. Grain prices thus recovered gradually from about 1896 onwards, helped by a general improvement in the economic situation. By the turn of the century the agricultural depression had spent its force, and the years leading up to the First World War were favourable to farmers in both North America and Europe.

In its early stages, the crisis affected mainly the grain market. Later on, techniques of refrigeration began to be applied on a commercial scale: the United States started to ship frozen meat in 1875, Australia in 1877, and shipments grew enormously from then on. However, production of livestock products was rising much less rapidly than that of grains, while with the gradual improvement in the standard of living the demand for livestock products was increasing comparatively fast. The result was that the decline in the prices of meat and other livestock products was never as large as that which occurred for grains. The experience of Britain is a useful guide, since the British market remained free:

the prices of livestock products began to fall only in 1885, some five years later than the beginning of the crisis for grains, and at the worst of the depression, in 1896, livestock prices were about three-quarters of their pre-depression level when crop prices were little more than a half.

This difference in the trend of prices as between crop and livestock products is of considerable significance. It means in the first place that the effects of the depression were not felt equally by arable farmers and by livestock producers. The former suffered directly from the reduced value of their produce. The latter also suffered a fall in value, but it came later and was not so serious; moreover, they *benefited* from the much greater fall in the price of feed grains.

It follows that the best policy for European countries during the Great Depression was to carry out a shift from crop production to livestock. Countries which were able and far-sighted enough to do this stood a much better chance of overcoming the crisis than those which, in the face of the new trends, persevered with former habits.

The difference in price trends also leads to a distinction between large and small farms in a given country. Returns per unit of land tend to be smaller with crops, so that profitable crop production requires a fairly large area: grain farming is therefore carried out mainly on the larger holdings. The smaller farms – peasant holdings in particular – generally depend for their livelihood on dairy cows, pigs and poultry; they may produce some grain and other fodder for the use of their own livestock, but they often need to purchase additional supplies. It was therefore the larger farms which were most affected by the crisis.

In some cases the distinction is also geographical. This was particularly so in Germany, with many large grain-producing estates in East Prussia. In England, arable farms were found mainly in East Anglia and parts of the south, while in the north and west livestock farming predominated. In France, the plains of the Paris basin were the principal grain-producing region.

These differences are important for an understanding of the Great Depression. At this time the farmers best able to make their views known and to exert an influence on official policy were the larger farmers, particularly when they belonged to a ruling aristocracy; the peasants were generally unorganised and unrepresented. This suggests that the traditional view of the Great Depression may have exaggerated its severity. Too much emphasis was laid on the fall in grain prices and on the unfavourable consequences for arable farmers; not enough attention was paid to the much less difficult situation of livestock producers.

It also follows that when various European countries introduced measures of protection, they did so largely at the instigation of the bigger farmers and for their benefit, and that for the peasantry the advantages of protection were much less evident.

The protectionist revival

Already, before European agriculture felt the blast of overseas competition, there were more general pressures for a return to protection after the free trade interlude initiated in 1860. One cause was the Franco-Prussian War of 1870 and the nationalist sentiment to which it gave rise in both countries. Another, closely associated with this, was the increasing concern with industrial development throughout the Continent; demands by industrialists for protection became more insistent, demonstrating that they at least had never been converted to free trade. Public opinion became increasingly willing to listen to them. These demands were accentuated as prices began a long downward trend, starting in 1873 and lasting till about 1896, with only a short respite from 1880 to 1882; this depression kept wages and profits low and reduced purchasing power. The protectionist movement, especially in Germany, was able to draw inspiration and justification from Friedrich List's school of nationalist economics, with its stress on economic development through protection. Thus, when European agriculture was awakened to the challenge of overseas competition, there was ample scope for powerful protectionist alliances between industrial and agricultural interests.

In France the reversal of policy was marked by the tariff of 1881, which introduced moderate protection for industry and made large increases in the duties on livestock and livestock products. In 1885 and 1887, the duties on all major agricultural products were very substantially raised. The Méline Tariff of 1892 gave increased protection to both industry and agriculture. These measures remained substantially unchanged up to the First World War: in 1910 a modernised tariff brought some increase in protection for industry, but no important change in agricultural duties.

Germany's reconversion to protection began with the tariff of 1879, which restored duties on various manufactures and also imposed moderate duties on agricultural products. In 1885 and 1887 there were further and substantial increases in the grain duties. This protectionist policy was the work of Bismarck: after his dismissal in 1890, Germany for a while followed the opposite course. Germany's need by then was to obtain expanded markets for her industrial exports, and in a series of commercial treaties Bismarck's successors made limited reductions in the agricultural duties in return for advantages for German exports. But this led to violent opposition from the farming sector and in particular from the powerful Prussian landowners. It also gave rise to violent controversy as to the desirability of Germany becoming an increasingly industrial nation. Satisfaction was finally given to the agricultural interests with the tariff of 1902.

Italy re-established protection in 1878, mainly with the object of protecting her infant industries, but the duties were moderate because of the desire not to provoke retaliation against Italy's agricultural exports. Various revisions were made in subsequent years. The main change in policy occurred with the tariff of 1887, which raised the duties on both industrial and agricultural products. By then industrialists had become increasingly sensitive to foreign competition, while grain-producing farmers were suffering from the universal fall in prices. The producers of export commodities – wine, olive oil, fruit, vegetables – had little competition to fear and did not want foreign countries to discriminate against their exports, but as in other countries the large grain producers were the most influential. One of the first results of the 1887 tariff was in fact sharp retaliation by France; there followed a tariff war between the two countries lasting till 1892, and normal trade relations were not restored till 1899. This brought great hardship to agricultural exporting interests in Italy. Meanwhile the grain producers succeeded in having the duty on wheat raised by successive stages from 3 lire per 100 kg in 1887 to 7.50 lire in 1898. The protectionist revival in Italy was accompanied by a tendency for the school of national economics on the German model to prevail over economists of the Manchester School: from 1875 the former made its voice heard in the *Giornale degli Economisti,* while an Adam Smith Society was formed and published the *Economista,* but became less and less effective.

Once Belgium became firmly established as an independent nation (her secession from the Netherlands took place in 1830), she realised that, as an industrial nation heavily dependent on foreign trade, her interests lay in the greatest possible freedom of trade. But here too the agricultural crisis caused a revival of protectionist agitation and led in 1887 to the imposition of tariffs on livestock and meat. The protection involved remained moderate. In 1895 the duties on a number of agricultural products were raised, but grains (other than oats) remained free.

Economic policy in Switzerland was originally dominated by the liberalism of the Manchester School, and the Constitution of 1874 did not allow the Confederation to intervene in economic matters. Switzerland's trade policy was to obtain advantages for her exports of manufactured goods by offering an open market in agricultural products. Swiss farmers themselves were for a time inclined to favour free trade, because of their interest in exporting cheese and breeding cattle. But when the crisis broke, they too began to demand protection. Though in 1884 and 1887 some duties were imposed on manufactured products, no immediate satisfaction was given to the farmers, except for an increase in the subsidies granted for various purposes. Protection for agriculture was first introduced with the tariff of 1891. Like the Belgian tariff, this left grains largely unprotected: grains were only a small part of total Swiss agricultural output but they were a large import item, and duties would have weighed heavily on consumers and on users of feed grains. Another feature of the tariff was that the large dairy sector could not be given direct protection, because cheese exports formed an important outlet for this sector and the price of cheese on export markets largely determined the whole price structure for dairy products. The tariff thus concentrated protection on beef cattle and meat, and it was hoped that this would give some indirect help to the dairy sector by diverting part of the milk supplies to feeding more calves. The agricultural duties were raised in the 1902 tariff and modified by treaties in 1905.

From the late 1870s onwards, measures of protection for both agriculture and industry were adopted also by Austria-Hungary, Sweden, Spain and Portugal, while Russia and the United States sheltered their growing industries behind prohibitive tariff walls; in the United States the McKinley Tariff of 1890 extended protection to agriculture, with increased duties on grains and new duties on some other products.

It is not easy to give a general idea of the degree of protection for agriculture attained in the course of the Great Depression. Practically all tariff rates were 'specific' (i.e. in terms of weight or quantity), and to express them in *ad valorem* terms involves difficulties in selecting appropriate price data. An attempt was made by Liepmann to compare tariff levels in European countries: the results for foodstuffs in 1913 are shown in Table 1.3. Both the basic material available and the method adopted leave much to be desired, and the results are not to be regarded as precise. However, it seems reasonable to conclude that in most of the countries which had adopted protection for agriculture, the degree of protection before the First World War lay somewhere between 20% and 30%.

Table 1.3: Average 'potential' tariff levels for foodstuffs, as a percentage of export prices in European countries, 1913*

Country	Tariff level
France	29
Germany	22
Italy	22
Belgium	26
Switzerland	15
Austria-Hungary	29
Sweden †	24
Finland	49

* The method used was to select thirty-eight foodstuffs, said to represent a 'fair sample' of goods in European trade; to calculate for each country the incidence of duty on each of these foodstuffs, on the basis of 'normal prices' – i.e. the export prices of the leading European exporting countries in the year in question; and to work out an average of the various duties for each country. The average was unweighted, so it took no account of the varying importance of different items in the imports of each country: indeed each country was not necessarily an importer of all the thirty-eight foodstuffs. This difficulty was recognised by the author, who accordingly called his results 'potential' tariff levels.
† Fruit and vegetables not included.
Source: Liepmann (1938).

Laissez-faire, Free Trade and adaptation

A few European countries were left holding determinedly to Free Trade principles, the most important of these being Britain.

Agriculture in Britain was no less subject to overseas competition than that of other European countries. In fact a large sector of British agriculture was ruined in the course of the depression; many agricultural workers moved to the towns and much arable land was turned over to grass. 'High farming' gave way to the utmost economy in operations and even to neglect. Yet the British Government refused the slightest degree of protection. Moreover, there was no effective move for protection among the farmers themselves. It was only after the end of the agricultural depression, from about 1903 to 1905, that a vigorous campaign for 'Tariff Reform' was started, but this was based on the desire to secure preferences for Empire trade rather than on the wish to protect British agriculture (indeed, the conflict between the two aims was never satisfactorily reconciled).

Denmark too held firmly to Free Trade, but while in Britain the policy was one of pure laissez-faire, Denmark's reaction to overseas competition was to carry out a fundamental transformation of its agriculture. There was a large shift from crops to livestock production, and Danish exports of livestock products became established on the British and German markets.

Developments in the Netherlands were similar to those in Denmark. The importance of foreign trade to the economy had led to a long-standing attachment to Free Trade principles. When the unprotected Dutch market came to be swamped with foreign goods to which other markets were increasingly closed, and when unemployment began to rise in consequence, a protectionist movement made itself heard. The government nevertheless withstood these demands. The National Agricultural Council declared that proposals to raise the grain duties would do no good to agriculture, as it would remove the incentive to work and lead to many protective measures which would raise prices of agricultural inputs; it also considered that the situation in the Netherlands under free trade was no worse than in countries which had already adopted protection. This energetic refusal put an end to protectionist claims for agriculture: the farmers thereupon accepted the Free Trade policy, set up associations to improve the processing of their products and made use of the training institutes set up by the government. A justification for this policy could be found in the subsequent increase in exports of livestock products, fruit and vegetables.

Factors influencing the choice of policies

What was responsible for the wide variety of policies followed during the Great Depression? Several factors played a role. Perhaps the first was the general attitude in each country towards economic policy and in particular to trade. Britain, with its early start in the Industrial Revolution, had a clear advantage in maximum freedom of trade. Small countries like Denmark, Belgium and the Netherlands, with a shortage of natural resources, were heavily dependent on trade and favoured free trade for that reason. France and Germany, on the other hand, wanted to develop their industries and needed protection to do so; with their extensive domestic resources they could afford to restrict trade (at least up to the point when, in the case of Germany, markets had to be found for industrial exports).

Economic theory played a part, but probably only a small part, in shaping public attitudes. There is no doubt that List was the dominant influence on academic thinking in Germany, as Smith and Ricardo were in Britain, and their respective theories provided powerful arguments for the opposing camps of Protectionists and Free Traders. But Kindleberger, in a study of the different responses to the agricultural depression rightly observed:

It might be fair to say that the economists of Britain and the national economists of Germany provided the rationale for the action taken rather than its impetus. And the relative unimportance of this function may be indicated by the action of France and Italy, taken in the absence of any distinctive rationale. [Kindleberger (1951) page 37]

The extent of public willingness to support agriculture was another important factor. Prevailing opinions were composed of elements of both reasoning and sentiment. The resort to agricultural protectionism in both France and Germany owed much to a widespread belief in the virtues of rural life and in the disadvantages of a high degree of industrialisation. In France this sentiment was expressed in an extreme (but vague) form by Jules Méline, particularly in his book *Le retour à la terre*. In Germany, discussion on this point took the form of an academic debate on the *Agrar- oder Industriestaat* issue, and the arguments of Professor Adolf Wagner, the main advocate of the *Agrarstaat*, carried great weight: he stressed the social disadvantages of the rise in the urban population and the risks – which were not only strategic – of excessive dependence on food imports.

On the other hand, the main argument working against public willingness to protect agriculture was the effect on food prices. In Britain, the successful campaign against the Corn Laws had left a deeply-engrained hostility to any form of tax on foodstuffs: 'cheap bread' remained a powerful rallying-cry. In other countries, this was an argument frequently used by Free Traders, but with little success.

The strategic argument – the need to ensure self-sufficiency in food supplies in time of war – was part of the general case for an *Agrarstaat* and was frequently advanced in almost all countries as a justification for protection. This argument certainly played a role, but perhaps less than is sometimes thought: it was an additional reason for statesmen to satisfy pressure from agricultural interests, rather than an independent factor influencing policy. Perhaps it manifested itself most clearly in reverse: there is no doubt that a vital factor in Britain's indifference to the fate of her agriculture was her undisputed mastery of the seas and her possession of rich food-producing colonies.

These factors outside agriculture influenced the possibilities for forming protective alliances between industrial and agricultural interests. At critical stages in France, Germany and Italy, such alliances were responsible for restoring protection to both industry and agriculture. This unity could be maintained only so long as the common interest in obtaining protection was stronger than conflicts of interest in other respects. Some such conflict between the two sectors was always inevitable. Protection for industry was liable to raise the cost of farm tools and equipment, possibly also of farm wages. Protection for agriculture might, by raising food prices, cause increased labour costs for industry; it might also raise the price of raw materials. This last problem was usually not serious, since agricultural raw materials for industry were generally produced overseas and not by European agriculture; but an important exception was wool, and it is significant that, even in the countries where agricultural protection went furthest, wool was generally left entirely free from duty, and the result of competition from Australia, South America and South Africa was a drastic fall in the sheep population of Europe. Conflict was inevitable also when industry was concerned with enlarging its export markets: in Germany this problem became acute in the 1890s, and the fact that it was resolved in favour of agriculture demonstrates the political strength of the farmers.

The influence of the farming community was to a large extent a question of the relative size of the agricultural population: the contrast between Britain and the other countries needs no comment. It was also determined by the extent to which general democratic evolution had deprived the rural aristocracy of its power. In Britain the political reforms starting with the 1832 Reform Act had weakened the landed classes; moreover, the division of interests and outlook between landlords and tenants, and also between arable and livestock farmers, prevented the formation of any coherent agricultural pressure group. In Denmark a series of significant reforms had created an independent and well-educated class of farmers with medium-sized farms. In the Netherlands, the Reformation, the struggle for national freedom, the age-long combat against the sea and the importance of trade, all combined to produce an independent, energetic and commercially-minded farming population. But in countries where there remained a dominant class of large landowners (often interested primarily in the price of grain), this group was able to present itself as the representative of the whole agricultural interest, regardless of the real needs of other groups in agriculture, and to exert considerable influence over the course of official policy. Thus in Germany the Junkers remained a coherent group with vestiges of feudal power and an entrenched position in the state; in France the campaign for protection was led by the aristocratic *Société des Agriculteurs de France;* and in Italy by the big grain producers.

Consequences of policies followed

Where protection was adopted, it succeeded in restraining the fall in prices, though not in stopping it. Figure 1.1 shows the trend of wheat prices in the United States, England, France and Germany, using a five-year moving average to eliminate annual fluctuations. It also indicates the rate of duty on wheat in France and Germany. This shows clearly that till the mid-1880s wheat prices in all four countries followed a similar downward course. Then, as greatly increased tariffs were imposed in France and Germany, prices in these two countries parted company from the 'world' price: the French and German prices in fact remained above the US and English prices by roughly the amount of the tariffs, or by nearly a third.

Shielded from the worst of the depression, farmers in countries where protection was adopted probably suffered a smaller loss of income than those in Britain. In the absence of adequate statistics, it is not easy to assess the extent of this difference. The situation in Britain may not have been as bad as is often thought.

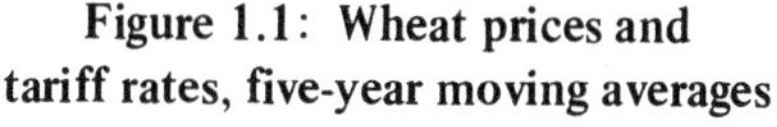

Figure 1.1: Wheat prices and tariff rates, five-year moving averages

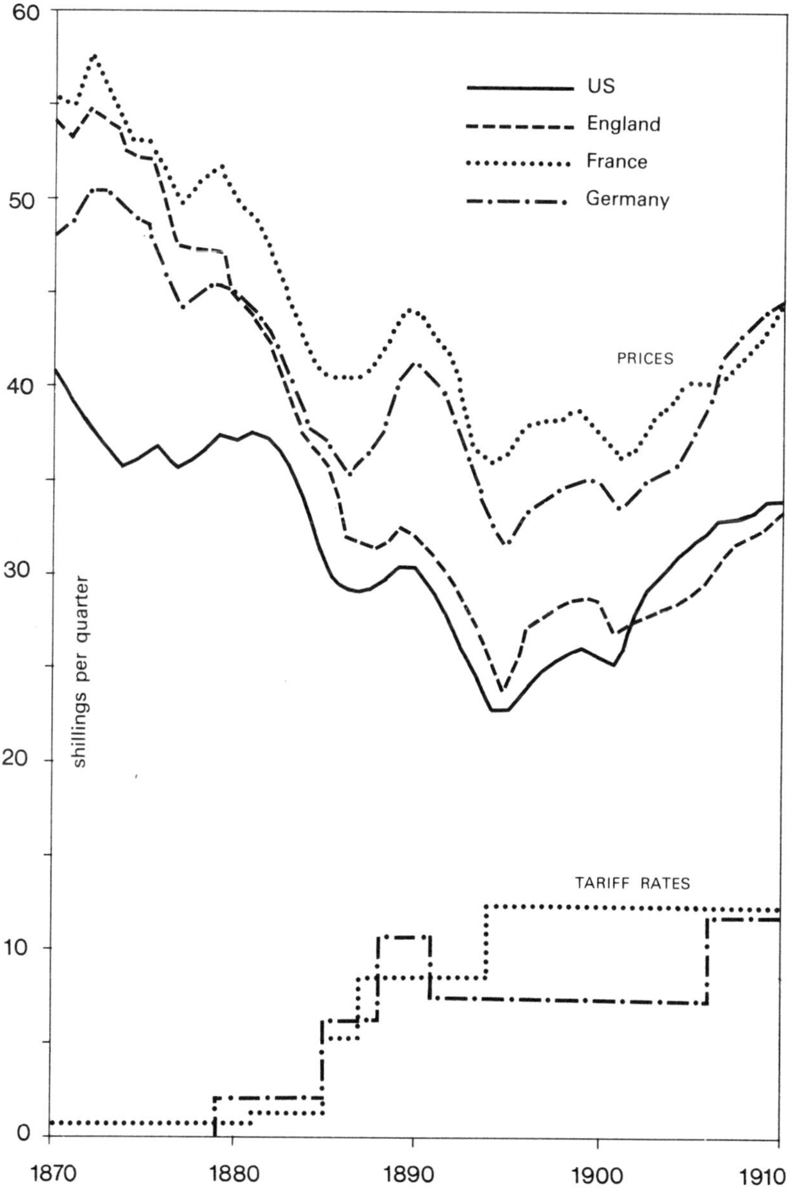

Source: Board of Trade (1909) *British and Foreign Trade and Industry*. Data for later years compiled from national sources.

It has already been pointed out that the fall in prices affected mainly the big arable farmers, while the smaller livestock producers did not do so badly. Conversely, in France and Germany the benefit of protection went mainly to the large grain growers, and gave much less benefit to the livestock-producing peasantry – perhaps it even harmed them by keeping up the price of bread and the cost of purchased feedingstuffs, as well as by maintaining land values. Such estimates as can be made for France and Germany suggest that in both cases only a minority of the total agricultural population stood to gain from protection on grains. It is true that livestock products were also protected – in Germany by 'sanitary' regulations more than by tariffs – but here overseas competition never exerted such a strong influence, and though precise indications are not easy to obtain it seems unlikely that the effect of protection was so marked.

Denmark and the Netherlands showed the possibility of meeting the crisis by adapting the pattern of agricultural production to the new situation, and thereby laid the basis for a prosperous agriculture making an important contribution to export earnings and to general economic progress. In France and Germany, though the livestock sector did expand, no such adaptation took place: the area under wheat actually expanded during the period, and in Germany the cultivation of rye, under the influence of various protective measures, rose quite out of line with trends in demand at home and abroad.

Equally serious, the preoccupation with tariff policy diverted attention from more constructive measures. In Denmark, the farmers themselves took the initiative in making agriculture more efficient, in particular by building their remarkable co-operative marketing system. In Germany, the co-operative movement made some progress, particularly the Raiffeisen credit co-operatives, and there was substantial technical progress. The need to consolidate fragmented holdings was realised and several states passed legislation to this effect, without achieving very much. In France *syndicats* and *mutuelles* were promoted by rival political groups with little initiative being shown by the peasants themselves; there was no attempt at farm consolidation. In general, farmers in France and Germany came to rely increasingly on tariff protection, and their governments did little to encourage them to improve their position through higher productivity, better marketing and so forth. The habits of mind thus created proved difficult to shake off.

The movement of manpower off the land seems to have been comparatively slow in France and Germany. In Britain the population occupied in agriculture, even though it was already relatively small by the 1880s, fell substantially between 1881 and 1901 – this at a time when total population was rising fast (Table 1.4). In France, according to the available data (not very reliable), the population occupied in agriculture appears to have *risen,* though there are indications that the active *male* population at least began to decline in the mid-1890s. In Germany there was a reduction in agricultural employment, involving in particular emigration from the rural areas of the east to the areas of industrial growth. In both France and Germany, the number of farmers remained approximately unchanged or even increased: the movement concerned farm workers, domestic servants and members of farm families. These differences between Britain, France and Germany probably reflect, above all, differences in the rate of industrialisation. Nevertheless, it seems likely that if France and Germany had not adopted agricultural protection, or if Britain had, the results in terms of labour movement would have been different.

Table 1.4: Occupied population and total population, in agriculture and in all sectors, in Great Britain, France and Germany

		Occupied population			Total population		
		Agri-culture	All sectors	Agri-culture as % of total	Agri-culture	All sectors	Agri-culture as % of total
		millions			*millions*		
Great Britain	1881	1.50	12.80	12	. .	29.7	. .
	1901	1.33	16.31	8	. .	37.0	. .
% change:	1881–1901	–12%	+28%		. .	+25%	
France*	1881	7.89	16.54	48	18.2	37.7	48
	1891	. .	. .	. .	17.4	38.3	45
	1906	8.86	20.72	43	. .	. .	. .
% change:	1881–91	. .	. .		–5%	+2%	
	1881–1906	+12%	+25%		. .	. .	
Germany†	1882	10.5	21.2	50	19.2	45.2	43
	1907	9.9	28.1	35	17.7	55.8	32
% change:	1882–1907	–6%	+33%		–8%	+23%	

* 'Agriculture' includes forestry and fishing.
† 'Agriculture' includes forestry.
Sources: See corresponding tables in country chapters.

As to the effects of protection on consumers, some indications can be obtained from studies published by the British Board of Trade between 1908 and 1910, which obtained comparative data on retail prices (Table 1.5). The price of white bread was some 15% higher in France than in England, and that of wheat flour very much higher in France and Germany than in England; the authors of the reports commented that in Germany bread was mostly made from a mixture of wheat and rye flours, and that hardly any wheat flour seemed to be used in households in the French industrial towns. In Belgium, where there was no duty on bread grain, the prices were comparable to England. For sugar, beef and mutton, British consumers were clearly benefiting from cheap imported supplies, consuming more at lower prices; pork was consumed particularly in Germany, but at a higher price than in the UK; bacon was an important item of consumption only the UK. Horsemeat was important in France and Belgium, so was *charcuterie* in France and *Wurst* in Germany. For butter, curiously, these data indicate that the price in England, in spite of the advantages of New Zealand supplies, was somewhat higher than in France, though slightly lower than in Germany. The price of potatoes – an item not much affected by international

Table 1.5: Predominant retail prices, 1905

	England and Wales	France	Germany	Belgium	
				estimate 1908	1905 *
	pence per lb.	*Indices: England = 100†*			
White bread	1.1–1.4	115	..	95	85
Wheat flour	8–10	153	140	107	90
Sugar	2	144	119	150	140
Potatoes	0.4–0.5	100	88	92	120
Milk	3–4**	71	75	64	..
Beef	7.5–8.5†† 5–6***	109	122	96	90
Mutton	7.5–9†† 4–5***	131	137	110	..
Pork	7.5–8.5	116	123	106	95
Bacon	7–9	..	123	98	95
Butter	12–13†† 14†††	94	105	98	100
Cheese	7	..	..	121	115

* These are very approximate figures, based on indications provided in the source as to the degree of change between 1905 and 1908.
† Calculated taking the arithmetic average of price ranges.
** per quart (1.136 litres).
†† British or home-killed.
*** Foreign or colonial.
††† Danish.
Source: Board of Trade, 1908, 1909 and 1910.

trade – was relatively low in Germany and the consumption there relatively high. It was only for milk, a commodity in which trade was still very localised, that British prices were relatively high.

The Board of Trade commented that if an English workman, with an average family, were to go to France or Germany and attempt to maintain there his customary mode of living, he would find his expenditure on food (and on rent and fuel) substantially increased. On the other hand, a French working-class family, if it were to go to England and keep the same consumption pattern as at home, would spend somewhat less; a German family keeping its consumption pattern would not gain so much, because it would not have the benefit of such low prices for potatoes and milk of which it normally had high consumption.

However, consumption patterns were adjusted to differences in relative prices, so that the impact of price differences on total food expenditure was to some extent offset. Table 1.6 indicates that when families with the same income are compared, food actually accounted for a smaller share of expenditure in France and Germany than in the UK (partly owing no doubt to the smaller size of family). The Board of Trade recognised that a French working-class family appeared to have a more varied diet than a British family with the same income, and also to eat rather more: *per capita* consumption of bread, meat, fish, eggs, butter and other fats, and potatoes, was distinctly larger, consumption of vegetables and fruit very much larger and that of milk slightly larger. In fact

Table 1.6: Pattern of expenditure in households earning 30–35 shillings per week, 1905

	UK	France	Germany	Belgium (1908)
Average number of persons per household	5.2	3.9	4.5	5.0
Percentage of income spent on food	65	58	59	62
Distribution of expenditure on food, in percentages:				
Bread and flour	15.8	17.9	16.1	19.2
Meat and fish	28.3	30.3	31.8	26.7
Fresh milk	6.1	4.4	7.4	3.9
Butter, oils, fats	11.4	10.8	14.2	15.9
Potatoes	4.2	3.6	4.6	5.8
Other vegetables and fruit	4.0	7.2	4.1	4.2
Tea, coffee (including chicory), cocoa	6.5	5.0	4.6	4.6
Sugar	4.3	2.7	2.2	1.5
Other foods (including meals away from home)	18.9	18.1	15.0	18.2
Total	100.0	100.0	100.0	100.0

Source: Board of Trade, 1908, 1909 and 1910.

sugar was the only item of which consumption was much less in France than in England.

The Board of Trade, however, estimated that average wages were substantially lower in France, Germany and Belgium than in England (respectively 75%, 83% and 63% of the English levels). Comparison of families at equivalent income levels may thus be misleading as to average *per capita* consumption levels[1].

The effects of protection on the cost of living were thus probably not so obvious to those concerned, partly because of differences between countries in consumption patterns and also because the general price trend at the end of the nineteenth century was downwards, so that the effect of protection was to restrain the fall rather than to cause an increase in food prices. Protection thus did not give rise to an obvious burden: it would be truer to say that consumers

[1] The Board of Trade surveys covered thirty cities and towns in France, thirty-three in Germany and fifteen in Belgium. They were carried out and reported upon with remarkable thoroughness, and constitute a mine of information which has been curiously neglected by economists and historians. However, P. Saunier, ((1975) *L'évolution du coût de l'alimentation depuis le début du siècle.* INRA. Paris) subjected the methods and results of these surveys to a critical analysis. He found that they accurately reflected living conditions of working-class households in the UK, France and Germany (he did not study the report on Belgium). The main defect of the surveys appeared to be overestimation of the average family size in the UK and ambiguity as to the situation of adults other than parents. The higher wage class seemed to be over-represented in France, though this may be justified by a relatively high proportion of working wives. (The household budget data were derived from surveys carried out mainly through trade unions and the Board recognised that they might not be representative). Consequently, though the reports give figures of average consumption per head, these seem insufficiently reliable for inclusion here.

in France and Germany were largely deprived of the benefits that could have come from cheap imported food and that changes in consumption patterns which might have taken place were impeded.

*

* *

The challenge of overseas competition had thus produced different responses. In most countries, including in particular France and Germany, there was a defensive reaction, taking the form of greatly increased tariff protection: their farmers were shielded from the worst effects of the crisis, but the incentive to adjust was diminished. Britain held to laissez-faire and an adjustment was forced on agriculture in the form of contraction and a decline in the arable sector, with adverse social effects at the time. At a later period, however, the relatively large British farms were able to achieve comparatively high labour productivity and incomes comparable on average with other sectors. In Denmark and the Netherlands there was a positive response, characterised by the absence of defensive measures and a deliberate adaptation and improvement of agricultural production and marketing; not only their farm sectors but their entire economies benefited.

To a large extent these different reactions were determined by earlier social and economic developments. In France and Germany, while the mass of the peasantry remained backward, large farmers and landowners, with a particular interest in cereals, still had political influence; they were able to join forces with industrialists in obtaining protection, or even – in the case of Germany – to prevail over the manufacturers when interests diverged. In Britain, the urban population and the industrial interest were already economically and politically preponderant. In Denmark a series of remarkable reforms conducted by an autocratic but enlightened government had created a class of independent farmers: the foresight of Crown-Prince Frederik and his ministers appears as a crucial turning point in Danish agrarian history (as well as in Denmark's whole evolution) and contrasts with the weakness or immobility of monarchs in other European countries before the French Revolution.

If the protectionist response was thus largely predetermined, it was *not* inevitable that the countries which took this path should have carried out so little positive adjustment. It would have been possible, while supporting prices through tariffs, to encourage technical progress, to consolidate fragmented farms, even to promote farm enlargement, to develop rural infrastructures, and so on. But in practice, protectionist attitudes inhibited such developments: adjustment, being made less urgent, was constantly deferred. The legacy, in the twentieth century, was a heavy one.

Bibliography

The feudal and manorial system was vividly described in Bloch's history, first published in 1939, latest edition 1968; it has been translated into English. Slicher van Bath, whose work was translated from the Dutch in 1963, partly covered the same field but paid more attention to economic developments. The agrarian reforms of the eighteenth and nineteenth centuries throughout Europe have more recently been analysed in masterly fashion by an American historian, Blum (1978).

Concerning the Great Depression, the work by Ashworth (1952) provided useful general background. A thorough contemporary account of protectionist policies was made by Poinsard (1893).

GENERAL

Ashley, P. (1920) *Modern Tariff History.* 3rd edn. London.

Ashworth, W. (1952) *A Short History of the International Economy, 1850–1950.* London, New York, Toronto: Longmans.

Birnie, A. (1961) *An Economic History of Europe, 1760–1939.* London: Methuen.

Bloch, M. (1968) *La société féodale.* 5th, single-volume edn. Paris: Albin Michel. (In English: *The Rise of Feudal Society* and abridged version in *Agrarian Life in the Middle Ages.* Cambridge Economic History, Volume I.)

Blum, J. (1978) *The End of the Old Order in Rural Europe.* Princeton University Press.

Board of Trade (UK) (1903 and 1909) *British and Foreign Trade and Industry* Cd. 1761 and Cd. 4954.

Board of Trade (UK) (1905) *Cost of Living of the Working Classes.* Cd. 3864.

Board of Trade (UK) (1908) *Cost of Living in German Towns.* Cd. 4032.

Board of Trade (UK) (1909) *Cost of Living in French Towns.* Cd. 4512.

Board of Trade (UK) (1910) *Cost of Living in Belgian Towns.* Cd. 5065.

Clough, S. B., and Cole, C. W. (1952) *Economic History of Europe.* Boston: Heath.

Coppock, D. (September 1961) 'The causes of the Great Depression, 1873–96'. *Manchester School of Economic and Social Studies.*

Crawford, R. F. (March 1895) 'An inquiry into wheat prices and wheat supply'. *Journal of the Royal Statistical Society.*

Farnsworth, H. (1934) *The decline and recovery of wheat prices in the 'nineties* ('Wheat Studies of the Food Research Institute'). Stanford University.

Földes, B. (Juni 1905) 'Die Getreidepreise im 19. Jahrhundert'. *Jahrbücher für Nationalökonomie und Statistik.*

Grunzel, J. (1916) *Economic Protectionism.* Oxford.

Kindlebergër, C. P. (February 1961) 'Group behaviour and international trade'. *Journal of Political Economy.*

Liepmann, H. (1938) *Tariff Levels.* London: Allen & Unwin.

Madison, A. (1962) 'Growth and fluctuations in the world economy'. *Quarterly*

Review, Banca Nazionale del Lavoro, Rome. No. 61.

Malenbaum, W. (1953) 'The World Wheat Economy'. *Harvard Economic Studies*, vol. **XCII**, Cambridge, Mass.

Poinsard, L. (1893) *Libre èchange et protection.* Paris.

Rudé, G. (1964) *Revolutionary Europe 1783–1815.* Fontana History of Europe. London & Glasgow: Collins.

Slicher van Bath (1963) *The Agrarian History of Western Europe, A.D. 500–1850.* London: Arnold. (Translated from Dutch.)

INDIVIDUAL COUNTRIES

(For the UK, France, Germany and Denmark, see the bibliographies at the end of the relevant chapters.)

Belgium

Leener, G. de (1934) *La politique commerciale de la Belgique* (publication de l'Institut Universitaire de Hautes Études Internationales, Genève). Paris: Sirey.

Italy

Bandini, M. (1937) *Agricoltura e Crisi.* Firenze: Barbera.

Herlihy, D. (April, 1959) 'The history of the rural seigneury in Italy, 751–1200'. *Agricultural History*, **33** (2), 58–71.

Porri, V. (1934) *La politique commerciale de l'Italie* (publication de l'Institut Universitaire de Hautes Etudes Internationales, Geneve). Paris: Sirey.

Smith, D. M. (1959) *Italy – A Modern History.* University of Michigan.

Netherlands

Buning, E. de Cock (1936) *Die Aussenhandelspolitik der Niederlande seit dem Weltkriege.* Berner wirtschaftswissenschaftliche Abhandlungen, 17.

Scandinavia

International Labour Office (January, 1960) 'Agricultural policy in Scandinavian countries'. *International Labour Review.*

Switzerland

Bickel, W. (1961) *Landwirtschaft und Landwirtschaftspolitik der Schweiz.* Bern: Unionsdruckerei.

Landmann, J. (1928) *Die Agrarpolitik des schweizerischen Industriestaates.* Kieler Vorträge, 26, Jena.

Neuhaus, J. (1948) *Die Entwicklung der bundesstaatlichen Agrarpolitik seit 1848.* Turbenthal: Furrers Erben.

Chapter 2

The United Kingdom

From early feudal times up to the nineteenth century, British agriculture underwent profound transformations. A long process of enclosures caused the medieval open fields and strips to disappear, giving way to a structure of relatively large, consolidated farms. Economic and social developments created, besides the large landowners, a class of prosperous tenant farmers and some owner-occupiers, with another class of hired farm labourers. The 'peasant' virtually disappeared. In the context of general technological and industrial development, farming productivity was greatly increased, while agriculture's share in the national income and in the total active population were much reduced. Britain became dependent on imports for much of its food supplies.

The enclosures and the age of 'high farming'

Anglo-Saxon settlement consisted primarily of villages, with open fields; medieval lords and religious houses controlled manors not unlike those of the Carolingian empire. The Norman Conquest of 1066 instituted a strictly regulated feudal society and generalised the manorial system in England. In the typical manor on arable land, the village was surrounded by open fields, two or three in number, their size depending on the population of the village. These were divided into furlongs; each peasant held strips in each furlong and in each field, thus having a fair share of good land and bad. The strips ran the length of the furlong.[1] The village would also have meadowland, marked into separate lots from which each peasant could take a crop of hay before the land was returned to communal grazing; also common rough grazing, wasteland, woodland, etc.

The peasantry after the Norman Conquest included various categories: freemen who were virtually rent-paying tenants without other commitments; 'villeins' holding perhaps thirty acres of land plus grazing rights in return for various services to the lord of the manor; cottars with much less land; and bondmen (*servi*).

This picture began to change after the Black Death in the middle of the fourteenth century, when many villages were wiped out, and in many regions the population became insufficient to maintain traditional arable husbandry. Many landlords, including the monastic houses, hastened the process by evicting the remaining population and converting arable fields to pastures: in such cases the strips disappeared and the vast open fields gave way to hedged enclosures, grazed by sheep. In some cases, open-field farmers seem to have agreed to enclose the fields and rearrange their strips to their mutual benefit in view of the advantages in management. This process was easier in regions where common pastures were abundant, for in other areas, common rights and the right to pasture animals on open-field stubble were jealously guarded.

By the early sixteenth century, serfdom had practically disappeared and labour services had generally been commuted into rents or been replaced by hired work. Manorial courts were progressively replaced by Justices of the Peace,

[1] An 'acre' corresponded in principle to a day's ploughing and was reckoned to be 220 yards (a furlong) by 22 yards, but in practice the size varied (see Higgs, 1964).

giving added status to rural gentry. A far-reaching change in land ownership occurred with Henry VIII's dissolution of the monasteries in 1536–9, for the monasteries owned over a quarter of landed property in England. Though some of this land was kept by the Crown and some was sold to noble families, a large part was sold or leased to smaller gentry.

The process of enclosure was stimulated by the profitability of sheep farming for wool. Forcible enclosures by landlords, pushing peasants off the land, even became a matter of official concern – in 1517 Wolsey's Commission of Inquiry into depopulating enclosures actually ordered some landlords to restore land to tillage and even to rebuild houses that had been destroyed. Under the Restoration Monarchy of 1660, however, landlords were left free to carry out enclosures and the process continued, particularly in the Midlands and the central southern counties. By the beginning of the eighteenth century, farming was carried out in open fields over much of England: probably about half of the arable land had by then been enclosed.

From the mid-eighteenth century, further enclosures – aiming now at a general improvement of the agrarian structure – took place mainly under private Acts of Parliament, a procedure whereby special commissioners were appointed for each area concerned. In the following hundred years, the process of enclosure was systematically completed, nearly 3000 parishes being affected. The Enclosure Commissioners sought to form squarish fields, varying according to the size of the farm itself from five up to about sixty acres. Hedges were planted around these fields and the countryside took on its characteristically 'English' appearance. The Enclosure Commissioners also replanned the roads, many new roads coming into existence as a result of their work.

The description so far relates primarily to the arable areas of England. Other parts of the country were different in several respects. In northern England and southern Scotland, where land was plentiful, the field arrangement frequently consisted of regularly cultivated 'infield' and less fertile but extensive 'outfield' of which only part would be cultivated at a time. In hill country, isolated farmsteads were more typical than villages. In the Highlands of Scotland, hereditary chieftains held sway over their clansmen and over large but unproductive mountain territories.

Important differences arose also in respect of relations between landlords and tenants. In England, the tenants installed on the enclosed land generally enjoyed good relations with their landlords and – even though there was no statutory control of leases – had reasonable security of tenure: in practice, tenants' sons often inherited the leases. In Ireland, on the other hand, exploitation of the native peasantry by absentee English landlords was a sad history and a major cause of the eventual revolt against English rule. Conditions in Wales were not much better: landowners, whether of Welsh or English origin, were generally on poor terms with their tenants. The latter had no security, little incentive to improve their farms, and suffered in many cases from acute poverty; linguistic and religious issues sharpened the conflicts. In the Highlands of Scotland, the clan system was broken up after the 1745 Jacobite rising and the chieftains had to adapt to new ways. Surplus population was forced off the land into the cities or overseas, and the big estates were given over to sheep, cattle, deer and game; a dwindling number of Gaelic-speaking crofters eked out a living from their small-holdings and from fishing.

The Scottish Lowlands enjoyed better conditions. Many farmers owned small or medium-sized holdings; tenants benefited from reasonably long leases (nineteen years was common, but sometimes leases covered three lives). The early spread of the banking system encouraged farming enterprise, while the children of farmers and farm labourers benefited from an advanced system of education.

Bearing in mind these regional differences, it is nevertheless reasonable to say that by the mid-nineteenth century the more productive lands of Great Britain were owned by large landlords and farmed by their tenants in units almost always large enough to permit efficient management.[1] The ownership of the estates had usually been in the same family for generations: personal ties between landlords and tenants were often strong. Many landlords brought in capital and commercial spirit. Chambers and Mingay, in their study of the agricultural revolution of this period (Chambers, 1966, page 201) comment that 'the English landed interest had long accepted the idea that land was an agent of production and not merely an instrument of feudal and social prestige'. (The payment of tithes to the Church, a final vestige of feudalism, was commuted into 'rent-charges' in England and Wales by an Act of 1836; Scotland had settled this question already at the Reformation.)

Relatively small, owner-occupied farms existed too: it can be seen from Table 2.1 that holdings of less than twenty acres were quite numerous but they accounted for only a small proportion of the agricultural area, two-thirds of which was in farms of more than 100 acres. As the custom was to hand the farm in its entirety from father to eldest son, there was little of the tendency to divide up holdings which plagued agriculture in other countries.

The result of the transformation was to create an agriculture able to take full advantage of the technological changes which came in the wake of the Industrial Revolution. Improved methods of cultivation, in particular the famous 'Norfolk' four-course rotation, were introduced, together with new crops – turnips, clovers, legumes. Increased fodder production made it possible to keep more livestock, and the extra manure in turn raised the fertility of the soil. The practice of 'marling' (the addition of clay to thin sandy soils) enabled heathland (in Norfolk, for example) to be reclaimed and enclosed. Livestock breeds were steadily improved. These developments laid the foundation for the period of

[1] Various attempts were made later in the century to discover who owned the land: the results indicated a marked concentration in very few hands. Thus, calculations by John Bateman (1883) *Great Landowners of England and Wales.* 4th edn., London, page 515, indicated that about four thousand titled families held just over half the land in England and Wales; the Church and universities were also substantial landowners; about 250 000 independent farmers owned only some 37% of the land. These data are not very reliable, but they are not inconsistent with the figures in Table 2.1 which are based on land *use* and not ownership.

Table 2.1: Agricultural holdings in Great Britain in 1870, by size

Size of holding		Number of holdings	Estimated area covered
acres	*hectares* (approx.)	*thousands*	*thousand acres*
under 5	under 2	136	407
5–20	2–8	150	2 252
20–50	8–20	86	3 010
50–100	20–40	64	4 800
over 100	over 40	93	19 940
Total		529	30 409

Source: Agricultural Returns of Great Britain, 1870.

'high farming' which characterised the early decades of Queen Victoria's reign: an age of carefully controlled livestock breeding, calculated feeding, scientific soil treatment, supplemented by a growing range of mechanical devices and the spreading use of steam power. New techniques of drainage were a great benefit to farmers on heavy soils. It was an age of intensive farming, aiming at maximum food output to meet the growing demands of the urban population. British agriculture, in the mid-nineteenth century, was recognised to be the best in the world.

The Industrial Revolution brought other changes which fundamentally affected the situation of agriculture. Growing labour needs in industry promoted the movement of people away from the land, so that already by the nineteenth century only a small proportion of the population earned their living from agriculture. Also, the growth in the demand for food was such that it could not be fully satisfied by British agriculture. Until the middle of the eighteenth century, Britain had been a net exporter of grain, but from then on imports began to exceed exports in years of bad harvest, until by the nineteenth century Britain's dependence on grain imports was established. This brought into the foreground the issue of the Corn Laws.

The repeal of the Corn Laws

'Corn Laws' had existed in England from the early Middle Ages. Their original purpose was to maintain fair prices by preventing speculation and monopolistic practices: they regulated internal trade and restricted exports. From 1660 onwards, imports too were regulated by a sliding scale of import duties, in which the duty was 2s. a quarter when home prices were at or under 44s., but only 4d. when home prices were at a higher level. However, so long as Britain was normally a net exporter of grain, the Corn Laws were of minor importance as a measure of import restriction.

By the end of the Napoleonic Wars, conditions were different. The Corn Laws became a protective measure, shutting out imports and raising domestic prices. Under a law of 1815, imports of foreign wheat were admitted free of duty when the price of British wheat was at or above 80s. per quarter; otherwise imports were prohibited. Similar provisions were adopted for other grains. In 1822 the system was modified: the price limit at which imports were prohibited was lowered to 70s. per quarter and above this level a sliding scale of duties was applied. A further change was made in 1828: there was no longer any prohibition, but a more complicated sliding scale was introduced.

The average annual price of wheat, which had been above 80s. during most of the Napoleonic Wars, dropped sharply around 1820 and lay between 40s. and 70s. in subsequent decades. At this price level, imports were either excluded or subjected to near-prohibitive duties for most of the time.

The Corn Laws were notorious chiefly for accentuating fluctuations in supplies and prices. Traders had an interest in holding supplies off the market till the price rose sufficiently for imports to be admitted or for the duty to be reduced: there would then be a flood of imports, depressing prices once again.

Given the miserable conditions of much of the urban working class at this time, the slightest rise in the price of bread was a great evil, and to have relatively low prices at one period was poor compensation for prohibitively high prices at another. Agitation began after the passage of the new law in 1815: it continued during the 1820s, eased off between 1831 and 1836 when harvests were large and prices low, but intensified as prices rose from 1837 on. In 1838 the Anti-Corn Law League was formed, under the leadership of Richard Cobden and John Bright, and began a vigorous campaign. It issued a regular publication and numerous pamphlets and held public meetings, stressing the need for 'cheap bread' and appealing to the interests of the urban population. In 1842 it took an active part in the elections: Cobden himself obtained a seat in Parliament. In the same year Sir Robert Peel modified the sliding scale and eased the degree of import restriction; this, combined with a good harvest, took the edge off the agitation for a while. But the harvest of 1845 was disastrous, and in Ireland the failure of the potato crop owing to blight caused famine. Peel's efforts at reform had so far been opposed by almost the whole of his Tory Cabinet and though he was now convinced of the need for total abolition of the Corn Laws, he still lacked support. He resigned, but after attempts to form another government had failed, he was called back to office with enhanced authority. In January 1846 he announced that from 1 February 1849, grain would be admitted subject only to a small registration duty (1s. per quarter on wheat) and that in the meantime the duties would be substantially reduced. After a vigorous Parliamentary debate, the Corn Bill received royal assent in June. The Anti-Corn Law League, having achieved its object, dissolved itself the following month.

The success of the campaign for abolition of the Corn Laws established firmly in Britain the Free Trade ideal, and for nearly a century helped to ensure the failure of any attempt to restore protection for agriculture. The outcome owed much to the efforts of the Anti-Corn Law League and its leaders, Cobden and Bright. Their basic theme of cheap bread was simple and easily understood: it appealed directly to the working classes and to humanitarians concerned at the distress of the urban population. Agricultural labourers too saw their interest in cheaper food, and the League even enlisted the support of tenant farmers (though with doubtful success) on the grounds that their rents were fixed on the basis of high prices, so that they suffered from the fluctuations caused by the Corn Laws.

Cobden and Bright, together with the majority of the League's leadership, belonged to the manufacturing class, which was interested in cheap bread in order to save wages. Essentially the repeal of the Corn Laws represented a victory of the manufacturers over the farmers and a concession by the landowning aristocracy, and reflected the shift in the balance of the population and of political power. The Reform Act of 1832 had by no means broken the power of the landed aristocracy, but through the extension of the franchise and the redistribution of Parliamentary seats it had added 217 000 voters, mostly in the towns, to the previous electorate of only 435 000.

Lord Ernle, in his standard work on English farming history, commented as follows:

> Down to the middle of the eighteenth century, the great preponderance of the nation had been interested in prices both as consumers and producers of corn. Now the proportions were completely altered and the majority had permanently shifted. The new manufacturing class was rapidly growing; the mass of open-field farmers had become agricultural labourers, whose real wages rose with the cheapness or fell with the dearness of bread. On the other hand, the interests of producers of corn were now represented by a comparatively small and dwindling class of landowners and farmers, who in recent years had enormously raised their own standard of living. Numerically small, but politically powerful, this class was convinced that the war-prices yielded only reasonable profits. The great majority of the population was convinced to the contrary. [Ernle (1962) page 272]

It should be added, however, that some of the large landowners accepted the repeal of the Corn Laws: some of them drew incomes from coal-mines, ground rents and government stock, so that the division of interests between town and country was in fact not quite so sharp. The repeal of the Corn Laws was the most controversial of a series of tariff reforms which made Britain a Free Trade nation. The process had begun in 1822, when Canning simplified the tariff and reduced some duties; it was pursued by Peel in 1842, 1845 and 1846 with the removal of many duties on manufactures as well as those on meat, cattle, potatoes, vegetables and other products; it was carried still further by Gladstone in 1853 with more removals and reductions of duty; and it culminated in the Anglo-French Treaty of Commerce of 1860 (the negotiations for which were conducted on the British side by Cobden). The reductions and abolitions of duty promised to France in this treaty were carried through by Gladstone as an independent tariff reform and extended to all other countries. After this, there remained only a few duties, mostly for revenue purposes: these were on sugar (abolished in 1874), manufactured tobacco, wines, spirits and hops (abolished in 1862), together with the small registration duties on grains and flour which were finally removed in 1869.

The Great Depression

For some thirty years after the repeal of the Corn Laws, British agriculture had little difficulty in surviving in its unprotected state. Though grain prices fell sharply from about 1848 to 1850, they recovered from 1853 onwards. The interruption of exports by the United States during the American Civil War (1861–4) gave a further stimulus to grain prices. British farmers benefited also from a series of good harvests. The cultivated area expanded, land values rose, and 'high farming' reached its peak. Seen in comparison with later years, this period was to appear as a 'Golden Age' for British agriculture.

Trouble began with a general economic depression in the late 1870s. In addition, from 1875 to 1879 a series of wet summers caused bad harvests: the crop of 1879 was one of the poorest ever recorded. Formerly, small harvests had generally been accompanied by higher prices, but now a new factor was introduced. Since 1873 grain imports from the United States had been rising (Table 2.2): in 1879 they increased to such an extent that prices fell, and British farmers suffered severe losses. At this period there was also a series of livestock epidemics: rinderpest in 1877, liver-rot in sheep in 1879, foot-and-mouth in 1883. This combination of disasters ruined many farmers and compelled landlords to remit or reduce rents.

Table 2.2: Imports of major agricultural products into the United Kingdom, annual averages

	Unit	1865–9	1870–74	1875–9	1880–84	1885–9	1890–94	1895–9	1900–04	1905–09
Wheat and wheat flour*	*million cwt.*	35	47	61	74	77	92	98	109	113
Barley		7	10	12	13	18	20	21	23	19
Oats		8	11	12	13	15	15	16	18	15
Maize		14	19	34	29	31	35	52	48	43
Beef†		0.2	0.3	0.6	1.0	1.0	2.0	3.0	4.2	5.6
Mutton†		..	..	..	..	.8	1.2	3.1	3.6	4.3
Pork†		0.2	0.3	0.4	0.4	0.4	0.4	0.7	0.9	0.7
Butter**		1.2	1.3	1.7	2.3	1.6***	2.2	3.1	3.9	4.2
Cheese		0.9	1.2	1.7	1.8	1.9	2.1	2.3	2.6	2.4
Eggs	*thousand millions*	0.4	0.5	0.8	0.9	1.1	1.2	1.2	2.4	2.2
Wool††	*million lb.*	147	184	201	221	284	311	383	350	397

* Wheat flour in grain equivalent.
† Both fresh and refrigerated, and, in the case of pork, salted.
** Including margarine till 1885.
†† Net imports.
*** Series not fully comparable before and after this date.
Source: Statistical Abstract for the United Kingdom.

In the 1880s, though imports continued to rise and prices to fall, farmers got some relief from better seasons. But a second crisis occurred when from 1891 to 1894 a succession of cold summers, drought and other difficulties caused bad harvests, and imports from the United States, after a period of relative stability, began to increase again. By the mid-1890s grain prices were the lowest on record: the price of wheat dropped to less than half the pre-depression level. The effects on arable farmers were all the more severe because the troubles of the previous years had eliminated their financial reserves.

Throughout the depression, however, livestock prices were relatively well maintained. Though imports were increasing steadily, helped by the new techniques of refrigeration, a decline in prices began only in the mid-1880s, some five years after the beginning of the crisis for grains. Subsequently, the fall in livestock prices was never so severe: in 1896 Sauerbeck's index (Table 2.3) reached its lowest point for both crop and livestock products, but while the former stood at 53 the latter was at 73 (1867–77 = 100). Over the whole period of depression, from 1880 to 1900, crop prices averaged 68% of the pre-depression level, livestock prices 86%.[1]

After 1897 the prices of most agricultural commodities ceased to fall; gradually they recovered, and British agriculture emerged from the depression.

Table 2.3: Price indices of foodstuffs on British markets: 1867–77 = 100

	1880–84	1885–9	1890–94	1895–9	1900–04	1905–09
Crop products	82	66	64	59	62	67
of which: Wheat	78	58	54	51	50	58
Barley	82	69	68	62	62	65
Oats	83	70	73	63	69	70
Maize	84	67	67	52	68	76
Livestock products	101	84	82	77	85	88
of which: Beef*	99	81	80	79	85	84
Mutton*	109	92	87	87	91	111
Pork	99	83	86	77	86	89
Bacon	100	87	87	74	82	88
Butter	98	83	83	77	82	88
All foodstuffs†	83	70	69	63	71	75

* 'Prime' quality: prices of 'middling' quality suffered a bigger price fall (up to 10–15% more at the worst of the depression).

† Including sugar, coffee and tea.

Source: Sauerbeck's Index, *Journal of the Royal Statistical Society*, various dates.

Contemporary views of the Great Depression

It is easier with hindsight to understand what was happening during this period: a later section in this chapter will describe the view of the Great Depression that emerged from later historical research. Those actually involved were in many cases confused as to the causes and nature of the problems and there was a tendency to overestimate the importance of the problems in the arable sector. A Royal Commission was appointed in 1879 to investigate the agricultural situation, as the result of pressure by 'Squire' Chaplin, who was the recognised spokesman for agriculture in the House of Commons; he was a firm believer in the crucial importance of corn. The leading members of the Commission belonged to the landowning aristocracy and gentry; it collected evidence mainly from large farmers and the arable counties of England. The Commission nevertheless con-

[1] Several of the series used by Sauerbeck relate to imported commodities. Price developments for home-produced commodities may not have been exactly the same. Also, the weighting of the index for all foodstuffs did not correspond to the pattern of British agricultural output (see Fletcher, 1961).

cluded that agricultural distress had prevailed over the whole country. It did note that the northern counties of England had suffered relatively little, but it failed to investigate the causes of this difference. It made little distinction between the trends of crop and livestock prices, and saw no implications in the low prices of animal fodder.

The Commission was equally undiscerning about the factors responsible for the depression. Bad seasons seemed to have been the main cause. It considered that foreign competition was also having serious effects, lowering prices even in years of poor harvests, and thought that this competition was likely to continue. Yet the Commission did not see in this a fundamental change in the pattern of world trade, necessitating a corresponding adaptation in British agriculture. It did not even consider protection, simply stating that the low prices caused by imports had to be accepted. As a piece of economic analysis, this report was dismally inadequate, and inevitably its recommendations bore little relevance to the basic causes of distress.

The ineptness of this Royal Commission may be contrasted with the insight of Sir James Caird, whose book *The Landed Interest and the Supply of Food* received its fourth edition in 1880, while the Commission was sitting. In this he stated that 'a great change in the agricultural position is impending' as a result of the competition of rich virgin soils in America and the cheapening of transport. This meant that British agriculture should turn to products which could not stand long transport or storage, such as milk, early wheat, vegetables, potatoes, sugarbeet and hay. Sir James saw, moreover, that the consumption pattern was shifting as the working class raised its living standards: bread was giving way to meat, vegetables and other high-quality produce. He concluded that:

> Our agriculture must adapt itself to the change, freely accepting the good it brings, and skilfully using the advantages which greater proximity to the best market must always command. [Caird (1880) page 175]

A second Royal Commission on agriculture, appointed in 1894, had a greater variety of members than the first, and though most of the witnesses it heard were again the larger farmers, there was more representation from the north and west of England. A conflict of opinion appeared in the Commission's second report in 1896 (a first progress report was issued in 1894): though all the members of the Commission recognised that the fall in prices had been the chief cause of depression, the majority – consisting of the landed gentry – confined itself thereafter to a discussion of rates, land tax and State loans to agriculture. However, a minority of three, including the chairman, declared that the depression had been far more serious in the eastern and southern counties of England than elsewhere in Britain, and attributed this difference to four factors:

(a) The east and south had suffered from particularly unfavourable seasons, starting in 1892;
(b) These regions had also been the most affected because they were the chief wheat-growing areas;
(c) They had comparatively few small farms, and small farms had done better

because they depended less on hired labour and because of their concentration on dairying;
(d) Burdens in the form of tithes, land tax and local rates were as a rule much heavier in the east and south, having been determined in relation to the former prosperity of these regions.

In the Commission's final report, in 1897, the conflict was largely resolved and a more careful analysis was made. The depression was attributed mainly to the fall in prices, and it was realised that this fall had been most severe for grain – particularly wheat – and for wool. It was thus not surprising that the worst trouble had been felt in the arable counties, while areas suitable for dairying, market gardening and poultry-raising had escaped comparatively lightly. The Commission's recommendations were limited to the relief of taxes, tithes, railway rates and so on, but it did not pretend that these would be a complete remedy. It recognised that foreign competition would cause a further reduction in the arable area, but did not propose to do anything about it, declaring:

> The grave situation we have described, affecting no inconsiderable part of Great Britain, is due to a long-continued fall in prices. This fall is attributed by the great majority of witnesses to foreign competition, and, as previously pointed out, we have not been able to find any promise, in the near future at all events, of a material relaxation of the pressure of this competition upon the British producer.
>
> So far, then, as the maintenance of this competition involves the continued depreciation of agricultural values, we must look forward to a further reduction of the area of British land susceptible of profitable arable cultivation, together with a corresponding contraction of our production and a diminution of our rural population. [Royal Commission (1897) page 159]

A purer expression of laissez-faire would be hard to find. A supplementary report by ten members of the Commission noted that many farmers would like a return to protection, but that 'several of those who are among the warmest advocates of protection told us they did not regard the adoption of that policy as within the pale of practical politics'.

Free Trade, Fair Trade, Empire and Tariff Reform

Following the report of the first Royal Commission, the government relieved farmers of part of the burden of local taxation, engaged upon a reform of land tenure and took action against animal diseases. In 1889 the Board of Agriculture was created. But otherwise the various governments in power during the Great Depression, Conservative as well as Liberal, adhered strictly to laissez-faire. The only measure of import restriction was the prohibition, in 1892, of imports of livestock from the Continent; this measure, originally applied for reasons of animal health, afterwards became permanent.

Throughout the period of the Great Depression, there was nevertheless much controversy on the tariff issue in general. A constant barrage of Free Trade propaganda was kept up by the Cobden Club, founded in 1866 to encourage 'the growth and diffusion of those economic and political principles with which Cobden's name is associated'. The Club held great sway in the 1860s and 1870s, and though its extreme Free Trade position gradually lost favour it continued to be influential for many years after. Its ideas were generally close to those of the Liberal party.

A reaction against unilateral Free Trade began with a slackening in economic activity in 1868. It was pointed out that even Adam Smith had thought that it might be good policy to use duties in retaliation against restrictions by other countries. This movement died down temporarily after 1870, as the economic situation improved, but it gained force later in the 1870s as a more serious depression began to be felt. 'Fair Trade' then became the motto for those discontented with current policy. The principle was to place home and foreign producers on an equal footing by eliminating 'artificial' differences (for example, export bounties were to be countered by equivalent import duties), but not to level out 'natural' differences in production. A National Fair Trade League was formed, and its policy included the following points:

(a) Commercial treaties to be made subject to revision at one year's notice;
(b) No duties on imported raw materials;
(c) Adequate duties on manufactured goods from countries refusing to admit British manufactures on an equitable basis;
(d) 'Very moderate' duties on foodstuffs, but not on the produce of Empire countries admitting British manufactures in 'reasonably free interchange'.

With the growth of imports of grain from the United States, the Fair Trade League provided a platform for advocates of protection for agriculture. Its influence grew after 1885 when the Liberal Cabinet fell and Lord Salisbury, who had already made speeches on behalf of Fair Trade, became the leader of the new government. One of his first actions was to appoint a Royal Commission to inquire into the causes of industrial depression. This Commission issued a report in 1886, in which a minority put forward Fair Trade arguments. Further, in 1887 the Congress of Conservative Associations declared itself in favour of Fair Trade. In the government, however, the Conservatives had joined forces with the Liberal Unionists, most of whom were convinced Free Traders. Lord Salisbury was therefore compelled to disown the Fair Trade movement, which ceased for a while to be an effective political force.

In the 1890s Fair Trade had a new lease of life, as the United States and various Continental countries reinforced their tariffs and a desire for retaliation was increasingly felt. However, the movement was hampered by the reluctance of any political party to take it up openly: there was no question of the Liberals doing so, and the Conservatives still owed their majority to the alliance with the Liberal Unionists.

Around the turn of the century, a stronger protectionist force arose. One cause of this was a lessening of confidence in Britain's command of the seas and of food supplies: in the Boer War of 1899–1902 Britain found herself without friends on the Continent, and Germany was starting to build a powerful navy.

Moreover, industry in both Germany and the United States was making rapid progress under protectionist policies, and one-sided free trade began to seem absurd to many people.

The main motive, however, was the question of the Empire, which brought a new dimension to the tariff issue. The first Colonial Conference had been held in London in 1887, and it marked the beginning of a move to closer union. The idea of a commercial union between the countries of the Empire, with preferential duties, gradually gained ground and was supported by the Fair Trade League. In 1902 the fourth Colonial Conference passed a resolution favouring the principle of Imperial Preference. This made a great impression on Joseph Chamberlain, who had been appointed to the Colonial Office in 1895 and was looking for a way of reinforcing imperial unity. In 1903, in a speech at Birmingham, he proclaimed his secession from Free Trade and his belief in Imperial Preference, as well as in the need for retaliation through tariffs against protection by other countries. Later Chamberlain made a similar statement in the House of Commons, but he received no support from the leaders of the government. Balfour, who had become Conservative Prime Minister in 1902, tried to avoid a split by putting forward a compromise proposal, to the effect that the Government should be given power to force down foreign tariffs by retaliatory duties, but without setting up a general tariff and without putting duties on foodstuffs. This however satisfied neither of the conflicting groups, and soon afterwards both Chamberlain and the Free Traders in the Cabinet resigned.

Chamberlain now unleashed a large-scale tariff campaign, backed by a newly-formed Tariff Reform League and a Tariff Commission. Though Chamberlain began the campaign primarily for the sake of unifying the Empire, the need to get a broad basis of support by appealing to industrial interests in Britain caused the emphasis to be put increasingly on the protectionist aspect. But so far as agriculture was concerned, the reluctance to propose high duties on foodstuffs and the desire to give preference to Empire produce made it impossible to formulate any effective policy of protection. The so-called Tariff Commission produced a report on agriculture in 1906, which stressed the dangers of Britain's dependence on imported food and tried to prove that the small size of British agriculture restricted the market for home industry. It proposed that there should be import duties on agricultural products, and that the revenue from these duties should be used to assist agriculture in various ways. But as preference was to be given to the Empire, which by then was an important competitor for British agriculture, it seems doubtful whether British agriculture would have benefited much from these proposals.

In any case, the tariff campaign failed entirely. The Liberal Opposition, reunited over this issue, made the most of the popular outcry against taxes on food. Balfour, on his side, managed to unite the Conservatives with another compromise formula more to the satisfaction of Chamberlain. But in the general election of 1906, which was fought almost entirely on the tariff issue, the Liberals gained an overall majority of eighty-four seats. Protectionism in general, and above all tariffs on foodstuffs, remained a political taboo until the 1930s.

Factors in the victory of Free Trade

In the last quarter of the nineteenth century, a large and influential sector of British agriculture was virtually ruined and little was done to help it. In other countries during this period, the farmers – particularly the large arable farmers – campaigned vigorously for protection, and in general their governments were persuaded to give them satisfaction. In Britain, nor only did the government make no important move, but among farmers themselves there was little protectionist agitation.

A vital factor in explaining this is, of course, the balance of power as between the urban and the rural population. With its early start in industrialisation, Britain had a particularly small part of its population in agriculture. In addition, the political strength of the landed classes had been undermined by the Reform Act of 1832 and thereafter steadily declined; meanwhile the power of the industrial population, to which 'cheap bread' remained a war-cry, had constantly increased. Public-opinion was still very much in the sway of the Free Trade ethic – an ethic based not so much upon economic reasoning as upon the aftermath of feeling resulting from the repeal of the Corn Laws. This point is well expressed by Clapham:

> There were plenty of elderly men of affairs in the 'eighties and 'nineties to whom corn law repeal was a vivid and blessed memory; many younger men for whom it had been canonised. So, just because agricultural depression first presented itself as a problem in wheat and wheaten bread, the political chance of its being handled by way of tariffs was almost infinitely small. [Clapham (1932) page 77]

There was, moreover, no chance of a protectionist alliance between farmers and industrialists: the latter attached far too much importance to cheap food and cheap raw materials. There was no strong movement for protection in British industry which, as yet, had relatively little fear of competition and, on the other hand, had a very clear interest in maintaining export markets.

The strategic arguments for maintaining a large agriculture, which played a certain role in Continental countries nervous about their food supplies, had little impact in Britain with its extensive food-producing Empire and its undisputed mastery of the seas. It was only after the Great Depression that the growing power of Germany began to give cause for worry on this score.

There were several reasons why British farmers themselves did not present a strong protectionist front. One was that they realised how small would be the prospects of success. Another was that they were remarkably slow to become aware of the real nature of the threat from imports: it has been seen above that the difficulties were at first attributed mainly to bad weather. Also, the landlord-tenant system made possible a sharing of the burden: for thousands of tenant farmers the first shock of falling prices was tempered by abatements in their rent. Many landlords, as has already been pointed out, were not dependent on income from their land alone.

The prevalence of the landlord-tenant system made it more difficult to form a united agricultural front. So did the division of interests between crop and livestock producers, which has already been discussed. While in other countries the large arable farmers were powerful enough to draw smaller producers into their movement, or simply to ignore their existence, in England the small livestock farmers of the north and west seem to have been too independent and too well aware of their real interests to fall into such a trap.

As a result, no effective agricultural pressure group was active during this period. The main agricultural organisation was the Central Chamber of Agriculture, an association of numerous local Chambers of Agriculture, Farmers' Clubs and Farmers' Associations; it was founded in 1865 and led mainly by large landowners. It did not take the tariff issue seriously until 1892, when it convened a national Agricultural Conference, at which the split between the arable farmers and the livestock producers was evident. A Farmers' Alliance, started in 1879 to represent the interests of tenant farmers, also split on the question of protection and gradually died out. A National Agricultural Union was formed in 1893 in an attempt to combine landlords, farmers and workers in a single body; but its work overlapped that of the Central Chamber and it exerted little influence. The present National Farmers' Union was not formed until 1908.[1]

The effects of the depression

Calculations made by Bellerby (Table 2.4) indicated that at the trough of the depression, in the mid-1890s, net farm income was down by some 30% as compared with the pre-depression period. Rents and farmers' profits had borne the brunt of the crisis.

The population occupied in agriculture fell steadily in the latter half of the nineteenth century (Table 2.5); with the continuing rise of employment in other sectors, the active population in agriculture was reduced to a mere 8% of the total.

Table 2.4: Net farm income in the United Kingdom, and its distribution

	Net farm income	Net rent	Wages	Interest	Farmers' 'incentive income'
		£ million, annual averages			
1867–78	150	36	54	20	40
1879–83	122	33	49	17	24
1884–91	115	28	46	15	27
1892–96	108	23	43	14	27
1897–1905	115	22	43	14	36
1906–14	135	26	46	16	47
1892–96 as % of 1867–78	72	64	80	70	67

Source: Bellerby, J.R. (1956) *Agriculture and Industry.* London: Macmillan.

[1] The Central Chamber of Agriculture continued to exist even after the NFU had taken the lead and was finally wound up only in 1959.

Table 2.5: Population occupied in agriculture in Great Britain, in relation to total occupied population and to total population

	*Occupied population**			Total population of Great Britain
	Agriculture	All sectors	Agriculture as % of total	
	thousands			*thousands*
1861	1 913	10 463	18	23 128
1871	1 690	11 646	15	26 072
1881	1 500	12 795	12	29 710
1891	1 402	14 676	10	33 029
1901	1 325	16 312	8	37 000
1911	1 381	18 351	8	40 831

* Including persons temporarily unemployed.

Sources: Population occupied in agriculture from Taylor, F.D.W. (1955) 'Numbers in agriculture', *Farm Economist,* **VIII** (4). Total occupied population from Bellerby, J.R. (1958) 'The distribution of manpower in agriculture and industry', *Farm Economist,* **IX** (1). Total population from the *Statistical Abstracts.*

A striking change occurred in the pattern of agriculture: much arable land was returned to grass and the wheat acreage in particular was reduced by a half between 1870 and 1900 (Table 2.6). On the other hand there were more cattle and pigs; the number of sheep remained roughly the same.

Later historians contended that previous views of the Depression, including that given by Lord Ernle, concentrated too much on the problems of the arable sector and on the south-east of England, thereby in fact exaggerating the gravity of the crisis. Investigations by Fletcher, particularly into conditions in Lancashire, led him to observe that while the arable farmers, mainly in the east and south of England, suffered a steep fall in the price of their products, livestock farmers in the north and west not only experienced a relatively small decline in prices but benefited from the much greater fall in the cost of animal feedingstuffs. His calculations of gross agricultural output, reproduced in Table 2.7, showed that while the output of the arable sector fell in value (in current prices) from £104 million in 1867–9 to £62 million in 1894–1903, the output of livestock products rose from £127 million to £146 million.

Fletcher's work was pursued by others and a reasonably complete picture of events during the Depression emerged (see Perry, 1973 and 1974). In general, lowland arable farmers were more prone to failure than upland pastoralists. Areas that could supply the urban milk market, which was unaffected by competition from imports, had favourable opportunities. Horticulture could also expand in areas where natural conditions permitted and markets were close.

Perhaps more farmers should have taken advantage of such possibilities for adjustment. But for several years, few farmers realised that a permanent change had occurred: as has been observed in an earlier section, the causes of the Depression were not at first recognised. In many arable areas, moreover, conventional crop rotations limited the scope for adjustment; landlords were at first reluctant to encourage major changes, and a tenant who laid down arable land to grass risked obtaining no compensation at the end of his lease for the costs

Table 2.6: Crop area and livestock numbers in Great Britain

	1870	1880	1890	1900	1910
			million acres		
Area					
Wheat	3.5	2.9	2.4	1.8	1.8
Barley and oats	5.1	5.3	5.0	5.0	4.8
Green crops and roots	3.6	3.5	3.3	3.2	3.0
Other crops and fallow	1.6	1.6	1.2	0.9	1.0
Temporary grass	4.5	4.4	4.8	4.8	4.2
Total arable land	18.3	17.7	16.8	15.7	14.7
Permanent pasture	12.1	14.4	16.0	16.7	17.5
Total cultivated area	30.4	32.1	32.8	32.4	32.1
Livestock			*millions*		
Cattle	5.4	5.9	6.5	6.8	7.0
Pigs	2.2	2.0	2.7	2.4	2.4
Sheep	28.4	26.6	27.3	26.6	27.1

Source: Statistical Abstract for the United Kingdom.

incurred.

So farmers adjusted as best they could, cutting down on labour, reducing costs wherever possible, minimising maintenance and indefinitely postponing any further improvements. The proud period of high farming had come to a sad end. Some farmers went bankrupt: in East Anglia in the early 1880s, one farmer in 250 failed each year. Failures were at first comparatively rare in north-east England and Wales, but in the 1890s they were more widespread. Not a few of the farmers who failed in the previously rich counties, however, were replaced by men from Scotland, Wales and the West Country, used to a lower standard of living.

Landlords in arable regions were forced to grant rent reductions or see their tenants go out of business. In consequence they too cut down on expenditure for the improvement of real estate or 'landlords' capital'. This was particularly apparent in drainage work. An aspect of the Corn Law repeal which has often escaped attention was that Sir Robert Peel then announced a scheme whereby, in compensation for the loss of protection, landowners who wished to improve their property by drainage could obtain government loans at 3½% over 22 years. This scheme, administered by the Enclosure Commissioners, was an immediate success. Besides the government funds provided, public corporations were set up to raise money through the issue of stocks and relend it to landowners, not only for drainage but also for enclosure, fencing and buildings. Large amounts were spent for these purposes until the onset of the depression: improvements of all kinds then dwindled (see table in Orwin and Whetham, 1964, page 196).

The decline of the landed interest must be regarded, socially and politically, as a crucial feature of this period. For reasons that have already been discussed, the

Table 2.7: Gross agricultural output

	Total United Kingdom			England	
	1867–9	1870–76	1894–1903	1867–71	1894–8
			£ million		
Wheat	35	28	8	28	8
Barley	17	18	9	13	8
Oats	11	9	8	4	4
Potatoes	14	14	11	3	3
Hay, straw, fruit and vegetables	20	19	22	14	17
Other crops	7	8	4	3	2
Total crops	104	95	62	65	41
Beef	35	46	42	15	16
Mutton	26	31	25	15	13
Pigmeat	19	23	19	10	11
Horses	2	2	3	1	2
Milk	34	39	44	15	20
Wool	7	8	3	6	2
Poultry and eggs	5	7	10	4	7
Total livestock	127	155	146	64	71
Total agricultural output	231	250	208	130	112

Source: Fletcher (1961).

ancient landowning families did not use their remaining powers to protect their immediate farming interests. Besides the erosion of their farming revenue through the depression, a series of Agricultural Holding Acts prescribed proper compensation for tenants' improvements and limited the rights of landlords. A permanent blow to the landed interest was dealt by the introduction in 1896 of death duties on agricultural property: as the inheritors of the great estates found it necessary to sell off part of the property to pay the duties (the purchaser in many cases being the sitting tenant), this initiated the break-up of the estates and the steady shift of land to owner-occupancy which has continued at each generation.

In Ireland, Gladstone's efforts from 1868 onwards to give some justice to the oppressed peasantry were followed in 1903 by Balfour's Land Purchase Act whereby the British Government made available loans on easy terms to assist tenants to buy their holdings. By then most landlords were weary of the struggle and were willing to part with the land. Ireland thus became predominantly a country of small peasant proprietors. In Scotland, crofters[1] were given the status of hereditary tenants by an Act of 1886 and a Crofters' Commission was set up to fix rents; an Act of 1897 gave powers to compel landowners in crofting areas to surrender land for the creation of new crofts or the enlargement of existing ones.

In several other countries the growth of agricultural co-operatives was a major feature of the late nineteenth century. This was not nearly so significant in

[1] In other words, smallholders who generally combined farming with other activities such as fishing, forestry and crafts.

England. Some co-operative cheese and butter factories were set up in the Midlands and a few groups for bulk purchase of agricultural requisites were initiated. There was a period of co-operative activity after the turn of the century. Not all of these ventures were successful.

British agriculture thus emerged from the Depression reduced in size and in ambition. The reduction in numbers employed no doubt proved advantageous later on, when there were opportunities for mechanised large-scale farming. But the transition had been a painful one, and has been well summarised by Orwin and Whetham:

> The immediate effects of a long period of falling prices are almost uniformly disastrous for those whose profits take the first impact, in this case the arable farmers of Britain. Whatever benefits eventually emerged from the great agricultural depression, there remained the personal tragedies of thousands of respected and competent families, who saw their incomes diminish, their capital evaporate and themselves plunged from secure social positions into bankruptcy and destitution. From them, the trail of poverty afflicted the land, which was often left exhausted and weed-infested; their landowners, whose finances were often undermined by falling rents and rising costs of maintenance; and their workers, some of whom were forced, briefly or for longer periods, into the workhouse, as they were thrown out of employment. In these depressed arable districts, farmers, landowners and workers alike suffered, in the years after 1875, from economic forces which they could but imperfectly understand, but which deprived them of those benefits of economic progress enjoyed by other sections of the nation. [Orwin and Whetham (1964) pages 287–8]

Bibliography

Medieval patterns of settlement and their subsequent transformations are clearly set out by Higgs (1964) and Hoskins (1970). The most comprehensive history remains that of Lord Ernle, which has gone through several editions since it first appeared in 1912. More recent works supplement it and correct its emphasis in certain respects. Chambers and Mingay (1966) provide a clear account of the period of 'high farming'. The Great Depression itself has been the object of a number of studies since Fletcher's important article in 1961; Perry (1973) has provided a useful collection of studies on the subject, and has also investigated it in his own book (1974).

Ernle dealt with England: a particularly valuable contribution has been made by Orwin and Whetham (1964) as their work also takes into account Wales and Scotland; it is also a particularly vivid account of social conditions in the various classes of rural society.

Among writings contemporary with the Great Depression, the reports of the two Royal Commissions are significant. Most writing in Britain at the time naturally tended to reflect the authors' particular views and prejudices. The best early study was made by a German (Fuchs, 1905).

Armitage-Smith, G. (1903) *The Free-Trade Movement and Its Results.* 2nd edn. London.

Ashworth, W. (1960) *An Economic History of England, 1870–1939.* London: Methuen.

Caird, Sir James (1880) *The Landed Interest and the Supply of Food.* 4th edn. London.

Chambers, J. and Mingay, G. (1966) *The Agricultural Revolution 1750–1880.* London: Batsford.

Clapham, J.H. (1932) *An Economic History of Modern Britain,* Volumes II, III: Cambridge University Press.

Clark, G.K. (1951) 'The repeal of the Corn Laws and the politics of the forties'. *Economic History Review* (1).

Ensor, R.C.K. (1936) *England, 1870–1914* (Oxford History of England). Oxford.

Ernle, Lord (formerly R.E. Prothero) (1962) *English Farming, Past and Present.* 6th edn. with introduction to Part II by O.R. McGregor. London: Heinemann.

Fay, C.R. (1932) *The Corn Laws and Social England.* Cambridge.

Fletcher, T.W. (April, 1961) 'The Great Depression of English Agriculture, 1873–1896'. *Economic History Review.*

Fuchs, C.J. (1905) *The Trade Policy of Great Britain and Her Colonies since 1860.* London. (First published in German, Leipzig, 1893).

Green, J.L. (1911) *Agriculture and Tariff Reform.* 2nd edn. London.

Higgs, J. (1964) *The Land.* London: Vista.

Hoskins, W.G. (1970) *The Making of the English Landscape.* Pelican.

Leadam, J.S. (1887) *What Protection Does for the Farmer and Labourer.* 3rd edn. London: Cobden Club.

Matthews, A.H.H. (1915) *Fifty Years of Agricultural Politics, Being the History of the Central Chamber of Agriculture, 1865–1915.* London.
Mongredien, A. (1880) *History of the Free-Trade Movement in England.* London.
Nicholson, J.S. (1904) *The History of the English Corn Laws.* London.
Orwin, C.S. (1949) *A History of English Farming.* London: Nelson.
Orwin, C. and Whetham, E. (1964) *History of British Agriculture 1846–1914.* London: Longmans.
Perry, P.J. *ed.* (1973) *British Agriculture 1875–1914.* London: Methuen.
Perry, P.J. (1974) *British Farming in the Great Depression – An Historical Geography.* Newton Abbott: David & Charles.
Rea, R. (August, 1908) Untitled paper in *Report of the Proceedings of the International Free Trade Congress.* London.
Robertson, J.M. (1928) *The Political Economy of Free Trade.* London.
Royal Commission on Agriculture (appointed 1879). Reports of 1881 (C. 2778) and 1882 (C. 3309).
Royal Commission on Agriculture (appointed 1893). Reports of 1896 (C. 7981) and 1897 (C. 8540).
Sheldon, J.P. (1893) *The Future of British Agriculture.* London.
Stearns, R.P. (April and June, 1932) 'Agricultural adaptation in England, 1875–1900'. *Agricultural History.*
Stenton, D. (1962) *English Society in the Early Middle Ages.* 3rd edn. Penguin.
Tariff Commission (1906 and 1914) *Reports of the Agricultural Committee.* London.
Tremayne, H. (1903) *Protection and the Farmer.* London.
Woodward, E.L. (1938) *The Age of Reform, 1815–1870* (Oxford History of England). Oxford.

Chapter 3

France

The French Revolution, while it swept away feudal privilege and established the peasantry in full ownership of their holdings, did not significantly alter the farm structure. In the mid-nineteenth century, French agriculture included some quite large farms, but also very many small and fragmented ones. Farming had been a despised occupation under the *Ancien Régime*, and little technological improvement had taken place: productivity remained low and the greater part of the population still worked on the land.

Seigniorial privileges, land tenure and the Revolution

The manorial system as described in Chapter 1 originally characterised northern and north-eastern France. Under Charlemagne, the *villa* – whose ancestry can be traced back to Gallo Roman times – was incorporated in the Frankish feudal system, the *beneficium* granted to vassals consisting frequently of a grant of land. The typical Carolingian *villa* on the arable northern plains, with its village settlement, open fields and strips, established a pattern which remained virtually unchanged over the following centuries.

The feudal *seigneurie* was characterised also by the numerous servitudes which have been described briefly in Chapter 1. In return for their *tenure* of land, peasants were required to work on the seigneur's reserve and to pay him *cens* ('rent' would not be the correct term at this stage) in cash or more usually *champart* in kind. Tithe (*la dîme*) was due originally to the Church, but often the right to tithe was acquired by the landlord, whether lay or ecclesiastic. The *taille* was an irregular but heavy tax. Over the centuries, numerous other charges came to be added to the peasants' burdens: charges (*banalités*) for the compulsory use of the seigneur's mill, baking oven, brewery, wine-press, etc. were among the most onerous. Hunting and fishing rights were another much-resented seigniorial privilege.

Serfdom was never very common in France: it disappeared early in Ile-de-France, Normandy and Brittany, and there were only a few areas – the Nivernais, northern Bourgogne, Champagne, Lorraine – where hereditary serfs formed the majority of peasants. The distinction between serfs and other peasants, however, tended to become obscured: thus serfs could hold land and have other possessions, while ordinary *vilains* were also subject to restrictions on their personal freedom, except where village charters eventually defined the respective rights and obligations of seigneurs and peasants and permitted the latter to leave the *seigneurie*.

It is, however, difficult to generalise about France. Though most of the territory was brought under the control of feudal landlords, and though the ruling class tried to maintain the principle *'nulle terre sans seigneur'*, there were always some free peasants, particularly in the south-west and centre of the country. Moreover, as the population increased from about the year 1000, new land was brought into cultivation, sometimes on the initiative of seigneurs who founded new villages with relatively attractive statutes to attract settlers, sometimes on the individual initiative of peasants who created independent holdings with enclosed land. The Cistercians, establishing their monasteries in remote areas and often on infertile land, rejected tithes and other seigniorial dues, and ran their *granges* through the work of lay-brothers (*conversi*).

Moreover, the pattern of two or three open fields characterised mainly the arable plains of the *Bassin Parisien* and Alsace-Lorraine. In areas where pasture was plentiful and livestock farming predominated, as in Britanny, the Cotentin peninsula, the *Massif Central* or the Basque country, different patterns prevailed with fields often rectangular, enclosed by hedges or dykes. In place of village settlements, tiny hamlets might be found, consisting of a few farmhouses grouped together. On poor land in Brittany and the *Massif Central,* a system of 'infield' and 'outfield' was sometimes found, similar to that existing in parts of Britain. The south of France, always subject to different influences, had its own patterns adapted to the terrain and never as regular as in the north.

Tenancy (*fermage*) and share-cropping (*mētayage*) originated from about the twelfth century, as many landlords found it more profitable to have their land farmed in this way than through the traditional method of compulsory labour services on the reserves. In the north, the new class of *fermiers* prospered, working farms of adequate size on good land; recourse to hired labour also developed, giving employment to peasants with little or no land. In the south, however, share-cropping prevailed and the *mētayer* was typically a small and poor peasant, still bound to his landlord by semi-feudal ties.

The *seigneurie* nevertheless persisted as the dominant form of agrarian organisation, even after the disruptions caused in the fourteenth and fifteenth centuries by plague and warfare. The recovery of population from the sixteenth century meant increasing pressure on the land and increasing fragmentation of plots. Unlike England, France did not have the advantage of early industrial development to relieve rural population pressure. State taxes were added to seigniorial charges. In the reign of the *Roi Soleil,* the peasant world was more than ever scorned for its backwardness and ignorance. Noble landlords stayed at the Court, their interest in their land limited to the rents and hunting rights it provided.

In the eighteenth century, as new crops and new technology became available, some reform-minded landowners (influenced by the Physiocratic School) became well aware of the disadvantages of the open-field and strip system. The English enclosures and English farming techniques were looked on with envy by such people. There were numerous treatises on agricultural techniques: one would-be reformer, Duhamel du Monceau, put the problem thus:

> It has often been justly said that the establishment of temporary pastures is the surest way to increase the product of the land, but how can such pastures be obtained in countries where land is so divided up among the inhabitants that most plots are only a few perches in length? – in such cases more time is spent turning the plough round or moving it from one of these tiny fields to another than in the ploughing itself. This is a considerable inconvenience; but there is an even greater one, which is that all the owners are obliged to follow the same method of cultivation. If for good reason, a peasant wishes to sow

> one of these little fields to grain or to temporary pasture, while the neighbouring fields are open to livestock, his little portion of land will inevitably be overrun.... The English, who have recognised these inconveniences, have not hesitated to authorise forced exchanges.... This procedure, violent as it may seem, has succeeded quite well in England, but it seems to me that it would be impracticable in certain provinces of France ... where all the houses of a parish are grouped together: the land around these houses becomes so precious that one would do a great wrong in depriving anyone of it, however much land might be given in compensation further away. [Duhamel du Monceau (1762) pages 375–6]

Du Monceau also inveighed against the practice of allowing all the animals of the village to graze the open fields after harvest (*vaine-pâture*), which made it impossible for any farmer to introduce green fodder crops in place of fallowing, or to develop the new crops – turnips, potatoes, maize, sarrasin – then being advocated, with support from the authorities through the *intendants.*

But though some landowners, particularly in Normandy, did manage to consolidate their land, by purchase or by exchange with peasants, enclosure and technical innovation did not go far in France. Only a minority of landlords were interested and practically all peasants remained suspicious of change. Peasants daring enough to put up a fence around their property to keep off animals were liable to find it torn down overnight.

Conditions for the peasantry grew worse in the second half of the eighteenth century. Many landlords, including the Church, being anxious to take advantage of an expanding market, extended the area in their own use, taking over peasants' strips and common land; moreover, to keep pace with rising prices, they increased their exactions from the peasantry. There was growing abuse of seigniorial privilege in respect of hunting rights and so on.

Turgot, Louis XVI's *Contrôleur des Finances* and a well-known Physiocrat, had plans for developing France's agricultural potential which apparently included alleviating the burden of taxes and dues borne by the peasantry in order to stimulate new methods and higher productivity. A pamphlet written in 1776 by Boncerf, one of Turgot's officials in the Ministry of Finance, advocated cash redemption of seigniorial dues and services but the *parlement* of Paris ordered it to be suppressed, to the indignation of Voltaire among others. Support for reform was limited to a small band of intellectuals; Turgot himself was dismissed (for different reasons) in 1776 before he could put any of his plans into effect.

Peasant unrest was endemic. There was rioting in the countryside after a bad harvest in 1788, even before the Paris crowds stormed the Bastille on 14 July 1789. Subsequently, in the phenomenon known as *La Grande Peur,* armed bands of peasants attacked manor houses, sacking or burning them, and burning the registers in which peasants' charges and obligations were listed. The Assembly, after heated debate on the night of 4 August 1789, declared an end to feudal privilege. It tried initially to distinguish between rights which had been usurped or established unlawfully: labour services, tithes, *banalités,* hunting and fishing rights, what remained of serfdom and seigniorial jurisdiction were abolished, but payments related to the holding or transfer of land were declared lawful and had to be redeemed at quite high rates of compensation. The peasantry refused to pay any compensation at all and finally a law of 10 August 1792 declared that all charges were abolished without compensation unless the seigneur could produce evidence of his entitlement (which was usually difficult); further legislation on 17 August 1793 abolished all remaining seigniorial rights.

Peasants who previously had the use of land under a seigneur now found themselves owners in full. The Assembly moreover, in line with the general revolutionary principle of individual liberty, declared the right of every farmer to cultivate his land as he wished and to fence it off; *vaine-pâture* was in principle circumscribed; temporary pastures were to be respected. This facilitated the necessary breakaway from the constraints of co-ordinated cultivation. Still, however, not many French peasants fenced off their plots, which remained, as before, small and scattered, and the landscape of the plain areas retained its open-field aspect.

How much extra land did the peasantry acquire as a result of the Revolution? Church lands were declared *biens nationaux.* The initial purpose however was to gain revenue for the State and therefore the land was sold in large parcels to buyers with ready cash – *bourgeois* or wealthy farmers. From 1792 a new category of *biens nationaux* became available from land belonging to nobles who had fled the country. Much of this land was put up for sale in relatively small lots, but the terms of sale still made it difficult for small peasants to acquire it. At one stage a decree provided that a small piece of land (*un arpent,* about a quarter of a hectare) from *émigré's* estates should be given to peasants holding less than this amount, but this measure was scarcely implemented. Indeed the peasantry itself was divided: those who already had an adequate amount of land were reasonably satisfied once they had obtained the suppression of seigniorial privileges; those with little or no land demanded a share of the *biens nationaux,* but with limited success. Another controversial issue concerned common land, the ownership of which had long been contested between seigneurs and the *communes;* the revolutionary assemblies favoured the latter, but while well-to-do peasants wanted common land divided up, poor peasants naturally wanted to maintain their pasture and other rights. A decree of 10 June 1793 in fact authorised the division of common land, subject to approval by a third of the inhabitants of the *commune.*

The Revolution was thus by no means egalitarian in the countryside. Unfortunately there are no reliable figures for the distribution of land use or ownership in the country as a whole either before or after the Revolution. Putting together data from various sources, Blum (1978, bibliography to chapter 1) has concluded that before the Revolution the nobles held about 20% of the cultivated land in their own reserves, the Church about 10%, the *bourgeoisie* about 30%, while about 35% was farmed by peasants holding the land from a seigneur; there were however wide variations between regions. The most detailed study was made for the *département* of the Nord by Lefebvre (1924). He found that of the Church lands put up for sale about 17 600 peasant buyers obtained slightly more than half while about 5500 *bourgeois* buyers obtained slightly less than half; peasants had gained about 12% of the area of the *département* and the *bourgeoisie* likewise, while Church property had virtually disappeared and the nobility had lost at least 8% of the area. Lefebvre found that some previously landless peasants had become owners of one or two hectares; other had added to the size of their holdings. Considering however that five hectares at least were needed to support a family, it was clear that many peasant families, perhaps the majority, were still below the level of viability.

Napoleon used some of the undistributed *biens nationaux* to endow his new Imperial nobility; Louis XVIII in 1814 recognised these titles and bequests, and subsequently the restored Bourbon monarchy, while respecting the rights of those who had purchased Church and noble lands, helped their followers returned from exile to recover their property wherever possible. In some cases their estates had been bought on their behalf by agents or relatives and kept in expectation of their return; in others the land had not yet been sold; and in other cases again the new owners could be persuaded to sell the land back at a relatively low price. The general outcome is well summarised by Bloch:

> The great crisis which broke in 1789 had not meant the destruction of the large landed property as reconstituted in the immediately preceding centuries. Such noble and bourgeois proprietors as did not emigrate – far more numerous, even among the nobility, than is sometimes imagined – hung on to their possessions. Some émigrés were able to arrange for their properties to be bought back by relations, or had them restored by the Consulate or Empire... Even the sale of national property – lands belonging to the Church and to émigrés – did not inflict a really damaging blow. The procedure followed made it possible for the land to be sold in blocks, or even as complete estates; big tenant farmers became big proprietors, the bourgeoisie continued the patient and fruitful work put into the land by their predecessors, prosperous *laboureurs* added to their patrimony and made their decisive entry into the ranks of rural capitalists. [Bloch (1966) pages 242–3]

The point concerning the role of the *bourgeoisie* is significant in terms of later political developments. Lefebvre pointed out:

> [The *bourgeoisie*] had manoeuvred with great skill to rally to the new order very many peasants. By letting them acquire most of the land that became available, and by provoking in this way the splitting-up of many farms, it had allowed them to get satisfaction individually: through this personal interest it had separated them from the rural proletariat and had broken the collective action which in 1789 had seemed possible.... The *bourgeoisie* moreover, allied to what remained of the former nobility, now constituted a landed aristocracy which was powerful enough to hold in check, through its economic dictatorship, this rural democracy which it had in part created but which, in order to live, still had to ask for work or for land to rent. Its political power was safe for a long time to come. [Lefebvre (1924) pages 546–7]

The state of agriculture in the mid-nineteenth century

Thus at the Revolution seigniorial privileges had been abolished and some extra land had been acquired by the peasantry; but the pattern of plots still reflected the old open-field and strip system and the limited area owned by individual peasants was almost always divided between numerous tiny, scattered plots. Further sub-division between heirs at each generation was now given legal force by the *Code Napoléon*. The rural population, moreover, was growing until the middle of the nineteenth century, causing the number of farms to increase to a peak in the 1880s and leading to more and more fragmentation. In 1882, 5.7 million holdings were counted: even excluding from consideration 2.2 million with less than one hectare which may not have been genuine farm businesses, there were 2.6 million holdings with between one and ten hectares (Table 3.1). Moreover, the 5.7 million holdings included no less than 125 million separate plots, making an average of twenty-two per holding; the average area of each plot was a mere 0.39 hectares. At the other end of the scale, there was a minority of large holdings of forty hectares and upwards; because of their large average size, this group accounted for nearly half the agricultural area.

On the numerous small holdings, progress was difficult. Structural conditions were not the only obstacle: the peasantry was largely uneducated (indeed all too often illiterate) and lacked the capital resources necessary for improvements. In cases where holdings were let to tenants, leases were often short – rarely for more than ten years – and the tenant had no guarantee of receiving compensation for improvements he made. The backwardness and poverty of the French peasantry at this time are well described in the following passage from Augé-Laribé's outstanding history of French agricultural policy:

> At this time [around 1880], newspapers and public speakers often referred to the 'remote rural areas'. It was an expression which conjured up terrae incognitae, forgotten, godforsaken villages where the most basic necessities were lacking and where it was not even certain that the French language would be understood. In fact many country districts were very far from any town and indeed from modern civilisation. Though the towns were linked by excellent roads and railways, it was a much more difficult matter to reach the villages and penetrate to all the farms. [Augé-Laribé (1950) page 55]

Weber (1977) provided a vivid description of large parts of the French countryside in the latter half of the nineteenth century: mostly subsistence farming using traditional methods, with low productivity, little contact with the market or the outside world in general; low educational level and widespread illiteracy; primitive living conditions.

French agriculture was indeed largely untouched by the other major influence which had affected farming in Britain – the Industrial Revolution. France's Industrial Revolution began much later than that of Britain; by 1861 the rural population (persons in villages of less than 2000 inhabitants) still amounted to 26.6 million out of a total of 37.4 million, and 19.9 million of the rural population depended on agriculture for a living. There was as yet little opportunity for

Table 3.1: Agricultural holdings in 1882, by size

Size *ha*	Number *thousands*	Area covered *thousand ha*	Average area per holding *ha*
0–1	2168	1.1	0.5
1–5	1866	5.6	3.0
5–10	769	5.8	7.5
10–20	431	6.5	15.0
20–30	198	5.0	25.0
30–40	98	3.4	35.0
over 40	142	22.3	156.7
Total	5672	49.6	8.7

Source: Statistique Agricole de la France – Enquête Décennale de 1882.

impoverished peasants to better themselves by seeking work in the towns, and those who did leave agriculture were impelled by their poverty rather than attracted by any firm prospect of improvement. Moreover, there was relatively little technological impulse to pass from other sectors to agriculture, even if the state of communications and the educational level of the peasantry had been such as to facilitate this process. In another passage Augé-Laribé commented on the unprogressiveness of French agriculture:

> The agricultural population, almost in its entirety and in practically every region, did not want to make progress. Its aim was self-preservation: it wanted to maintain itself and its environment unchanged. It did not realise that not to advance is to retreat, that what remains stagnant and does not adapt, in a world in evolution, risks its own extinction. [Augé-Laribé (1950) page 58]

The more recent history of French agriculture edited by Duby and Wallon stressed the failure of the rural population to take advantage of profitable prices in order to invest:

> Agricultural prosperity from 1850 to 1870 was finally an illusion. Techniques were not sufficiently modernised, for the available capital arising from the use of the land was not reinvested to improve it. Those who had capital used it for other purposes: buying shares on the stock market, or buying land. Those who did not have capital could get it only at rates approaching usury, rarely less than 8% and often 12% or more, because there were no adequate credit facilities. [Duby et Wallon (1975–6) *Histoire de la France rurale* vol. 3, page 253 © 1976 Editions du Seuil]

But a time was reached when drastic changes forced themselves on French agriculture. Augé-Laribé continued:

> But now economic progress was speeding up. Steamships had reduced the cost of ocean transport. Railways were ready to take from the ports the grain which had crossed the seas. All the continents were linked by trade. The railway companies gave preferential rates to foreign goods. Techniques of cultivation, which were not new but had remained unknown or neglected, reached a point where no one could afford not to adopt them. Chemical fertilisers forced

themselves on the attention of even the most backward farmers. Highly efficient commercial organisations spanned the world. The development of other sectors and the competition of agriculture in foreign lands compelled the French peasants to join a race to which they were not accustomed. They complained about all this agitation. They were short of breath. It was the crisis. Protectionist doctors, anxious to please the patient, prescribed rest, caution, avoidance of draughts; no healing medicines, but soothing drugs. When finally the farmers tried to recover, to exert the strength they gained through their association, it was far too late. Their competitors had outdistanced them. [Augé-Laribé (1950) page 59]

Free trade – the 1860 Anglo-French treaty

Throughout most of French history before 1860, protectionist tendencies dominated. Under Napoleon's 'Continental System' British goods were prohibited or subject to high duties. French industry thus grew accustomed to high protection, and vested interests were formed which impeded subsequent attempts at tariff reform. Under the Restoration monarchy of 1814–30, manufactured goods from any origin were liable to be prohibited. Also, for the first time in France, protection was extended to agriculture. The big landowners made common cause with the manufacturers and demanded high import duties, together with freedom to export; the Bourbon monarchy wanted to restore a rural aristocracy and was ready to satisfy these demands. In 1819 a sliding scale of duties on grain, similar to the British Corn Laws, was introduced.

Under the July monarchy of 1830–48, a Free Trade Association was formed by Frédéric Bastiat and Michel Chevalier, along lines similar to Cobden's Anti-Corn Law League. But this movement gathered little public support. The Parliament of the Second Republic after 1848 was still opposed to tariff reforms; in 1850, by a large majority, it rejected proposals to remove protective duties on food and raw materials. The duties on wheat were maintained in principle, though they were suspended after bad harvests on several occasions.

With the Second Empire in 1852, the Free Trade movement found a new champion in Napoleon III, who was impressed by Peel's reforms in Britain. In 1856 he presented to Parliament a further proposal for freeing trade, but this too was rejected; indeed, the government had to promise not to revive the tariff issue for another five years at least.

Napoleon III and the Free Traders were thus forced to look for other ways to carry out their purposes. As the Emperor had the right to conclude treaties with other nations without consulting Parliament, the device of a commercial treaty with Britain provided a solution. In addition, Napoleon was anxious to establish closer relations with Britain and to reduce France's political isolation. The Anglo-French Treaty of Commerce of 1860 was therefore negotiated and signed in secret, and presented to Parliament as a *fait accompli.*

The concessions given by Britain to France in this treaty included reductions in the duties on wines and spirits. Much greater changes were made by France: the tariffs on manufactured goods were reduced to moderate levels, and the duties on nearly all foodstuffs and raw materials were removed. Abolition of the sliding scale of duties on grain followed in 1861.

Between 1861 and 1867, France concluded ten other commercial treaties: with Belgium, the *Zollverein,* Italy, Switzerland, Sweden and Norway, the Hanseatic League, the Netherlands, Spain, Portugal and Austria. Each of these treaties included the 'most favoured nation' clause, so that the relatively low duties resulting from the treaties were extended to the greater part of European trade.

The protectionist revival and the agricultural depression

This edifice of free trade in France was subjected to its first strains after the Franco-Prussian War of 1870. France's defeat led to a revival of nationalist sentiment and to a heavy burden of debt. The government of the Third Republic, under Thiers, tried to raise import duties in order to obtain more revenue, but this required the revision of the commercial treaties, to which Britain in particular would not agree. In addition, French industrialists objected to Thiers's intention of imposing duties on raw materials. The failure of his customs policy was one of the reasons for Thiers's resignation in 1873.

The next government was less protectionist, but as the commercial treaties were due to expire around 1878 a revision of the tariff appeared necessary. When the preparation for this began in 1875, a strong protectionist movement began to make itself felt both inside and outside Parliament. The impulse came from the iron and textile industries rather than from agriculture, which was not yet in any special difficulty. The industrialists however needed to make an ally of agriculture in order to win their battle. In this they succeeded, helped by a bad harvest in 1879 which caused discontent among the farmers. The principal agricultural organisation, the *Société des Agriculteurs de France,* adopted the protectionist cause at its General Assembly of 1879: it demanded equality of treatment with industry, pleading the competitive advantage of agriculture in new countries and the relatively heavy burden of taxation on agriculture.

In 1881 the new tariff was adopted. The rates on manufactures were higher than before but still moderate; in the next few years many of the increases were renounced by the government in a new series of commercial treaties. For agriculture, there were large increases in the duties on livestock and livestock products, but wheat remained subject to only a nominal duty and other grains were still exempt (Table 3.2).

The protectionist revival thus began before French agriculture got into serious difficulties, and for reasons largely independent of agriculture. But from 1879 on, the effects of overseas competition in grains made themselves increasingly felt. As in Britain, a succession of poor harvests, of which 1879 was the worst, encouraged the growth in imports. French agriculture no longer benefited from higher prices in these years of short supply; in 1882 and 1883, when domestic production recovered, imports were maintained and prices fell drastically.

French agriculture in its backward state was quite unable to compete with the flood of cheap imports. Further, the importance of peasant proprietorship (it was estimated that 80% of the agricultural area was in the hands of owner-occupiers in 1882) meant that there was little possibility for passing on part of the loss of income to landlords, as in England.

The agricultural crisis thus made protection a vital issue. The *Société des Agriculteurs de France* was disappointed with the 1881 tariff and intensified its campaign. It continued to receive the support of industrialists, who took the opportunity of demonstrating their supposedly disinterested sympathy with the problems of the farmers. In 1885 agriculture received substantial measures of

Table 3.2: Tariffs on major agricultural products, 1881–1906

	1881 7 May	1885 28 and 30 Mar.	1887 29 Mar. 5 April	1892 7 Jan.	1894 27 Feb.	1898 5 April	1899 1 Feb.	1903 31 July	1906 18 July 21 Nov.
	francs per 100 kg. except where otherwise indicated								
Wheat	0.60	3	5		7				
Oats	nil	1.50	3						
Barley	nil	1.50		3					
Flour	1.20	6	8	8–12*	11–16*				
Butter	13			6–13†					
Cheese (hard)	8			15–25†					12–35
Eggs	10			6–10†					
Beef (fresh)	3	7	12	25				35–50†	
Pork (fresh)	3	7	12					25–40†	
Bullocks	15**	25**	38**	10				20–30†	
Cows	8**	12**	20**	10				20–30†	
Pigs	3**	6**		8		8–12		15–25†	
Wine††	5†††			7–12†‡			12–25‡‡		12–35

Note: (a) From 1861 to 1881, duties were nil or negligible (wine excepted).
(b) Temporary changes in the duties are not indicated.

* Variations according to gluten content.
† 'Minimum' and 'general' tariffs from this date on.
** Francs per head.
†† Francs per hectolitre.
††† Duty imposed on 8 July 1871.
‡ Under 11° proof.
‡‡ Not exceeding 12° proof from this date on.

Source: Sirey, *Recueil général des lois et arrêts.*

protection: the duty on wheat was raised to 3 francs per 100 kg, duties of 1.50 francs were imposed on feed grains and the livestock duties were increased (Table 3.2). In 1887 most of the agricultural duties were further increased, that on wheat rising to 5 francs, which was about a quarter of the prevailing price.

The reinforcement of protection – the Méline Tariff 1892

Even after these tariff increases, a vigorous fight for further protection was pursued by both agriculture and industry. The liberal element in Parliament had been gradually reduced in successive elections. In the 1889 election campaign, candidates were requested to sign a programme in favour of agriculture, and the names of both those who accepted and those who refused were made public. This election was fought mainly on the constitutional issue raised by *Boulangisme*, and many candidates were reluctant to lose the rural vote over the tariff question. The election gave a protectionist majority: the successful deputies who had signed the pledge were immediately reminded of it, and a unified Agricultural Group was formed in Parliament, consisting of 301 deputies. The Cabinet was at first headed by the liberal Tirard, but in March 1890 he was overthrown: a government with decidedly protectionist leanings took office in which the Ministers of Commerce and of Agriculture were both members of the Agricultural Group.

The tariff question soon came to the boil: the commercial treaties were due to expire again in 1892 and a new tariff was necessary. In October 1890 the Government put forward its tariff proposals, which were submitted to a Tariff Commission.

The subsequent discussions were dominated by the personality of Jules Méline, a former Minister of Agriculture with a long-standing commitment to protection. He was president of both the Agricultural Group and the Tariff Commission. The Commission itself had a strong protectionist majority, and in its report in March 1891 it recommended duties generally higher than those the government had proposed. There followed a long and vigorous debate in the Chamber and in the Senate. Méline's main arguments were that other countries had raised their tariffs, and that agriculture deserved to have equal treatment with industry as well as protection against overseas competition. He claimed to be acting in the interests of consumers, since the development of home production would tend to reduce prices.

The Free Traders in Parliament, led by Léon Say (grandson of the famous economist, and a former Minister of Finance), vigorously opposed the Tariff Bill, which was finally passed only in January 1892. It represented almost a complete success for the protectionists, earning its title of the 'Méline Tariff'. It reinforced the whole structure of agricultural protection, imposing higher duties on barley, livestock, meat, cheese, wine, beer and hops, and putting new duties on certain goods which formerly entered free, including maize, rice, vegetables and potatoes (Table 3.2). The duties on agricultural products generally had an incidence ranging from 10% to 25%; those on industrial products were mostly over 25%, sometimes as much as 60%. The Méline Tariff was reputed to be the stiffest in the world, with the exception of Russia and the United States. However, most agricultural raw materials – wool, skins, cotton, flax, etc. – continued to enter free: here the interests of manufacturers prevailed over those of the farmers.

One of the innovations of the tariff was the institution of a 'general' rate, intended to be the normal one, and a 'minimum' rate, representing the lowest to which the government could go in negotiations with other countries. This arrangement was another victory for the protectionists, who opposed the freedom of the government to reduce duties in commercial treaties. For grains and livestock, a single rate was laid down, which could not be changed by treaty.

In principle, imports from French overseas territories remained free of duty, with exceptions for the sake of revenue (on coffee, tea, sugar, etc.). Since most of the products of the overseas territories did not then compete directly with those of French agriculture, this exemption originally raised no great problem. At a later date, rising imports of wine and wheat from North Africa proved troublesome.

Later in 1892, France (like Britain) prohibited imports of cattle for reasons of animal health. This prohibition was maintained till 1903, when it was replaced by duties which were almost equally prohibitive; the system of general and minimum duties was extended to meat and livestock at this time.

After 1892, French agriculture entered on a second period of acute depression, as a result mainly of a renewed increase in imports. In 1894 the duty on wheat was further raised to 7 francs per 100 kg; the domestic price at this time averaged around 22 francs.

In 1897 a law was passed which was to play an important role at a later period. Entitled *loi de cadenas,* it enabled the government to raise the duty on major agricultural products without waiting for the agreement of Parliament, the object being to prevent increased imports in the period pending approval.

The 1892 tariff, with the increased duty on wheat of 1894 and those on livestock of 1903, remained substantially unchanged up to the First World War, in spite of the improvement which took place in the agricultural situation after 1900. A revision of the tariff was carried out in 1910: this however primarily concerned manufactured goods and there was no important change in the agricultural duties. The report presenting the Tariff Bill stated:

> We believe that agriculture has everything to gain from simply maintaining the customs arrangements for wheat, barley, rye and maize. This system has proved itself: it has given excellent results which are confirmed and consolidated every day. [*Rapport général de M. Morel,* Chambre des Députés, 11 juillet 1908]

Yet Méline, presenting the Bill of duties on grain as Minister of Agriculture in 1885, had declared:

> I do not hesitate to say that, the day when the price of wheat recovers, in conditions which enable French farmers to withstand competition and to remain in business, you may revise this duty or even abolish it, and French agriculture will then accept the sacrifice. [Chambre des Députés, 10 février 1885]

Protectionist theories

France's main reaction to the depression brought about by overseas competition was thus to restore to agriculture a substantial measure of protection. This policy was carried out much more in response to practical and immediate needs than in accordance with any carefully conceived doctrine. France had neither a Free Trade school like that of England, nor an influential body of scientific protectionists like that of Germany. Such influence as economists exerted, however, was mainly in the protectionist direction.

Paul-Louis Cauwès, Professor and Dean of the Law Faculty in Paris, was an eminent economist who argued that protection should be given to assist a nation's economic development. He thought that agriculture was entitled to protection when it needed it, and approved the resort to protection during the depression. It has been suggested that Cauwès provided a systematic theoretical background for the Méline Tariff. In fact this appears doubtful: Meredith, writing in 1904 on protection in France, seems nearer the truth when he said that the role played by scientific protectionists was much less important in France than in Germany. Meredith considered that the 'mercantile protectionists' were the dominant influence: their arguments were not very different from those of the mercantilists of a former period, for the basis of their belief was that imports diminished the sum of employment at home, and that a country would be more prosperous, the smaller its imports in relation to its exports. This view was corroborated by Haight in his history of French commercial policy:

> In many respects the protectionism of 1880–1913 resembled the mercantilism of the seventeenth century. Both aimed at independence of foreign products, saw an inherent goodness in the output of industry, and sought national greatness in productive capacity rather than in the satisfaction of needs. [Haight (1941) page 58]

The title of 'mercantile protectionists' might reasonably be applied to a succession of French politicians during the second half of the century, including Méline and others responsible for the return to protectionism. Protection was part of the nationalist response to the defeat of 1870, and part of the Republican reaction to the previous monarchist regime. (From 1879 the Republicans held the majority in both the *Sénat* and the *Chambre.*)

On the other side there was a mixed bag of academic Free Traders and businessmen with an interest in free trade, together with some moderate protectionists whose aim was to restrain the excesses of the mercantilists. Careful academic argument was not much in evidence in this group either. The handful of academic Free Traders was never able to get its theories accepted by any political party. Meredith (1904, page 40) commented acidly that 'no important party in France ... has ever been educated in that part of the theory of international trade which is supported by the weight of economic authority all over the world'. Indeed there was a tendency deliberately to avoid abstract arguments because of the unpopularity of 'economists'. Owing to this lack of basic principles, the Free Traders were unwilling to contest the mercantilists' arguments, and contented themselves in the first instance by trying to show that the situation under near-Free Trade was not so bad as the protectionists made out; when protection was restored, they emphasised the danger of increased costs, of foodstuffs in particular.

The balance of advantage in this argument was with the protectionists. Meredith's diagnosis seems correct:

> The weakness of the position of the Liberals became particularly plain when the demand for agricultural Protection began. The reformers in the 'sixties had not protected agriculture, first, because they wanted to reduce Protection to a minimum, and second, because the agriculturists at that time were, many of them, in favour of Free Trade. Wheat and meat were imported into France in small quantities only.... But when with the cheapening of ocean transport and the opening up of new countries the great fall in the prices of agricultural produce began, the Liberals had no logical reply to make to the demand for Protection. Many of them, indeed, objected to 'food taxes', and voted in accordance with this sentiment. But the formula – 'compensatory duties wherever needed' – certainly did cover the case of agricultural produce just as well as the case of manufactures. We need not be surprised that Liberal agriculturists who had been bred upon this formula went over with a clear conscience to the mercantilists as soon as they saw their business interfered with by foreign competition. [Meredith (1904) page 42]

Controversy began again with the revision of the tariff in 1910, when the effects of the Méline Tariff were widely discussed. In 1908 Edmond Théry, the director of *L'Économiste européen,* wrote a defence of protectionism (with a preface by Cauwès): this was a much more solidly-based work than other apologies for protectionism. On the other hand, a vigorous campaign against protection was carried on by certain economists, particularly in the *Journal des Économistes,* whose editor, Yves Guyot, wrote numerous articles with titles such as 'La cherté et le protectionnisme.' A *Ligue de Libre Échange* was formed in 1911 to fight against the increased cost of living, 'to obtain a reduction of tariffs and in the first instance the conclusion of commercial treaties'. The debate however remained on a pragmatic level, as Golob pointed out:

> In general, the debates were similar in character to those of 1891. The Liberals disclaimed any doctrinaire notions of free trade, the protectionists insisted that they were not systematic protectionists ... [The Méline Tariff] was attacked as having raised the cost of living of the French masses, and defended as having saved French agriculture from the destructive competition which threatened it. [Golob (1944) page 216]

Politics and the peasantry

The main influence on policy thus was not so much the application of any body of principles as the immediate economic realities, and the consequent agitation of leaders of both industry and agriculture for protection. The role of the *Société des Agriculteurs de France* has already been mentioned and needs further attention.

This body had been founded in the 1860s, and by 1890 it had about ten thousand members. It was very definitely not representative of all French agriculture, but was essentially a club of distinguished landowners, members of the old nobility or conservative upper *bourgeoisie* who had acquired estates (but often continued to live in Paris). The agitation by this élite for protection and the emphasis which they laid on wheat, earned them among Free Traders the sobriquet of *'les marquis du pain cher'*. Yet they were largely successful in maintaining the myth of the single *classe paysanne*, of peasant unity. It was in the interests of politicians to preserve this myth, while preventing the peasantry from becoming an effective political force, for the peasants held massive electoral power, had they been aware of it: at least half of the electoral districts for the Chamber of Deputies were predominantly rural, and in the Senate rural preponderance was even greater.

After the Waldeck-Rousseau law of 1884 had legalised associations for the defence of economic interests, farmers' syndicates were set up throughout the country – mostly, however, by local crusaders from the landed aristocracy or the *bourgeoisie*, impelled partly by the spirit of *noblesse oblige*, partly by the desire to insulate the peasantry from the spread of Republican ideology; the Catholic clergy encouraged the movement. These syndicates had little political activity and concentrated on the joint purchase of fertilisers and other agricultural requisites, together with, in some cases, collective sales of farm produce and other activities. Their number grew to 648 in 1890, with a membership of 234 000. Under the patronage of the *Société des Agriculteurs de France*, the *Union Centrale des Syndicats des Agriculteurs de France* was founded in 1886 and located in the premises of the *Société des Agriculteurs* in the rue d'Athènes. By 1914 it claimed to have over a million members in about 5000 syndicates.

Other forces were at work among the rural population, as Wright (1964) and Barral (1968) pointed out. On the Republican side, Gambetta had recognised the potential political importance of the rural vote: in 1880 he founded the *Société nationale de l'encouragement à l'agriculture* as a Republican counterpart to the *Société des Agriculteurs*, and when the Republicans came to power in 1881 he created for the first time a separate ministry of Agriculture (agriculture had previously been linked with trade). The Republican administration was uneasy at the activities of the right-wing agrarian syndicates, and gradually the Republican movement developed in the countryside. Its leaders, often radical deputies, were of middle-class origins – lawyers, doctors, veterinary surgeons, tradesmen, residing in small towns and owners of some land. They were helped by the Republican administration, in particular by the technical agricultural services in the *départements;* and while the right-wing syndicates had the support of the local *curé*, the Republican movement could usually count on the village schoolmaster. The development of Republican co-operatives was relatively slow until the *loi Viger* of 1900 gave an impulse to the formation of *mutuelles* for group insurance. The rural credit movement, largely inspired by the *Raiffeisen* example in Germany, was also primarily led by the Republican side. In 1910 the Republican groups federated at the national level in the *Fédération Nationale de la*

Mutualité et de la Coopération Agricoles, in the boulevard St Germain.

On both sides, however, traditional values were upheld: there was nothing revolutionary about the Republican approach to agricultural issues. Both political groups proclaimed the virtues of small farm ownership (though in the minds of right-wing leaders this was certainly not intended to put in question their own political predominance); neither set out to solve fundamental economic or social problems. The 'co-operative' movement was led from outside: the peasants themselves played virtually no part in the organisation and few of them had the education that would have enabled them to do so.

The Socialist party challenged the myth of peasant unity: according to Jean Jaurès, the *bourgeois* parties were indifferent to peasant needs and their real goal was capitalist farming; but such Socialist attacks had little effect beyond provoking redoubled insistence on the value of individual farm ownership, together with voluntary association in co-operatives. From another angle, traditional attitudes were also attacked by a vigorous young Breton priest, Félix Trochu, who founded syndicates for tenants distinct from those for landowners.

Jules Méline

Jules Méline, though a Republican, significantly furthered the aims of the élite *Société des Agriculteurs de France.* Méline entered Parliament in 1872, and was Minister of Agriculture from 1883 to 1885, introducing in 1885 the new duties on grain. His role in the formulation of the 1892 tariff has already been described. From 1896 to 1898 he was Prime Minister, and demonstrated his attachment to agriculture by taking the Ministry of Agriculture at the same time. His fall in 1898 was inglorious, being a result of his attitude over the Dreyfus Affair (perhaps his most celebrated remark was *'Il n'y a pas d'affaire Dreyfus'*). He returned to the Minstry of Agriculture during the First World War.

To a certain section of agricultural opinion in France, Méline has remained their greatest Minister of Agriculture. A testimony written in 1928 by Lachapelle declared:

> As Minister of Agriculture, he sought to promote by all possible means the production which enables us to live, to maintain on the land the mass of the rural population which constitutes the essential element in the nation's prosperity. For this purpose he took every occasion to defend the interests of agriculture and to reconcile them with those of industry. [Lachapelle (1928) page 209]

However Augé-Laribé took a different view, accusing Méline of having lacked any constructive agricultural policy:

> In matters of progress he favoured slowness, in reforms timidity and in public order the respect of established rights.... Decidedly the venerable M. Méline, in the list of physicians of the agricultural crisis, is entitled only to appear among the charlatans. [Augé-Laribé (1950) pages 69 and 108-9]

Mêline's basic philosphy was set out in his own book, of which the sixth edition was published in 1912. The title itself was eloquent: *Le retour à la terre et la surproduction industrielle.* The theme was simple: following the movement of labour from the land, there was a state of overproduction and unemployment in industry, which only a return to the land could cure. He wrote:

> There now remains only one field of action and of expansion capable of absorbing all the forces now unemployed ... this is the land, the land which is the nursemaid of humanity, fertile and everlasting. [Méline (1912) page 97]

Barral (1968) suggested that Augé-Laribé might have been somewhat unfair to Méline. He pointed out that Méline was aware that behind protectionist barriers it was necessary to undertake basic reforms. In his first speech after becoming Minister of Agriculture in 1883 Mêline had put forward an overall programme; and in a speech to the *Chambre des Députés* during the debate on the tariff issue on 11 February 1885 (J. O. page 190) he declared: 'There are many things to do for agriculture besides what we are doing now, to improve our production and to increase our yields... Our agriculture must improve its methods, perfect its techniques, and become scientific'. Rural education and credit were among his main ideas: universal education was at last made possible by the reforms of Jules Ferry during the 1880s, while farm credit developed on the basis of a law introduced by Mêline in 1894 and particularly after the *Banque de France* in 1897 provided substantial funds. According to Barral, Mêline also wanted to tackle problems of land distribution, but in this field he does not seem to have put his ideas into practice.

Conservative elements on the right, however, saw little need for such reforms and gave no encouragement. Tariff protection and reductions in certain taxes remained their principal objective.

French agriculture under protection

Wheat was the product with which the protectionists were most concerned, and for which the effects of protection appear most clearly. The duties imposed in 1885 and 1887 probably contributed to the recovery in prices which took place in the late 1880s, and though they did not prevent a renewed fall in the early 1890s, they ensured that throughout the period from 1885 on, the French price remained above the free market price by approximately the amount of the duty (see figure 1.1). The growth of imports, already restrained by the duties of 1885 and 1887, gave way to a decline after the further increase in the duty in 1894; as Table 3.3 shows, imports were actually less after 1900 than they had been in the 1860s (in Britain, meanwhile, they had trebled). As a result, wheat cultivation in France was scarcely discouraged: the area under cultivation even tended to rise up to 1890, and though it afterwards fell slightly, rising yields resulted in increased production up to the First World War. The area of rye fell throughout this period; that of oats expanded, while barley fell.[1]

Table 3.3: Wheat: trade, production, etc., in annual averages

	Imports*	Exports*	Production	Area	Average value
		thousand tons		*thousand ha*	*fr. per 100 kg*
1861–65	354	142	7594	6896	26
1866–70	430	141	7378	6979	30
1871–75	632	242	7586	6801	31
1876–80	1311	154	7070	6900	29
1881–85	1110	24	8541	6991	25
1886–90	1026	13	8348	7000	23
1891–95	1363	24	8101	6762	22
1896–1900	613	37	8918	6844	21
1901–05	285	28	8906	6575	21
1906–10	318	51	8920	6562	23

* Including flour in wheat equivalent.

Source: Statistique Agricole Annuelle, 1947 (*tableaux rétrospectifs*).

There was an increase in imports of cattle and meat in the late 1870s, but this was checked by the increased duties imposed from 1881 on. As a result prices were well maintained throughout the Depression. From 1892, when imports of cattle were prohibited for health reasons, there was a substantial increase in the number of cattle in France, and after 1900 France became self-sufficient in cattle and beef, with occasional export availabilities. The number of pigs also increased.

Wine was the most important product of French agriculture after grains. In the late 1860s and again from 1879, the vineyards were devastated by phylloxera. The fall in production led to a large increase in imports. After an increase in the import duty in 1892, imports from foreign countries were reduced, and production revived. However, foreign wines were diverted to other markets where they competed with French exports; also, wine from Algeria became established on the French market. The sale of French wine both at home and abroad became more difficult: exports failed to recover to the level of the 1870s (Table 3.4).

In a useful study of the effects of the 1892 tariff, Dijol commented as follows:

> Though it is true that, thanks to protection, French wine production has been put on its feet again, it is also a fact that under the protection there has been an artificial expansion of the vineyards, leading to overproduction and to difficulties in finding markets. [Dijol (1911) page 78]

[1] There is considerable difficulty in obtaining consistent and reliable series of crop areas and livestock numbers in France during this period. One of the problems is that the regular annual data (where available) do not seem to bear comparison with the results of the censuses carried out in 1862, 1882 and 1892. It therefore seems preferable to avoid detailed tables, which could be misleading unless carefully interpreted.

Table 3.4: Wine: trade and production, in annual averages

	Imports	Exports	Production
		million hectolitres	
1861–70	0.2	2.6	52.9
1871–80	1.5	3.2	49.2
1881–90	9.7	2.4	29.7
1891–1900	7.7	1.8	39.9
1901–10	5.9	2.2	51.8

Source: As Table 3.3.

Indeed, following large harvests, wine prices collapsed in 1901 and again in 1907, provoking disturbances among the winegrowers of the Midi of the kind which have presented French policy-makers with an intractable problem ever since.

The production of sugar from sugar-beet was encouraged not so much by tariff protection as by export subsidies, which were granted in one form or another from 1884. Various exporting countries began to compete, each being forced to increase the rate of subsidy, until at the Brussels Conference of 1902 they agreed to refrain from subsidies. After this, however, France had difficulty in competing and its exports fell off. Dijol again pointed out:

> The sugar industry, artificially developed through direct and indirect bounties, but now left to fend for itself, is suffering from all the drawbacks of protection after having enjoyed all its advantages. [Ibid., page 89]

Dijol's general conclusion on the effects of protection is of interest:

> [The tariff] has given excellent results when it has promoted the normal development of production, but it has had no effect or even undesirable effects in cases where the productive resources of the country have been directed into a type of activity for which France is less well suited than other countries, or in cases where protection has led to an excessive increase in production. [Ibid., page 337]

As examples of favourable effects, Dijol lists wheat and other grains, cattle and dairy products; cases of unfavourable effects include sugarbeet, vines and flax.

As in England, livestock producers seem to have suffered much less from the depression than crop producers: though the prices of livestock products fell, the reduction was smaller and did not last for so long. As the urban demand increased, growth in production was able to offset the fall in price, and the area under fodder crops also expanded in consequence. Many dairies were established, including co-operatives. Normandy and other livestock-producing areas thus seem to have been relatively untouched by the crisis.

The case of sheep was different. As has been pointed out above, wool was left free from duty, in deference to the interests of the manufacturers; the result was a big increase in imports from the middle of the century on, and a prolonged fall in prices. The number of sheep in France was reduced from 24 million in 1852 to 17 million in 1910, and the fall would have been even bigger but for increased utilisation for meat. The report on the agricultural census of 1882 declared:

> Formerly, wool was the most important product of our sheep. Today, with abundant supplies of wool arriving from overseas countries, it has become in France little more than a by-product. While the value of sheep sold for meat amounts each year to nearly 200 million francs, their wool is worth scarcely 78 million francs. [Ministère de l'Agriculture (1887) page 247]

Yet this decimation of the sheep flocks took place without any public outcry, while the difficulties of the arable farmers were made a national issue.

Other items left virtually unprotected were flax, hemp and oilseeds: in these cases, too, production fell drastically and imports increased.

There is little reliable information on the overall trend in farm incomes during this period. Table 3.5 suggests general progress in the agricultural situation up to 1882 and a worsening – but hardly a catastrophe – between 1882 and 1892. Subsequently there are no more data up to the First World War, but there is no doubt that in the period after 1900 there was a substantial improvement. Table 3.6 indicates a decline, though not a very drastic one, in the total agricultural population up to 1891, where this series ends; the active population in agriculture seems to have declined from 1866 to 1881, and then to have risen once again – the active *male* population however seems to have declined about 4% between 1896 and 1906. Not much faith can be put in these figures, but it does not seem that French agriculture suffered an exodus comparable to that of the British. It should of course be remembered that in Britain it was mainly the farm labourers who left agriculture: the French peasant, to whom the farm was a home as well as a place of work, could not so easily move to the towns. It seems in fact that the number of *chefs d'exploitation* remained fairly stable from 1882 to 1892, while the number of labourers and other *auxiliaires* fell.

Whatever the benefits of protection for the big farmers, and for grain producers in particular, its advantages for the peasantry are questionable. The small peasant usually sold only a part of his grain crop, and he often had to buy flour for bread-making. It seems likely that a high grain price benefited only the occupiers of holdings of ten hectares at least: such holdings amounted in 1882 to only about 870 000 out of the total of 5.7 million (or 3.5 million if the dwarf holdings under one hectare are excluded).

The most serious objection to protectionism was that it diverted attention away from the need for a constructive long-term policy for French agriculture. Most contemporary thinking, even among protectionists, was aware that aid through protection should be only temporary and that agriculture should adapt itself to the new situation. But agriculture was bound by tradition and inertia, and the various governments did little to bring about an improvement, apart from the useful encouragement given to rural credit and mutual insurance, the beginnings of which are described above. The development of a consistent policy of basic reform was hardly facilitated by the frequency of government changes: from 1881, when the Ministry of Agriculture was formed, to 1914, there were

Table 3.5: The economic situation of agriculture

	Unit	1852	1882	1892
Gross product of agriculture	*million fr.*	8061	13461	..
Net product of agriculture	"	..	1198	800
Average value of agricultural land	*fr. per ha*	1266	1686	..
Average value of first-class arable land	"	..	3442	2866
Average wage of male agricultural workers, not fed on the farm, in summer	*fr. per day*	2.77	3.11	2.94

Source: Statistique Agricole de la France: Enquêtes Décennales.

Table 3.6: Active and total population in agriculture, forestry and fishing, in relation to population in all sectors

	Active population			*Total population*		
	Agri-culture, etc.	All sectors	Agri-culture as % of total	Agri-culture, etc.	All sectors	Agri-culture as % of total
	thousands			*millions*		
1861	..	..	..	19.9	37.4	53
1866	8535	16643	51	19.6	38.1	51
1872	..	..	..	18.5	36.1	51
1876	..	..	..	19.0	36.9	52
1881	7890	16544	48	18.2	37.7	48
1886	..	..	..	17.7	38.2	46
1891	..	..	..	17.4	38.3	45
1896	8501	18935	45	..	..	..
1901	8244	19735	42	..	..	..
1906	8855	20721	43	..	..	..

Sources: For active population: 'Quelques aspects de l'évolution des populations actives dans les pays d'Europe occidentale', *Études et Conjoncture,* November 1954; for total population: *Statistique Agricole de la France: Enquêtes Décennales.*

no less than forty-two different governments and nineteen different Ministers of Agriculture. Joseph Ruau, who was Minister of Agriculture for a relatively long time (1905–10, in successive governments) based his policy on three principles: encouragement to free association in co-operatives, agricultural education, and maintenance of tariff protection. In 1906 he introduced a scheme of long-term, interest-free loans to co-operatives and *mutuelles;* and around this time most *départements* created winter schools in farming techniques, though attendance was limited to a few thousand pupils. More than this was required and progress was not helped by the fact that Finance Ministers were reluctant to provide the funds which would have been necessary for irrigation, drainage, access roads, research and training.

The structure of French agriculture remained almost unchanged. There was no attempt to deal with the problem of small and fragmented holdings – indeed, it was scarcely recognised as a problem. Nor was anything done to reform the system of land tenure and to give tenants some guarantee that by making improvements they would not be wasting their money (even in laissez-faire England some important action was taken in this respect); the status of share-croppers too remained unimproved.

Changes did take place, as rural France came increasingly into contact with the outside world. Weber (1977) considered that the main period of change was between 1880 and 1910, and identified three main causes. After 1881 the building of rural roads was officially promoted and these, together with the spread of railways, gradually brought hitherto remote and inaccessible regions into easy contact with urban markets and ways of life. Basic education began to make significant progress even in the countryside after Jules Ferry's reforms in the 1880s. Illiteracy was reduced and the use of the French language spread, while the teaching of French history and geography developed national sentiment. Also, military service, from the 1870 war onwards, proved an agency for change in attitudes, as many young peasants discovered for the first time other ways of life and higher standards of living. Ruttan (1978) pointed out that, by comparison with the United Kingdom or Germany, French agriculture was hampered in its development by the relatively slow growth of the economy as a whole: both the demand for farm produce and the opportunities for rural labour to move into urban employment grew relatively slowly. This, in his view, was the main cause of the slow rate of modernisation of French agriculture, rather than its defective structures. The history by Duby and Wallon emphasised the lack of productive investment in both agriculture and industry, attributing the lack of progress to:

> ... a timid entrepreneur class, a peasantry too attached to its roots in the soil, and also a *bourgeoisie* obtaining land rents which constituted a substantial part of the agricultural product and thereby reduced the capital available to the rural sector ... [Duby et Wallon (1975–6) *Histoire de la France rurale* Vol. 3, page 467 © 1976 Editions du Seuil]

The absence of dynamic growth was to a large extent a deliberate choice – Volume 4 of the Duby–Wallon history regarded the French case as a particular 'model' in which funds were put into overseas investment rather than domestic industrial expansion.

Perhaps the favourable natural conditions of French agriculture – a temperate climate and generally fertile soil – created the illusion that farming would remain a strong pillar of the economy and prevented realisation of the need for adjustment to modern conditions.

Food prices, the balance of trade and self-sufficiency

Méline and other protectionists sometimes tried to argue that protection did not raise the price of food to the consumer, since national production would be encouraged in the place of imports. They also tried, in retrospect, to show that tariffs had not raised prices by pointing to the fact that prices had continued to fall in most years; sometimes it was even stated that tariffs had *reduced* prices.

The truth was, of course, that prices of foodstuffs in France fell to some extent during the depression, but less than prices in Britain where there was no protection; the result was that by 1900 bread grain, meat and sugar cost more in France than in Britain, French consumers being deprived of access to relatively cheap imported supplies of these goods. As has been pointed out in chapter 1, French households seem nevertheless to have eaten more of most foodstuffs than British, in *per capita* terms, possibly because the heads of household generally had fewer mouths to feed.

During the period under consideration, the reduction of imports and the expansion of certain exports had the result that France became a net exporter instead of a net importer of food. Various protectionists claimed this as a success for protectionist policy. The limitations of this mercantilist approach were pointed out by Clapham:

> France remained self-supporting – at a price. The price was paid in many ways. Part of it in rather stagnant exports; for the nation that will not buy neither shall it sell. Part in the failure to develop home industries connected with the handling of imported food, and food export industries which required cheap materials. [Clapham (1945) page 183]

The need to promote self-sufficiency in wartime was also frequently advanced by protectionists. Thus Méline (1912, page 247) asked: 'Is there a greater danger for a nation than to have its food supplies in the hands of foreign countries and at their mercy?' and evoked the case of England as a frightening example. There is no doubt that protection maintained the output of certain items and increased the degree of self-sufficiency: by the outbreak of war France produced at least 90% of its requirements of all major crops except maize, and all its needs of meat and dairy products. But when the war came there was a big decline in production, due not only to the loss of part of the territory but also to falling yields, as a result of the shortage of labour, the lack of a prepared plan and the inflexibility of techniques. Large imports of wheat, sugar and meat were soon required.

Bibliography

The need for a comprehensive history of French agriculture has been abundantly met by the four-volume work edited by Duby and Wallon (1975–6); the third volume is particularly relevant to the present chapter. Bloch's pioneering work (which first appeared in 1931, republished 1952) on developments from feudal times up to and including the Revolution can however still be read with profit (it is also available in English). The most detailed account of the impact of the Revolution itself is still Lefebvre's study of the *département* of the Nord (1924) – the fact that this runs to 886 pages gives an idea of the complexity of the subject. (More recent work on other regions, particularly by Soboul or collected by him, has not been included in the bibliography as it is only marginally relevant to the present theme.)

An American author, Weber (1977), has provided a vivid account of conditions in rural France in the nineteenth century. Clapham's economic history (1945) of the period 1815–1914 described the general context. On the growth of agricultural protectionism, the obvious starting-point is Augé-Laribé's history (1950). This is completed by Wright (1964) who emphasises political movements and developments in farm organisations; Barral (1968) throws more light on the aims and activities of the various agricultural organisations.

A useful early study of protectionist measures was made by Meredith (1904); that by Dijol (1911) provided a more thorough analysis. Méline's policies in particular were further examined by Golob (1944). Méline's own book (which went through many editions, the sixth in 1912) is a significant work.

Amé, G. (1868) *Le libre-échange en Angleterre et en France* (Conférence faite à la Gare Saint-Jean, à Bordeaux). Paris.

Amé, L. (1876) *Étude sur les tarifs de douanes et sur les traités de commerce.* Paris.

Arnauné, A. (1911) *Le commerce extérieur et les tarifs de douane.* Paris.

Augé-Laribé, M. (1950) *La politique agricole de la France de 1880 à 1940.* Paris: Presses Universitaires Francaises.

Augier, C. et Marvaud, A. (1911) *La politique douanière de la France dans ses rapports avec celle des autres États.* Paris.

Barral, P. (1968) *Les agrariens français de Méline à Pisani.* Paris: Colin.

Beaurieux, N. (1909) *Les prix du blé en France au XIXe siècle.* Paris.

Bloch, M. (1952) *Les caractères originaux de l'histoire rurale française.* Nvlle. édn. Paris: Colin. (In English: *French Rural History – An Essay on its Basic Characteristics.* London: Routledge & Kegan Paul; 1966.)

Châle, J. (1901) *Transformation du problème douanier, relativement au blé.* Paris.

Clapham, J. H. (1945) *The Economic Development of France and Germany, 1815–1914.* 4th edn.: Cambridge University Press.

Dijol, M. (1911) *Situation économique de la France sous le régime protectionniste de 1892.* Paris.

Duby, G. et Wallon, A. (dir.) (1975–6) *Histoire de la France rurale.* Volumes 1–4. Paris: Seuil.

Duhamel du Monceau (1762) *Eléments d'agriculture.* Paris.

Dunham, A. L. (1930) *The Anglo-French Treaty of Commerce of 1860.* Michigan.

Golob, E. O. (1944) *The Méline Tariff: French Agriculture and Nationalist Economic Policy.* New York: Columbia University Press.

Guyot, Y. (octobre, 1911) 'La cherté et la protectionnisme', *Journal des Économistes.*

Haight, F. A. (1941) *A History of French Commercial Policies.* New York: Macmillan.

Imbart de la Tour, J. (1901) *La crise agricole en France et à l'étranger.* Nevers.

LaChapelle, G. (1928) *Le Ministère Méline.* Paris.

Lefebvre, G. (1924) *Les paysans du Nord pendant la Révolution française.* Paris.

Margadant, T. W. (July, 1979) 'French rural society in the nineteenth century'. *Agricultural History.* **53** (3), 644–51.

Méline, J. (1912) *Le retour à la terre et la surproduction industrielle,* 6e édn. Paris.

Meredith, H. O. (1904) *Protection in France.* London.

Ministère de l'Agriculture (1887) *Enquête décennale de 1882.* Nancy.

Ministère de l'Agriculture (1897) *Enquête décennale de 1892.* Paris.

Nogaro, B. et Moye, M. (1931) *Le régime douanier de la France.* Paris.

Paturel, G. (juin, 1911) 'Le protectionnisme et le coût de la vie', *Journal des Économistes.*

Risler, E. (1er février, 1885) 'La crise agricole en France et en Angleterre', *Revue des Deux Mondes.*

Ronce, P. (1900) *La crise agricole.* Paris.

Rosier, C. (1943) *La France Agricole.* Paris: Alsatia.

Ruttan, V. W. (1978) 'Structural retardation and the modernisation of French agriculture: a skeptical view', *Journal of Economic History.* **XXVIII** (3), 714–27.

Soboul, A. (1980) 'Les paysans "partageux" et la Révolution française', *L'histoire,* **26**, 30–37.

Théry, E. (1908) *Les progrès économiques de la France: Bilan du régime douanier de 1892.* Paris.

Weber, E. (1977) *Peasants into Frenchmen – the Modernisation of Rural France, 1870–1914.* London: Chatto & Windus.

Wright, G. (1964) *Rural Revolution in France – The Peasantry in the Twentieth Century.* Stanford University Press.

Chapter 4

Germany

Agrarian reforms in Germany, completed only in 1848, abolished feudal servitudes but – as in France – made little impact on farm structures. There was great diversity between the various parts of Germany, the most obvious contrast being between the regions east of the Elbe, occupied mainly by large estates in the direct management of powerful landlords (the Junkers), and the rest of Germany, where small peasant-owned holdings prevailed. On these small holdings, little technical progress had occurred. By the mid-nineteenth century, industrial development had not yet drawn large numbers from the countryside.

The slow emergence from feudal servitudes

The diversity of farm structures reflected different paths of evolution from the feudal society of the Middle Ages in the German territories of the Holy Roman Empire (which did not initially extend east of the Elbe). The situation in the late Middle Ages was well described in a history of the German peasantry by Franz:

> The peasant stood in a network of dependencies with numerous masters above him – even though in the Germany of former times the principle *'nulle terre sans seigneur'* did not apply to the same extent as in France. Most peasants had to pay to a landlord, the principal owner of their land, rent and payment in kind, regardless of the nature of their own title to their holdings, whether they held it in hereditary lease (*Erbleihe*) or only for their lifetime (*Leibfall*), or even just 'at will' (*Freistift*) in which case it could at any time be taken away from them.... Besides the landlords (*Grundherren*) there were the lords with right of jurisdiction (*Gerichtsherren*) ... and the beneficiaries of tithes (*Zehntherren*) (who by this time were not merely the Church).
>
> Besides these dependencies in relation to land and the law, there was the personal bond to a lord. In the previous centuries, many peasants had put themselves under the protection of a lord, above all the Church, and paid a protection fee; others were free, but even this freedom implied .. a tie to the lords who had settled the peasant on new land as a freeman (*Rodungsfreier*), whether this was the Emperor himself or the ruler of the State. Finally, there was the state of personal subservience.... The serf (*Leibeigener*) too could hold land in hereditary lease ... but his mobility was restricted – without the lord's permission he could not leave his position – and likewise his choice in marriage....
>
> A decisive role was naturally played by the jurisdictional lordship ... From this legal authority flowed peasants' obligations regarding public services and payments, and the construction of roads and castles, as well as regarding assistance to the law ... and regarding armed service, conscription and taxation. [Franz (1970) pages 35 and 42]

In the original German territories west of the Elbe, this situation remained broadly unchanged until the nineteenth century. The landed nobility remained in power and in possession of the land. In Protestant states following the Reformation, lands acquired from the Church appear mainly to have been added to the possessions of state rulers or nobility; in Catholic states, Church possessions remained considerable. On the whole, noble landowners in the western parts of Germany did not set out to farm their estates themselves: they were primarily interested in the rents they could obtain and often did not live on their estates but had posts in government service, the army or the Church. Even after the ravages of the Thirty Years' War, when many farms lay empty, the policy – supported by the state rulers who had an interest in having as many active peasants as possible for purposes of tax and military service – was to reinstate peasants on the holdings. In some of the states, over time, peasant tenure rights came to be more firmly secured and their obligations to the lords better defined and regulated. Some enlightened rulers in the eighteenth century freed serfs on Crown lands – for example in Baden, Hohenlohe and Bavaria – but the nobility rarely followed this example: indeed in Bavaria, the landed nobility gave up virtually nothing of their feudal privileges, and the labour services and payments (especially in kind) imposed on the peasantry remained excessive.

The lands east of the Elbe were conquered, in the thirteenth and fourteenth centuries, by the Knights of the Teutonic Order, paving the way for peasants from the overpopulated lands of the west. Pomerania, Brandenburg, Silesia and finally the Prussian lands were colonised. However, the Black Death of the mid-fourteenth century, and the Thirty Years' War in the seventeenth, gave the nobility the opportunity to take over peasant land (*Bauernlegen*) and extend their powers over the peasantry. As they sought to exploit the land themselves, producing especially cereals for export to the west, there emerged a characteristic type of estate under the direct management of the owner (*Gutsherrschaft* as opposed to the *Grundherrschaft* of the western territories).[1]

This was the varied situation when pressures for reform grew overpowering towards the end of the eighteenth century. The invasions by France after the Revolution brought about the first major changes. In the territories west of the Rhine which Napoleon annexed, in the Kingdom of Westphalia and the Grand Duchy of Berg which he carved out for his brother and brother-in-law, as well as in several other states of the Confederation of the Rhine which he controlled, peasants were freed from ancient servitudes and obtained possession of their holdings. After 1815, however, there was a period of reaction as the nobility sought to reestablish themselves and in most states of west and south Germany peasant freedoms were not fully secured until the revolts of 1848: particularly in Bavaria, the nobility and the Church hung on to their privileges and possessions till 1848 (see Lütge, 1963, for an account of the reform movements). After this date, however, the liberation of the peasants from feudal ties was complete and the new situation is again well described by Franz:

> Subjection to lords who owned the land and administered the law, bondage, the imposition of tithes – factors which for a thousand years had conditioned the lives of peasants and villages – now belonged to the past, together with

[1] German agricultural historians distinguish clearly between these two forms. There are no satisfactory English equivalents.

> the multitude of payments and obligations which had been exacted from the peasantry.... It is true that the duty of the lords to assist their vassals in times of need was also removed. The peasant stood alone, free, responsible only to himself, but also unprotected in a changed world. [Franz (1970) page 265]

The reforms, however, concerned essentially peasant freedoms and their right to own land. The financial burden of purchasing ownership of the land they farmed proved too much for many peasants. Some, but not all states, set up land banks to provide credit for this purpose, but even where such provision was made, some peasant families could not free themselves from debt repayments before the post-1918 inflation wiped them out. The reforms did not in themselves create a more rational farm structure. Redistribution of land was mainly confined to allocation of land from common pasture and forest – a process which reduced landless villagers to beggary – and the farming pattern continued to reflect the former open-field strips. In most of west and south Germany, the practice of equal distribution among heirs (*Realteilung*) tended to aggravate the degree of fragmentation: some states began consolidation early, as maps 2a and b in chapter 1 have shown, but the overall impact was limited.

Developments in the east, in the Kingdom of Prussia, were different. The Hohenzollern emperors had made ineffectual attempts to protect the peasantry, their efforts foundering on the resistance of the landlords. The Prussian defeat by Napoleon in 1806 and the humiliating terms of the treaty of Tilsit in 1807, by which Prussia lost almost half its territory, brought a wider realisation of the need to free the peasantry – it was said that 'a slave has no interest in the State', and 'where there is a free peasant, there is a brave people and a free land'. The Stein-Hardenberg law of 1807 (from the names of the Minister and the Chancellor responsible) freed peasants from the control of the landlords and from services and payments, and authorised them to own land. In return for possession, however, they had to cede to the landlord between a third and a half of their holdings. If this left them with too little land to feed their families, payment in money or in kind could be substituted. Many peasants, finding themselves unable to make a living and heavily in debt, subsequently resold the holdings they had acquired to their former landlords. A decree of 1811, moreover, limited the grant of freedom to peasants owning a team of draught animals (*spannfähige Bauern*); while previous legislation protecting the peasants' right of occupancy was abolished in 1816, so that landlords could impost severe terms in return for the grant of freedom or could even take over the holdings themselves. In the province of Mecklenburg and in parts of Pomerania (previously subject to Swedish rule), peasants had never had any significant protection, and major reforms were delayed until there were hardly any land-holding peasants left to free (Lütge, 1963). Thus in Germany east of the Elbe, the results of the 'reforms' were to reinforce the landed nobility in possession of their estates and to augment the class of landless labourers. When, later in the century, a policy of resettlement was adopted, the argument was advanced that about two million hectares, or a third of the land then in estates, had previously been peasant land.

(This was probably an overestimate – other calculations indicate a figure about half this size.)

The effects of these divergent developments are clearly apparent from Table 4.1: in 1895, in the provinces of the Kingdom of Prussia lying east of the Elbe, estates of more than 100 hectares accounted for 43% of the agricultural area; peasant smallholdings, though numerous, were insignificant in terms of area (some of the small farms had recently been created under the resettlement programme – see later section). In the rest of Germany, a more varied farm structure was found. Large estates were proportionately fewer in number, though they still accounted for a substantial part of the area. Leaving aside the numerous holdings under two hectares – many of which would not have been farmed full-time – farms between two and twenty hectares were a major element in the farm structure, accounting for about half the area. Many of these would have been adequate in size to support a family, given the techniques and standards of the period. In most cases, however, fragmentation into tiny strips reduced productivity.

Table 4.1: Agricultural holdings in 1895, by size

	Number of holdings		Area covered	
Size group	Prussia*	Rest of Germany	Prussia*	Rest of Germany
hectares	%	%	%	%
0–2	58	58	3	7
2–5	15	19	5	13
5–20	18	18	20	37
20–100	7	4	29	31
over 100	1	0.2	43	11
Total	100	100	100	100
	thousands		*thousand ha*	
	1 433	4 125	13 169	19 348

* Six eastern provinces.
Source: Statistisches Jahrbuch.

The smaller farms usually could not profitably cultivate grain for sale, but practised more intensive lines of farming: cows, pigs and poultry were their main sources of revenue. The large estates of the east, by contrast, produced mainly grain – rye, wheat and other cereals – and exported large quantities of this produce to the rest of Germany and to other countries until the 1880s.

Free trade and the return to protection

After the Napoleonic Wars, Prussia had reduced its tariff rates, and the moderate duties which resulted formed the basis of the *Zollverein* tariff of 1834. This included import duties on grain, but in 1853 these were abolished. The first major change in the *Zollverein* tariff was brought about by the treaty with France in 1862, which, together with subsequent treaties between the *Zollverein* and other countries, substantially reduced the rates of duty on manufactured goods. These treaties were carried through Parliament by Bismarck, who had become Minister-President in 1862. His motives were largely political: to secure

French support in Prussia's quarrels with Austria and Denmark. The Franco-German Peace Treaty of 1871 renewed trade relations for an indefinite period, establishing 'most favoured nation' treatment on a reciprocal basis.

At this time agricultural opinion in Germany was strongly in favour of free trade. The years from 1850 to 1870 were prosperous; Germany was a net exporter of grain and its agriculture did not need protection. Moreover, the farmers were opposed to protection for industry, since this might raise the cost of agricultural requisites and also lead to reprisals by Britain against German grain exports. Industrialists, on the other hand, wanted to obtain protection behind which their activity could develop, shielded from the competition of Britain's established industries. Friedrich List's 'National System of Political Economy' provided strong arguments for the protection of young industries (though not for agriculture, which in List's view should have sought greater specialisation and taken advantage of the opportunities for intensified output offered by an industrialising economy). The opposition of farmers was largely responsible for the failure of an attempt by the industrialists to retain duties on iron. The leaders of industry thus became aware of the need for agricultural support: as in France, they took the initiative in converting agriculture to protectionism. The task became easier as agriculture began to feel the effects of foreign competition. On the markets of Britain and France, German grain was beginning to lose ground to American supplies. With its growing population, Germany became an importer of wheat and rye, obtaining supplies from Russia, Austria-Hungary and the United States.

In 1876, both a Central Association of German Industrialists and an Association of Tax and Economic Reformers were founded, the latter being composed mainly of large landowners; in 1878 they joined forces in a campaign for a return to protection for both industry and agriculture. At the same time a strong protectionist group was formed in the *Reichstag.* Bismarck was looking for means of securing additional revenue for the imperial finances and therefore welcomed the idea of restoring import duties. A new tariff was prepared and became law in July 1879. It included moderate duties on grain and increased duties on livestock products (Table 4.2). (Imports of live animals were restricted, ostensibly for 'sanitary' reasons, under a law of 1880.)

The duties on grain were too low to prevent the fall in prices which occurred from 1879 on. German grain by now had been almost completely forced off its former export markets and was meeting intensified competition on the home market. Further protection was demanded.

In 1885, the *Reichstag* agreed to raise the duties on wheat and rye from 1 to 3 marks per 100 kg; the rates for other grain and for livestock were also increased. But prices continued to fall. Bismarck was by now a firm believer in import duties on foodstuffs both as a source of revenue and as a means of assisting agriculture, and in 1887 he proposed to raise the duty on wheat and rye to 6 marks. This was too much for many members of the *Reichstag,* but a rate of 5 marks was accepted, with increases for other grains as well. The price of wheat on German markets was then around 17 marks.

Table 4.2: Tariffs on major agricultural products, 1879–1906

Product	1879 15 July	1885 22 May	1887 21 Dec.	1891 6 Dec.*	1902 25 Dec.†	1906 1 March**
		marks per 100 kg except where otherwise indicated				
Wheat	1	3	5	3.50	7.50–5.50	5.50
Rye	1	3	5	3.50	7–5	5
Oats	1	1.50	4	2.80	7–5	5
Barley	0.50	1.50	2.25	2	malting: 7–4	4
					feeding: 7	1.30
Maize	0.50	1	2	1.60	5	3
Flour	2	7.50	10.50	7.30	18.75	10.20
Butter	20			16	30	20
Cheese (hard)	20			15	30	15
Eggs	3			2	6	2
Meat	12	20		15***	45	27
Bullocks	20††	30††		25.50††	18	8
Cows	6††	9††			18	8
Pigs	2.50††	6††		5††	18	9
Wine in barrel	24			20†††	24‡	20‡

Note: From 1865 to 1879, duties on grains were nil and those on livestock products were relatively low.

* Rates in treaties with Austria-Hungary and Italy.

† Rates provided for under tariff law, with minimum duties for wheat, rye, oats and malting barley.

** Rates actually applied on entry into force of 1902 law, following commercial treaties.

†† Marks per head.

*** 17 marks on pigmeat.

††† 10 marks on red wine.

‡ Higher rates on wine of 14° proof and over.

Source: Reichsgesetzblatt.

Commercial Treaties, 1891–4

After 1887, agricultural prices recovered for a while, helped by the increased duties but owing also to a lull in overseas competition. This upward movement was accentuated in 1890 and 1891 by bad harvests in Germany and elsewhere: food supplies ran short and there were public demonstrations against the increase in prices. The grain duties were temporarily suspended.

In 1890, Bismarck was dismissed by the Emperor and replaced by Count Caprivi. Caprivi, unlike Bismarck, had no agricultural ties: indeed he declared in a misguided moment that he possessed *'kein Ar und Halm'* (no acre or blade of straw). He came under pressure from industrialists, who no longer had interests in common with agriculture. On the contrary, they now wanted not only cheaper foodstuffs to keep down costs, but greater access to foreign markets: this could only be achieved through tariff concessions on foodstuffs. The elections of 1890, moreover, brought gains to the left-wing parties which opposed protection because of its influence on the cost of living. Thus a number of factors were combining to bring about a change in tariff policy, and a revision of the tariff was necessitated in any case by the fact that several of the existing commercial treaties were due to expire in 1892.

Caprivi thus sought to conclude new commercial treaties in which Germany would obtain advantages for its exports of manufactures in return for concessions on the agricultural tariff. The first treaty to be concluded was with Austria-Hungary; others, with Italy, Belgium and Switzerland, followed closely. The Austrian treaty lowered Germany's tariffs on wheat and rye from 5 to 3½ marks, with similar reductions in the duties on other grains. The treaty with Italy lowered German duties on wine, poultry, eggs and fruit.

These treaties were passed by the *Reichstag* only against stiff opposition. Shortly after they came into effect, grain prices suffered a sharp fall. This was due mainly to developments on the world market, but the tariff reductions were widely blamed. Discontent among farmers was intensified. In 1893, the Prussian landlords organised themselves in the *Bund der Landwirte* (Farmers' League),[1] which immediately demanded adequate tariff protection and fiercely opposed further commercial treaties. Also in 1893, new elections brought a recovery of the right-wing parties, many of whose candidates received effective support from the *Bund* (see below).

Nevertheless, a treaty was concluded with Rumania in 1893 which confirmed the extension to Rumania of the reduced agricultural tariffs contained in the Austrian treaty. Well aware of Rumania's importance as an exporter of grain, the *Bund der Landwirte* whipped up opposition, and the treaty was passed only by a narrow majority.

Still greater opposition was aroused by the Russian treaty of 1894, for Russia was an even more dangerous competitor than Rumania, and the treaty offered Russia the lower rates on agricultural products, as well as reductions in the duty on wood and free entry for certain other products. The German Emperor attached great importance to this treaty, because of his desire to preserve personal relations with the Czar. In a speech from the Throne he tried to placate the agriculturalists, promising to set up chambers of agriculture which would work to improve the credit system and relieve agricultural indebtedness (these were items in the programme of the *Bund der Landwirte*). Nevertheless the *Bund* petitioned the *Reichstag* to reject the Russian treaty. The conservative element in the *Reichstag* was now torn between its traditional loyalty to the Kaiser and his government and its equally traditional ties with the landed class. For the time being, the Kaiser won, and the Russian treaty was passed by a bigger majority than the treaty with Rumania. Through the 'most favoured nation' clause, the reduced rates were now applicable to a long list of countries, including the United States and other grain-exporting countries.

Caprivi's position, however, had been steadily undermined by the agricultural opposition in the *Reichstag* and even in his Cabinet. He had received support from the Kaiser for the Russian treaty, but otherwise the Kaiser found it intolerable that his government should have to rely on left-wing support for its commercial policy. Caprivi had been constantly sniped at by Bismarck in retirement. In October 1894 he resigned, and the leaders of the *Bund* had gained their first victory.

[1] In fact the word '*Landwirt*' suggests something more important than 'farmer' – 'landed proprietor' would be more exact. The choice of term reflects the nature of the organisation.

However, they found little consolation in Caprivi's successor. Prince Hohenlohe was a South German who had no sympathy with the Junkers and considered their demands excessive. He frequently made use of the argument that small holdings had no grain to sell and gained no benefit from the tariff.

In any case the damage, from the agriculturalists' point of view, was now done: the treaties were in force and could not be revised until the end of 1903. The *Bund* nevertheless set about organising opposition in readiness for that time and meantime supported vigorously a plan put forward by Count Kanitz (a Prussian landowner) for a government grain monopoly, which through import control would maintain certain minimum prices on the domestic market. These minimum prices would have been well above world market prices. The scheme was put forward several times in the *Reichstag* from 1894 to 1896, but to many members it appeared undesirable or impracticable, or both, and it was defeated on each occasion. The plan is nevertheless of considerable interest as the first attempt to set up a mechanism which at a later date was to become a vital part of agricultural support policies; it is discussed more fully in a later section.

Intensified protection – the 1902 tariff

As the time approached for the commercial treaties to be renewed, and in spite of a significant recovery in world agricultural markets, agitation by the agriculturalists was intensified. In its foreign policy Imperial Germany was aiming at expansion: it sought rearmament and a greater degree of economic self-sufficiency. The political climate thus became more favourable to agricultural protection. In 1900, when a proposal for increased naval expenditure came before the *Reichstag,* the agriculturalists seized the opportunity of bargaining their vote against a promise that the agricultural duties would be raised. Prince Hohenlohe was replaced in 1900 by Count von Bülow, who was more willing to grant increased protection to agriculture.

The preparation of the new tariff had begun in 1897. In 1901, a draft was produced which included substantially higher duties on all agricultural products as well as on many manufactured goods. The increases were not enough to satisfy the *Bund der Landwirte,* which demanded still higher duties on grains. The subsequent debate was long and fierce. Finally the government succeeded in keeping the increases within approximately the limits proposed in the draft. The new tariff was enacted in December 1902.

For most grains it instituted maximum and minimum rates, the latter representing the lowest concession which the government could make in commercial treaties. For wheat the duty was raised from 3.50 marks per 100 kg to a maximum of 7.50 marks and a minimum of 5.50 marks; the increases for other agricultural products were equally large (Table 4.2). There were no minima to the rates on livestock products, which suggested to livestock producers that if concessions had to be made in future treaties, their products were those most likely to suffer.

The government did in fact negotiate a new series of commercial treaties under which the minimum rates on grains became generally applicable. The duties on livestock products were cut by a third, a half or even more, and in several cases were finally no higher than after Caprivi's treaties. The new rates came into force in March 1906.

The livestock sector was however protected in other ways. Reference has already been made to the law of 1880 under which imports of livestock were controlled for 'sanitary' reasons. By 1889 the government had all but closed the borders to imports of live animals. Caprivi relaxed these restrictions, but they were reintroduced after his fall; by 1910 cattle and pigs could be imported only from specified countries and on restrictive conditions. Meat imports too came to be restricted. A law of 1900 prohibited imports of sausages, canned meat and meat with preservatives; imports of pickled or salted meat had to be in pieces of at least 4 kg; imports of meat (other than pickled or salted) had to consist of whole beef carcases or half pig carcases, could enter only at certain ports and on certain days, and were subject to high inspection fees. If the quality of imported meat was judged doubtful, it was destroyed, though domestic meat of similar quality could be sold.

'Agrar- oder Industriestaat?'

It has been seen above that, in the 1870s and 1880s, the industrialists in Germany sought an alliance with the farmers in order to re-establish protection. The alliance, however, did not last. By the 1890s, German industry was in a much stronger competitive position; its further development was hampered both by relatively high wage costs, resulting from high food prices, and by protectionism in other countries, which in the case of the United States at least was in part a reprisal against German agricultural protectionism and could only be broken down by concessions in the German agricultural tariff. This was Caprivi's basic preoccupation. Though Bismarck constantly criticised Caprivi's trade policy, it seems likely that he himself, had he remained in power, would have seen the need to take similar action.

At the same time, there was growing preoccupation with the respective roles of agriculture and industry in the nation. While Germany's total population was growing fast, the agricultural population was falling. It was widely felt that the relative decline of agriculture was undesirable, for social and strategic as well as economic reasons. This question gave rise to much controversy. It was a major political issue during the 1890s, underlying the tariff debates in the *Reichstag,* and it was also the subject of much academic argument.

Many persons feared that a nation exporting ever more manufactures and importing ever increasing amounts of food and raw materials would fall into a precarious situation. Widespread concern on this point seems to have begun around 1897, when Oldenberg stressed the dangers in a speech at the Evangelical-Social Congress in Leipzig. The argument reached its fullest development in a book by Professor Adolf Wagner, *Agrar- und Industriestaat* ('An Agricultural and

Industrial Nation'), of which the first edition, published in 1901, was quickly exhausted. It is worth summarising the analysis of this important work.

Wagner drew attention to the growing industrialisation and to the associated rise in the urban population. This rapid development he regarded as harmful, firstly because it created a degree of specialisation which, though it might have economic benefits, was socially undesirable: it 'makes the occupations of the people one-sided in the extreme, narrowing their outlook, limiting the scope of physical and mental education, removing all the advantages of comprehensive and varied national production' (Wagner, 1902, page 33). Secondly, the increased dependence on imports of food exposed the agricultural population to the vicissitudes of the world market and endangered their livelihood; the agricultural population was a traditional and essential part of the nation and should be preserved. Thirdly, this dependence on imported food endangered the nation's food supply: apart from the possibility of war, there was the risk of year-to-year fluctuations in the harvest, and in the long run it was not certain that world food production would always meet the needs of the rising world population. Further, increased food imports could only be paid for by increased exports of manufacturers, and the prospects for this were doubtful since many countries, including the United States and Russia, were also becoming more industrial. Wagner thus contended that:

> Adequate protection for agriculture, even if higher than the present level, is in the national interest: even if this means that the creation of an industrial nation is retarded, though not entirely prevented, it should thereby benefit the workers and the German economy as a whole, and perhaps also assist the demographic trend. The maintenance of a viable German agriculture signifies the preservation of the German people, both now and in the future. [Wagner (1902) pages 1–2]

Wagner defended himself against the charge that he wanted a purely agricultural nation (as did some who shared his point of view). The question was one of degree: of ensuring the right 'mixture' of agriculture and industry in the nation. Agricultural protection should not seek to raise prices, but to prevent excessive falls in price and to ensure 'normal, average prices'. As a practical proposition, Wagner considered that the agricultural duties should be considerably higher than those in force under the Caprivi duties, and that the increased rates of the 1901 proposals were about right.

It should be noted that Wagner's case had little to do with the need to protect agriculture in times of crisis: the original argument for protection had shifted to a much more sophisticated reasoning which saw protection as a basic long-term need.

Wagner's arguments, well presented and carrying considerable logical force, exerted much influence. They contained the seeds of most of the arguments which appear today in favour of support for agriculture.

A middle-of-the-road position was taken by the eminent economist Schmoller. In a comprehensive economic treatise in 1904, he included a chapter in which he criticised both the Free Traders and the Protectionists. The former, he said, over-emphasised the consumer interest and took too dogmatic a view:

> The main argument [of the Free Traders] is now as always the consumer's point of view: the complaint that protective duties raise the cost of goods. They overlook the fact that the interests of producers deserve equal attention. [Schmoller (1904) Part II, page 643]

But Schmoller considered that the arguments of the Protectionists were equally weak. No country could do without trade. The argument that all economic sectors should have equal protection defeated itself: none would gain if everything were dearer. Sectors in decline should not be protected so much as rising ones, and raw materials not so much as end-products. The argument that agriculture must have tariffs if industry had them was irrelevant: agriculture needed tariffs to help it over the crisis caused by foreign competition. There should be tariffs sufficient to ensure that the agricultural population could survive, but they should not cause much increase in food costs or remove the stimulus to technical improvement.

Several other economists, including Conrad and Dade, felt that the unsatisfactory state of agriculture made it necessary to maintain protection, but that this should be only a transitional measure and should not prevent necessary structural adaptations.

The Free Trade school included Brentano, Dietzel, Lotz and Schäffle. They favoured increased industrialisation of Germany and better international division of labour. They demanded that the tariff should be (gradually and cautiously) pulled down, if possible in the context of commercial treaties giving reciprocal concessions. A fairly full and clear statement of their position was given by Brentano (1901). He stressed the overall benefits of the division of labour and the importance of the law of comparative advantage, citing Adam Smith and Ricardo. He pointed out that even List and his followers were not in favour of protection as a permanent policy. German agriculture, in Brentano's opinion, had not made use of protection to facilitate a shift from grain, now unprofitable, to more economic lines of production: on the contrary, protection had been regarded as an inducement to expand production of grains. But no amount of protection could make German grains competitive with overseas production, because of the much higher cost of land in Germany.

The Free Traders made much of the argument that only a minority of the agricultural population – the large landowners – benefited from high grain prices. They demonstrated (after the 1902 tariff) that higher levels of protection were causing an inflation of land prices, further indebtedness and thus still more limitations on the capacity of small farmers for structural innovation. They also stressed the effects of protection on consumers: Brentano (1910) calculated that a duty on bread grains of 5.50 marks per 100 kg meant that the average worker, with an average family, had to work thirteen days in the year simply to pay the duty.

However, the Free Traders failed to take account of the difficulties in adapting established farming patterns to the new situation. Their arguments were ineffective against the political strength of the agricultural movement. Though there was some popular agitation against 'dear bread', this never became a decisive force. The Free Traders had a smallish representation in the *Reichstag* among the Radicals, but on the whole did not exert great influence. The most vociferous advocates of lower tariffs were the growing numbers of Socialist representatives, for whom cheap bread was a major political issue.

The Junkers and the *Bund der Landwirte*

The political action undertaken by the *Bund der Landwirte* from its foundation in 1893 onwards must be counted the decisive factor in bringing about the reinforcement of agricultural protection. The *Bund* was formed and led by the Prussian landlords ('Junkers') who – as has been seen in the first section of this chapter – had actually reinforced their position following the 'reforms' earlier in the century, and its primary objective was to secure better markets for the grain produced on the Prussian estates. It included, from about 1896 onwards, an increasing number of relatively small farmers from the western states – a fact which the leadership emphasised – but these members played little part in the decision-making organs of the *Bund.* It succeeded in imposing itself as the principal or even the sole voice of the farming profession. Numerous other farmers' organisations did exist in west and south Germany (cf. in particular Farr, 1978). Some of these co-operated with the *Bund,* some opposed it, but conflicts were on religious or doctrinal grounds, and even the 'Peasant Leagues' – rightly or wrongly – did not seriously oppose the *Bund* on the issue of grain tariffs. In fact the *Bund* won the allegiance of many small farmers by promising increased duties on livestock products, claiming that the interests of big landowners were identical with those of small peasants and stressing that only the solidarity of all farmers could achieve political results.

The *Bund der Landwirte* was a new type of pressure group. Its methods were at times unscrupulous. One contemporary writer declared that 'agrarian terrorism has become one of the most dangerous phenomena in Germany's political life' (Bürger, 1911). It organised demonstrations, distributed propaganda, issued manifestos; it ordered its members to boycott traders who did not give support, and was said to have threatened physical violence against opponents; it was vigorously anti-semitic. Its principal means of action, however, was political and was exerted mainly – but not exclusively – through the Conservative party. In elections the *Bund* gave support to candidates willing to subscribe to its programme; if no such candidate was found, the *Bund* might put up a candidate of its own. It kept a strong hold over all members of parliament elected with its support: any who departed from its line usually found their political career at an end. Already after the 1893 elections, 140 members of the *Reichstag* formed an economic group (*Wirtschaftliche Vereinigung*) inspired by the *Bund,* which played a crucial role in the subsequent debates on the commercial treaties. Puhle (1966) considered that the *Bund* had altered, for the worse, the character of the Conservative party as a whole and was also responsible for introducing political agitation outside the parliamentary procedure as a means of pressure. It thereby, in his view, bore heavy responsibility for later political developments.

The *Bund's* attacks were particularly directed at Caprivi, accused of being an exponent of 'un-German' liberalism; the *Bund* was mainly responsible for Caprivi's fall, though subsequently it could not impose a candidate of its own as Chancellor. The *Bund's* opportunity came with the new tariff of 1902 and with the expiry of Caprivi's treaties which enabled the new higher rates to be put into effect; the treaty with Russia was regarded as particularly significant. The Vicomte de Guichen aptly described the influence which the Junkers had exerted through the *Bund:*

> The members of the distinguished German aristocracy, a considerable number of agriculturalists ... were the most reliable supporters of the Throne, the descendants of the founders of the Empire. Would their Emperor take the risk of displeasing them, possibly of ruining them, in order to give satisfaction to the Russians? A cruel choice! This was in fact one of the most serious problems facing Germany, with an industry in full expansion asking only to conquer new markets, but with a great landed class insisting that there should be no departure, so far as agriculture was concerned, from a restrictive customs policy. [Guichen (1917) page 18]

After the First World War, the *Bund der Landwirte* was amalgamated with the *Deutscher Landbund* to form the *Reichslandbund,* which in 1929 joined with other farm organisations in the Nazi-dominated 'Green Front'.

The proposed grain monopoly

Reference has been made above to the proposals put forward from 1894 to 1897 by Count Kanitz, with the support of the *Bund der Landwirte,* for an import monopoly for grain. The proposal (subject to variations in its successive versions) was that all imports of grain should be made on the State's account, and that the price at which the grain was resold on the domestic market should be the average of prices in the previous forty years. The profits made when import prices were below this level should go to a reserve fund, to be used for subsidising imports on occasions when foreign prices rose above the price of resale. As prices in the mid-1890s were much lower than those which had prevailed during most of the previous forty years, the immediate result would have been to raise domestic prices very substantially above the world price and above their existing level.

The proposal aroused passionate interest in agricultural circles. Buchenberger, in a contemporary study of German agricultural policy, wrote:

> Seldom has an agricultural programme attracted so much attention as the idea of nationalising grain imports ... The proposal showed how, with apparently simple means, a certain and immediate remedy could be found for all the problems of the time – the intolerable situation of inadequate prices, falling

rents, increasing indebtedness and impoverishment – and lead the way to a Promised Land safe from all fluctuations in prices. [Buchenberger (1899) page 240]

The proposal nevertheless encountered strong objections. It was opposed by Count Caprivi and his successor Prince Hohenlohe. The latter declared it to be an insidious and dangerous step towards Socialism, while the Social Democratic leader, Bebel, retorted that a measure to enrich one class at the expense of the community, and in particular of the workers and the poor, was anything but social. The Junkers however paid little attention to consumer interests, and the Prussian Minister of Agriculture declared in the *Reichstag* that the consumer had 'no right to get goods delivered to him below the cost of production' (the cost of production he had in mind being that which prevailed on Prussian estates).

The practicability of the scheme was also questioned. It would be difficult to determine an equitable minimum price. Millers preferred the higher-quality imported grain and would not necessarily pay more for home-produced grain. There would be technical problems in regulating supplies so as to avoid price fluctuations. Both supporters and opponents of the Kanitz plan drew the conclusion that, to ensure successful operation, control would have to be extended to domestic supplies. The objections to such far-reaching intervention by the State were at the time too strong to be overcome; but the essentials of the Kanitz proposals were introduced by the National Socialists in 1933 and maintained after the Second World War by the German Federal Republic.

The import certificate system

Import certificates were an arrangement devised to take account of some special features of the German grain trade; they provide a further illustration of the extent to which the interests of the Prussian grain-producers were catered for. It was a long-standing practice for Russian wheat and rye to pass through east German ports, and there to be mixed with the local grain, the quality of which was inferior and which thereby obtained an increased outlet. When duties on grain were revised in 1879, arrangements were made to enable this transit trade to continue: the re-export of foreign grain, or of foreign grain mixed with German grain, gave the right to import a corresponding quantity of grain free of duty.

This arrangement, however, did not cover another aspect of the grain trade, which was that east Germany sent most of its produce abroad, while west Germany imported the greater part of its requirements. Transport from east German ports by sea to Scandinavia was often cheaper than transport by rail to west Germany; further, west German consumers were used to a higher-quality flour than could be obtained from Prussian grain.

The opportunity for extending the system arose immediately after the passage of the commercial treaty with Russia in 1894, the justification being that producers required compensation for the expected increase in Russian competition. Any exports of the major cereals, whether foreign or home-produced, now gave entitlement to a certificate which could be used to obtain exemption from duty on imports of grain. Moreover, the import certificate was negotiable. Certificates obtained through the export of east German grain could thus be used to import grain into west Germany. The exporter could sell the certificate for its face value less only a small deduction for interest and transfer charges: he therefore received the export price for his grain plus nearly the full amount of the duty, and he had an incentive to export so long as the differential between the world price and the domestic price did not exceed the amount of the duty. The arrangement worked in effect as an export subsidy.

Initially, the certificate was valid only for importing grain of the same type as that which had been exported. However, under the Bülow tariff of 1902 this limitation was removed: the certificate was valid for importing any kind of grain. This enabled the system to be used for exporting German rye to Scandinavia and importing various kinds of grain, including feed grains, into west Germany.

As a result of the import certificate system, German exports of wheat and rye, which had been negligible since the 1880s, made a spectacular recovery from 1894 onwards. Moreover, the price level in east Germany was raised above the world price by approximately the amount of the duty and brought closer to the price on west German markets.

Considerable encouragement was thereby given to the production and export of grain – the trends are further discussed in the following section – and the value of the import certificates granted rose correspondingly: from 7 million marks in 1894 to 22 million in 1900 and 121 million in 1910. These amounts represented a loss of revenue which would otherwise have been obtained from import duties and were equivalent to a direct subsidy to producers and exporters. Even when exports of one kind of grain exceeded imports, as happened regularly for rye and occasionally for oats from about 1907 onwards, the system still operated owing to the possibility of using the certificates for importing other types of grain. Proposals were made to abolish this interchangeability, so that there would be no use for the certificates once exports were greater than imports; excess production would then be discouraged. However, no changes were made up to the First World War, and the system was revived to play once again an important role from 1925 to 1930.

German agriculture under protection

The measures of protection adopted in Germany from 1879 on did not prevent the fall in the world price of grain from being reflected on German markets: prices fell from 1879 onwards, and, after some recovery around 1890 with short harvests, they reached their lowest point in the mid-1890s. There was then a recovery lasting up to the First World War (Table 4.3).

The duties on grain certainly tempered the fall in price (see figure 1.1 in chapter 1) and seemed for a while at least to have restrained the growth of imports. Imports of wheat and rye, after beginning to increase in 1876, were checked in 1880 and did not rise again till the 1890s, helped by the reduction of duties under Caprivi's treaties (Table 4.4). With the rising population and hence the rising demand for bread, imports of wheat then increased very fast indeed up to the war. Imports of barley, mainly to feed the increasing numbers of pigs in west Germany, also expanded rapidly.

Perhaps the most striking feature of German agriculture during this period is that there was no contraction in the area under grain: indeed, the area sown to rye and oats expanded in the 1890s and 1900s (Table 4.5). As yields were rising under the influence of improved techniques, production of all grain – rye and oats in particular – expanded rapidly, doubling in the course of the thirty years 1880–1910 (Table 4.6). The tariff system gave a big advantage to rye as compared with wheat, for the duty on rye under the Caprivi treaties was the same as

Table 4.3: Wholesale prices of major agricultural products

	Rye	Wheat	Oats*	Barley	Slaughter cattle	Butter
			marks per 100 kg			
1860–69	20	15	14	15	..	..
1870–74	24	18	16	17	..	..
1875–79	21	17	15	16	..	..
1880–84	16	20	14	18	..	..
1885–89	14	17	13	16	94	..
1890–94	16	18	15	17	110	..
1895–99	13	16	13	13	115	228
1900–04	14	16	13	13	124	231
1905–09	17	20	16	14*	141	238
1910–13	17	21	16	15*	160	261

Note: The data prior to 1880 refer to the whole of Prussia, and are taken from Brentano (1910). Subsequent data refer to Berlin, except that for barley till 1893 the figures refer to Magdebourg, and from 1894 for both barley and oats to Breslau; the source is the *Statistisches Jahrbuch für das Deutsche Reich.*

* Feeding barley only from 1906 onwards.

Table 4.4: Imports and exports of grain

	Rye		Wheat		Oats		Barley	
	Imports	Exports	Imports	Exports	Imports	Exports	Imports	Exports
				thousand tons				
1865–69	298	117	409	590	95	110	112	117
1870–74	626	125	365	432	175	110	119	100
1875–79	1082	155	819	660	304	124	355	194
1880–84	732	15	535	82	265	33	321	95
1885–89	737	3	450	6	181	8	480	30
1890–94	630	10	946	16	208	5	799	9
1895–99	866	87	1404	23	399	43	1056	23
1900–04	804	168	1891	162	420	139	1165	35
1905–09	490	407	2255	186	552	298	2078	5
1910–13	418	830	2419	363	564	445	3037	3

Source: Statistisches Jahrbuch für das Deutsche Reich.

Table 4.5: Area of major crops

	Rye	Wheat	Oats	Barley
		million hectares		
1880	5.9	1.8	3.7	1.6
1890	5.8	2.0	3.9	1.7
1900	6.0	2.0	4.1	1.7
1910	6.2	1.9	4.3	1.6

Source: As Table 4.4.

Table 4.6: Production of grain

	Rye	Wheat	Oats	Barley
		million tons		
1880–84	5.6	2.5	4.1	2.2
1885–89	5.8	2.6	4.5	2.2
1890–94	6.9	3.0	5.1	2.5
1895–99	8.4	3.5	6.3	2.8
1900–94	9.3	3.5	7.3	3.1
1905–09	10.2	3.7	8.2	3.2
1910–13	11.3	4.3	8.5	3.3

Source: As Table 4.4.

for wheat, and under the Bülow tariff it was only slightly less, whereas rye was much easier to grow on the soils and in the climate of east Germany.

The preference of consumers, however, was turning from rye bread to wheat. The increase in rye production, artificially stimulated by protection and above all by the import certificate system, was wholly out of line with the trend in demand. The result was that imports of rye fell after 1904, while exports increased enormously. Also under the influence of the import certificate system, Germany once again began to export wheat and oats on a substantial scale (Table 4.4).

As has been mentioned above, the question as to how many of the rural population actually benefited from tariffs on grain was frequently debated in Germany at this time, and various calculations were made – not always unbiased. Hunt (1974) and Farr (1978) refer to farm surveys which suggest that quite a high proportion of relatively small farms did sell some grain and that most were about self-sufficient in bread and fodder. The data, however, are fragmentary. It seems reasonable to suppose, as did some relatively objective studies at the time, that holdings of less than five hectares can rarely have benefited from high grain prices and that only those over twenty hectares got a substantial benefit. In 1895 there were 4 252 000 holdings under five hectares, 999 000 holdings between five and twenty hectares, and only 307 000 holdings over twenty hectares. The average number of persons per holdings was certainly higher on the larger farms, but the extra persons consisted mainly of hired labourers, whose benefit from increased farm prices must have been very indirect. The proportion of farmers deriving a clear benefit from increased grain prices therefore appears small indeed, in relation to the total agricultural population of some 18.5 million and to the total German population of 52 million.

It has been seen that the livestock sector received protection not so much through tariffs as through import restrictions imposed ostensibly for 'sanitary' reasons (see Hunt, 1974). Imports of live animals were held approximately constant from 1880 onwards. The volume of meat imports rose until about 1900, but was then reduced by implementation of the 1900 legislation relating to meat. Imports of butter and eggs, however, continued to rise (Table 4.7).

Behind these import barriers, and with the benefit of rising demand, prices of livestock products remained firm or even increased. The number of cattle rose steadily, that of pigs multiplied (Table 4.8). The number of sheep, however, fell even more drastically than in France: in Germany too, wool received no protection, in deference to the manufacturers, while German consumers preferred pigmeat to mutton.

Table 4.7: Imports of major livestock products

	Beef		Pigmeat	Butter	Eggs
			thousand tons		
1880–84		15*		5	17
1885–89		6		6	44
1890–94		24		8	64
1895–99		53		10	98
1900–04		45		22	123
1905–09	15		10	39	136
1910–13	25		..	52	135

* Including poultrymeat, meat extract, etc., up to 1883.

Source: As Table 4.4.

Table 4.8: Numbers of livestock

	Cattle	Pigs	Sheep
		millions	
1873	15.8	7.1	25.0
1883	15.8	9.2	19.2
1892	17.6	12.2	13.6
1900	18.9	16.8	9.7
1904	19.3	18.9	7.9
1913	21.0	25.7	5.5

Source: As Table 4.4.

The number of people living on the land, as well as the number working in agriculture, appear to have risen up to 1882 but to have fallen thereafter (Table 4.9). The main reduction in agricultural employment took place in the east: between 1882 and 1907, the agricultural labour force in the eastern provinces is estimated to have fallen by about 10% as compared with a national average of 6%. This reflected substantial emigration from the east to the rapidly growing industrial areas of the Ruhr, Berlin and Saxony; there was relatively little industrial development in the east, because of the lack of natural resources and also as a result of deliberate opposition to industrialisation by the Junkers and the

Bund der Landwirte, who feared the spread of socialism and also did not want the Polish proletariat to participate in industrialisation (see Tipton, 1974). All over Germany, the reduction in agricultural employment concerned primarily farm labourers and domestic servants: the number of farmers seems to have begun to decline only after the 1895 census.

This movement out of agriculture seems to have been caused primarily by the pull of industrial growth, only secondarily by 'push' factors in agriculture. However, the depression did undoubtedly cause a good deal of rural distress, despite protection, and many farmers got into debt.

Table 4.9: Population in agriculture and forestry, in relation to population in all sectors

	1871	1882	1895	1907
Active population			*millions*	
Agriculture and forestry	..	10.5	10.2	9.9
All sectors	..	21.2	24.0	28.1
			percent	
Agriculture as percentage of total	..	50	42	35
Total population			*millions*	
Agriculture and forestry	18.7	19.2	18.5	17.7
All sectors	39.4	45.2	51.8	55.8
			percent	
Agriculture as percentage of total	47	43	36	32

Source: Statistisches Jahrbuch für das Deutsche Reich, corrected as regards the active population by Tipton's calculations (Tipton, 1974) in which he adjusted the female participation rate for earlier years on the basis of the 1907 census.

Farm structures did not evolve significantly after 1848. It has been pointed out in the first section of this chapter that following the reforms up to 1848, the pattern of small fragmented holdings persisted in south and west Germany, the main difference being that now they were mainly owned by the peasants. A need to improve these structures was beginning to be felt: already around 1800 Albrecht Thaer had described the situation in Germany as 'barbaric' by comparison with that in England. There were cases of consolidation in some states from the early nineteenth century onwards, as well as some reallocation of common land. Progress was limited, however: a satisfactory rearrangement of plots required the consent of all owners of land in the area concerned. Somewhat more effective laws to promote consolidation were introduced by a number of states around the middle of the century, but still results were meagre. Some states also tried to curb the practice of *Realteilung* or equal division between heirs, which was at the root of the problem of fragmentation. Bavaria, however, was among the states where these efforts, though badly needed in many areas, did not succeed.

The main preoccupation in the structural field, in fact, was with the resettlement of small farmers in thinly-populated areas (*Innere Kolonisation*), particularly (under a law of 1886) in the Prussian provinces of Posen and west Prussia which had begun to feel the effects of emigration by farm workers to the industrialising regions of the rest of Germany or overseas. Small-to-medium holdings were created on a new basis, the *Rentengut,* enabling peasants with limited means to acquire possession over time through payments in money or in crops.

Technical progress was significant, especially on the big Prussian estates, but also in the rest of the country once the agrarian reforms had removed the obstacles to individual initiative. Albrecht Thaer's pioneering work on 'rational farming' earlier in the century remained influential. Crop rotations were improved once the constraints of the open-field system were lifted. England was the model for much technical progress, including drainage. New machinery came into use, including the steam plough. In agricultural chemistry Germany led the world, and artificial fertilisers came into wide use. Agricultural academies were created and agricultural research flourished. Basic education became compulsory and gradually the peasants learned to read and write. Mortgage credit institutions had been founded in a number of states under the agrarian reforms. Short-term credit developed later, but from the 1860s onwards the expanding network of *Raiffeisen* co-operative banks helped to solve the financial problems of the peasants and also acted as collective purchasers of agricultural requisites. Co-operative sales organisations – dairies in particular – made a start. Insurance against farming risks began to develop.

From 1890 to 1913, the German national income grew at a rate estimated at 2.9% annually, which was faster than Britain, France and several other European countries, but lower than Denmark, Sweden and the United States; German exports rose by some 5.1% annually – faster than almost any other country (calculations by Madison, 1962). It appears therefore that the protection given to agriculture, though it kept up food prices, encouraged excessive production of grains, retarded the shift of manpower to other sectors and impeded the conclusion of commercial treaties, was not an adverse factor strong enough to prevent continued economic growth. However, it certainly did not facilitate it; and the fact that more damage was not done can probably be attributed to Caprivi's commercial treaties and to the dynamism of German industry.

Bibliography

Several very thorough histories are available in German covering different periods. For the emergence from feudalism up to the land reforms of the early nineteenth century, reference has been made here mainly to Franz (1970) and Lütge (1963); some historians would prefer the analysis by Abel (in several works). (It was not possible at the time of writing to consult the recent book by Dipper, 1980.) Haushofer (1963) continued the story through the nineteenth century: his work is particularly useful for its account of the impact of technological developments prior to the Great Depression. Frauendorfer (1957) provided a general agricultural history up to the First World War. For the Great Depression in particular, an informative early study in English was that by Dawson (1904). Other valuable studies (by American historians) were those by Gerschenkron (1943) and Tirrel (1951): both these are helpful in bringing out the political manoeuvring of the period. The role of the *Bund der Landwirte* was further analysed by Puhle (1966); Hunt (1974) and Farr (1978) usefully complemented and qualified previous work on protection and the role of various farm organisations.

Reference has been made in the chapter to the writings of economists during the period studied. Wagner's work (1902, 2nd edn.) was particularly significant as a justification of protectionism; Schmoller (1904) took a moderate position and Brentano (1901) presented the Free Trade argument. Dietzel (1901) provided a useful account in English of the current controversies.

Abel, W. (1966) *Agrarkrise und Agrarkonjunktur.* 2nd edn. Hamburg.

Abel, W. (1967) *Geschichte der deutschen Landwirtschaft vom frühen Mittelalter bis zum 19. Jahrhundert.* 2nd edn. Stuttgart.

Abel, W. (1967) *Agrarpolitik.* Göttingen: Vandenhoeck & Ruprecht.

Beckmann, F. (1911) *Einfuhrscheinsysteme.* Karlsruhe.

Beckman, F. (1913) *Die Futtermittelzölle.* München.

Brentano, L. (1901) *Das Freihandelsargument.* Berlin-Schöneberg.

Brentano, L. (1910) *Die deutschen Getreidezölle.* Berlin.

Buchenberger, A. (1899) *Grundzüge der deutschen Agrarpolitik.* 2nd edn. Berlin.

Burger, C. (1911) *Die Agrardemagogie in Deutschland.* Lichterfeld.

Clapham, J. H. (1945) *The Economic Development of France and Germany, 1815–1914.* 4th edn.: Cambridge University Press.

Cobden Club (1910) *The Influence of Protection on Agriculture in Germany.* London.

Conrad, J. (1900) *Die Stellung der landwirtschaftlichen Zölle in den 1903 zu schliessenden Handelsverträgen Deutschlands.* Schriften des Vereins für Sozialpolitik **90.**

Dade, H. (1901) *Die Agrarzölle.* Schriften des Vereins für Sozialpolitik **91.**

Dawson, W. H. (1904) *Protection in Germany.* London.

Dietzel, H. (1901) 'Kornzoll und Sozialreform' *Volkswirtschaftliche Zeitfragen* **177/178.** Berlin.

Dietzel, H. (May, 1903) 'The German tariff controversy'. *Quarterly Journal of Economics.*

Dipper, C. (1980) *Die Bauernbefreiung in Deutschland 1790–1850.* Stuttgart: Kohlhammer.

Farr, I. (1978) 'Populism in the countryside: The Peasant Leagues in Bavaria in the 1890s'. In Evans, R. J. *Society and Politics in Wilhemine Germany.* London: Croom Helm. New York: Barnes & Noble.

Franz, G. (1970) *Geschichte des deutschen Bauerntums vom frühen Mittelalter bis zum 19. Jahrhundert.* Stuttgart: Eugen Elmer.

Frauendorfer, S. von (1957) *Ideengeschichte der Agrarwirtschaft und Agrarpolitik im deutschen Sprachgebiet – Band I: Von den Anfängen bis zum ersten Weltkrieg.* München: BLV Verlagsgesellschaft.

Gerschenkron, A. (1943) *Bread and Democracy in Germany.* University of California.

Gläsel, E. J. (1917) *Die Entwicklung der Preise landwirtschaftlicher Produkte und Produktionsmittel.* Berlin.

Guichen, Vicomte de (1917) *Le Problème agricole allemand.* Conférence faite le 5 octobre 1917 à la Société d'Economie Politique.

Haushofer, H. (1963) *Die deutsche Landwirtschaft im technischen Zeitalter.* Stuttgart: Eugen Ulmer.

Hunt, J. C. (1974) 'Peasants, grain tariffs and meat quotas: Imperial German protectionism re-examined'. *Central European History* **7**, 311–31.

Lotz, W. (1892) *Die Ideen der deutschen Handelspolitik, 1860 bis 1891.* Schriften des Vereins für Sozialpolitik **50.**

Lotz, W. (1900) 'Der Schutz der deutschen Landwirtschaft', *Volkswirtschaftliche Zeitfragen* **170/171.** Berlin.

Lotz, W. (1901) *Die Handelspolitik des deutschen Reiches unter Graf Caprivi und Fürst Hohenlohe, 1890–1900.* Schriften des Vereins für Sozialpolitik **92.**

Louis-Dop (1911) *La politique agraire en Allemagne.* Rome: Institut International de l'Agriculture.

Lütge, F. (1963) *Geschichte der deutschen Agrarverfassung vom frühen Mittelalter bis zum 19. Jahrhundert.* Stuttgart: Eugen Ulmer.

Macbride, H. (1945) 'Note on the economic basis of the Junker class'. In Viner, J. et al. *The United States in a Multi-National Economy.* New York. Council on Foreign Relations.

Madison, A. (June, 1962) 'Growth and fluctuations in the world economy, 1870–1960'. *Quarterly Review of the Banca Nazionale del Lavoro* **61.** Rome.

Puhle, H.-J. (1966) *Agrarische Interessenpolitik und preussischer Konservatismus im wilhelminischen Reich (1893–1914).* Hanover: Verlag für Literatur und Zeitgeschehen.

Ritter, K. (1925) *Die deutschen Agrarzölle.* Schriften des Vereins für Sozialpolitik **171.**

Schäffle, A. (1902) *Die agrarische Gefahr.* Berlin.

Schmoller, G. (1904) *Grundriss der allgemeinen Volkswirtschaftslehre* (zweiter Teil). Leipzig.

Schreiner, G. (1975) 'Die Entwicklung der deutschen Agrarstrukturpolitik von der Reichsgründung im Jahre 1871 bis zum Ende des Zweiten Weltkrieges'. *Berichte über Landwirtschaft* **53** (2), 219–315.

Sering, M. (1894) *Das Sinken der Getreidepreise und die Konkurrenz des Auslandes.* Berlin.

Tipton, F. B. (1974) 'Farm labor and power politics: Germany, 1850–1914'. *Journal of Economic History* **XXXIV** (4), 951–79.

Tirrel, S. (1951) *German Agrarian Politics after Bismarck's Fall.* Columbia University Press.

Treue, W. (1933) *Die Deutsche Landwirtschaft zur Zeit Caprivis und ihr Kampf gegen die Handelsverträge.* Berlin.

Wagner, A. (1902) *Agrar- und Industriestaat.* 2. Auflage. Jena.

Weihe, E. M. (1902) *Der Einfluss der deutschen Schutzzollpolitik auf die Entwicklung der Industrie und Landwirtschaft.* Heidelberg.

Chapter 5

Denmark

By the mid-nineteenth century the Danish countryside was already characterised by today's familiar pattern of farmsteads situated on compact blocks of land. Small by British standards, these farms were nevertheless large by comparison with most other Continental countries, besides being free of fragmentation. Danish agriculture had not yet undergone the transformation that was to turn it into a highly efficient sector specialising in the production of livestock products: cereals were still the main product. But social and economic conditions had been created which enabled Danish agriculture to adapt when the challenge came.

The liberation of the peasants, land reform and education

Only a hundred years further back, in the middle of the eighteenth century, a very different situation had prevailed. Practically all the land was then owned by the nobility, other large landowners, the Crown and the Church. Only a handful of peasants owned their holdings and even in these cases their ownership titles were often ambigious and uncertain. The mass of the peasantry farmed land that was not their own, under tenancy contracts that did not define their rights, paying rent and above all rendering burdensome compulsory services to their landlords. They lived in compact villages and the land was cultivated usually in open fields, the three-field system (winter grain, spring grain and pasture) being the most common. Moreover, the peasants were bound to their place of birth: 'adscription'[1] provided the landlords with cheap labour and by enabling them to choose who should go to the detested military service, gave them a weapon which they sometimes used unscrupulously. Finally, some landlords exerted legal authority over their tenants; all could impose punishment, often brutal.

It was in some ways one of the most repressive agrarian systems in western Europe, tempered only by a reasonably independent judiciary and by the possibility for peasants to petition the Crown directly. In Denmark, however, the system had not originated with the spread of feudalism in the early Middle Ages. Indeed, around the thirteenth century, some two-thirds of the land had probably belonged to the peasants and the peasantry had formed a Fourth Estate which participated in the election of the King. Progressively, however, the Danish peasantry had lost its land and its independence, largely through putting itself under the protection of the lords in troubled times.

Yet within a historically short space of time – a few decades at the end of the eighteenth and beginning of the nineteenth century – and without violence, the peasantry had regained possession of the land, bondage and compulsory labour service had been abolished, and a farm structure had been created which permitted Danish agriculture to become one of the most efficient in Europe.

How did such a significant reform come about? A major factor was the strength of the crown. In 1660, when Denmark had been humiliated by successive defeats at Swedish hands, losing its provinces in southern Sweden, and when the

[1] Historians prefer to render the Danish *Stavnsbaand* as 'adscription' rather than as 'bondage' or 'serfdom', as the Danish system was later in date than the feudal systems of other countries and somewhat different in nature.

nobility's prestige was thereby reduced, the Estates General had agreed to institute an absolute, instead of an elected, monarchy. Already in the early eighteenth century, the Crown made some efforts on behalf of the peasantry: in the 1760s it initiated reforms on its own estates. The other main element was the emergence of reform-minded leaders among the land-owning nobility itself. The currents of enlightenment then flowing in Europe may have had an influence, in particular those aspects of Physiocratic thought which urged the suppression of monopoly and privilege through firm government and a powerful monarch, but developments in Denmark had a strongly national character. Several of the nobility made their estates into models of reform.

Public discussion of the state of the countryside was stimulated by a Commission of Inquiry in 1757; another inquiry took place in 1767. In 1770–72 Johann Struensee, who temporarily gained control of the government, attempted numerous reforms (relating in particular to the freedom of the press and to the judicial system), but these had as yet no significant effect on agriculture and a brief period of reaction followed. In 1784 the young Crown-Prince Frederik was able to take over control from his mentally unbalanced father, and appointed a new government of reform-minded leaders, headed by Count Andreas Peter Bernstorff, with Count Christian Ditlev Reventlow in charge of agriculture. (Both Bernstorff and Reventlow were themselves enlightened landowners.) A 'small' Commission of inquiry into agriculture sat in 1784; then in 1786 the 'big' Agricultural Commission was instituted and in the succeeding years carried through major reforms. Besides Reventlow and other nobles, this Commission included several lawyers, in particular Christian Colbjørnsen who proved particularly active for reform.

Growing realisation of the inefficiencies of the existing system no doubt facilitated the specific reforms which were undertaken. With growing rural population, 'adscription' (*Stavnsbaand*) was becoming less necessary to landlords as a means of ensuring their labour supply, while the need for more labour in the towns made the system seem increasingly anachronistic. Many landlords were also aware of the relative inefficiency of compulsory labour services (*Hoveri*). There was increasing unrest among the peasantry, particularly provoked by the *Stavnsbaand* and *Hoveri* systems.

The *Stavnsbaand* was abolished in 1788 – to end immediately for all peasants under fourteen or over thirty-six, by 1800 in all other cases. *Hoveri*, already regulated to some extent since 1769, was reformed by a decree of 1791; then in 1799 a further decree provided for labour services to be commuted into money payments. Enclosure of the scattered open-field strips into consolidated plots (*Udskiftning*) was promoted under a series of laws starting in 1758–60 and became significant after major legislation in 1792; the initiative was taken by landowners, by village communities and by individual freeholders. Following abolition of the *Stavnsbaand* and *Hoveri*, landowners had a greater interest in selling land to their former tenants. From 1788 onwards, peasants were able to obtain loans from a government credit bank at low interest rates and with repayment of capital over twenty-five years, helping them to purchase land and to establish farmsteads on their land. Tenants were given much greater security of tenure by the provision that leases had to be for a minimum of fifty years or for life. Even cottagers, who previously had no share in the common fields but only some rights of common pasture, received grants of two or three hectares. Finally, legislation further encouraged the growth of a strong society of small farmers by prohibiting the merging of peasant holdings into larger farm units and by preventing the subdivision of farms into units too small to support a family.

By about 1810, most of the open fields had been enclosed, many peasants moving out of the villages into new farmsteads. A large part – between a third and a half – of the land farmed by peasants had become their own property and in the course of the nineteenth century the proportion of freehold farms steadily increased. The process was helped not only by the state loans but by the rise in product prices provoked by the Napoleonic Wars, which gave the peasantry extra cash for purchases and improvements. The maintenance of a favourable structure was assisted by the custom whereby the son (or daughter) best fitted to do so inherited the farm and compensated the other heirs in cash or by a mortgage.[1]

Most important of all, perhaps, was the change in attitudes brought about by the reforms, by the removal of restrictions and by the new opportunities created for individual initiative. 'Step by step' – comments Ilsøe (1940, page 161) – 'the whole social structure, which had rested on compulsion, was replaced by a new one, based on freedom'.

These reforms in agriculture formed part of the general movement to effective democracy in Denmark. Various political reforms in the first half of the nineteenth century brought about a democratic structure in which the influence of the remaining large landowners was substantially curbed, and in which the mass of the peasantry had a voice and an influence unknown anywhere else in Europe. It is true that throughout the Great Depression of the 1880s and 1890s the Conservative party held office, and this was to some extent the party of the remaining big agriculturalists. But the *Venstre* (left) party, which began as a rural group and gradually became a people's party, held a big majority in the *Folketing* (lower house) for most of the time; its internal divisions prevented it from forming a government until 1901, but it was able to exercise an effective check. Bearing in mind the disproportionate influence exerted in France and Germany by a dominating class of large landlords, the virtual absence of any such class in Denmark is a significant feature.

Rural education began early in Denmark. An Elementary Education Act of 1814 made schooling compulsory between the ages of six and fourteen. The famous Folk High Schools of Bishop Grundtvig began in the 1840s: by 1874 there were fifty such schools, with over three thousand pupils. The Folk High Schools aimed at providing education for students aged eighteen and upwards, mostly from farm families; the subjects taught were largely technical, but some

[1] A practice different from that in Norway, for example, where (as in England) the eldest son had preference.

academic courses were included. From 1868 the State provided grants to poor students; by 1900 about one student in three received a government grant.

There was also an increase in the number of agricultural colleges, which had about a thousand students by 1900. The Royal Agricultural Society was active in spreading the latest knowledge of farming techniques; from 1845 on it held assemblies every few years. The Royal Veterinary and Agricultural College was instituted in 1858.

Thus by the Depression, the Danish agricultural population had a high standard of both general education and technical agricultural knowledge. As a result, the Danish farmer was more willing and better able to embark on adaptations than his counterparts in other countries.

The adaptation to crisis

Before the advent of cheap overseas grain, Denmark herself was a major grain exporter. Indeed, grain was her most important export, though she also sent some butter, cheese, eggs and other livestock products to Britain and to Germany.

The fall in grain prices hit Denmark, as other countries, in the late 1870s; in the 1880s the decline for wheat in particular became acute; prices reached their lowest point in the mid-1890s, and there was a revival from about 1900 (Table 5.1). Livestock products were much less affected: for these, transport costs remained relatively high and restrained the growth in foreign competition. The

Table 5.1: Prices of major agricultural products in Copenhagen

	Wheat	Rye	Barley	Oats	Butter	Pig-meat	Eggs	Fat cattle
	kroner per hectolitre				*kroner per 100 kg*			
1866–70	14	10	8	5	140	96	..	..
1871–75	14	10	9	6	166	104	..	..
1876–80	13	10	9	6	170	106	..	..
1881–85	11	9	8	5	180	110	79	110
1886–90	9	7	7	5	178	94	79	96
1891–95	8	7	7	5	188	96	88	99
1896–1900	8	7	7	5	190	88	91	87
1901–05	9	7	7	5	192	104	102	98
1906–10	10	8	8	5	204	116	110	100

Sources: Statistisk Aarbog; and Danish Council of Agriculture (1935) for eggs and cattle.

prices of meat and livestock fell somewhat up to the end of the century, but the decline was much less acute than in the case of grain. A substantial improvement in the quality of Danish butter enabled its price to rise.

This difference in price trends gave an advantage to livestock production both directly, through the price of the product, and indirectly, in that feed grains and other feedingstuffs could be imported very cheaply. Danish farmers saw the significance of this, and instead of supporting the price of grain through import duties they embarked on a fundamental transformation of their agriculture.

The challenge of overseas grain production had already been foreseen in the 1860s by enlightened agricultural leaders such as Edward Tesdorpf, the president of the Royal Agricultural Society from 1860 to 1888: Tesdorpf was a large landowner who introduced greatly improved farming methods on his own land, and used grain and fodder as the basis for intensive dairying. As a result of the encouragement given by such men to livestock production, the transformation of Danish agriculture was already well under way when the crisis came. The number of cattle – dairy cows in particular – was rising steadily; pigs were multiplying even more rapidly (Table 5.2). Only sheep, in common with the trend throughout western Europe, were falling in numbers under the influence of overseas competition in wool. At the same time, livestock yields were rising enormously with improved methods: between 1871 and 1914 the average annual output of milk per cow rose from 1300 kg to 2700 kg. The output of livestock products thus rose in leaps and bounds: production of milk, butter, eggs and pigmeat trebled or more in this period (Table 5.3).

Table 5.2: Numbers of livestock

	Dairy cows	Other cattle	Pigs	Sheep	Poultry
			thousands		
1861	757	364	304	1749	..
1871	808	431	442	1842	..
1881	899	571	527	1549	4070
1893	1011	695	829	1247	5856
1903	1089	751	1457	877	11555
1914	1310	1153	2497	515	15154

Sources: Landbrugsforhold i Danmark siden midten af det 19. Aarhundrede: Statistisk Tabelvaerk, Femte Raekke, Litra C Nr. 4. København 1911. For 1914: *Statistisk Aarbog.*

Table 5.3: Production of major livestock products

	Butter	Milk	Eggs	Pork and bacon	Beef
			thousand tons		
1871	36	1082	8	39	..
1881	49	1391	9	46	59
1893	75	1945	15	72	68
1903	107	2613	30	127	92
1914	143	3574	48	217	123

Sources: Von Arnim (1951) and Skrubbeltrang (1953).

The adaptation was reflected in the use of agricultural land, which was increasingly devoted to the needs of livestock: the area under feed grains increased and there was a very big expansion in root crops (especially mangolds and swedes). The relatively small area under wheat was reduced still further, while the area of rye fluctuated but on the whole did not rise (Table 5.4).

Table 5.4: Utilisation of agricultural land

	1876	1888	1896	1907	1912
			thousand hectares		
Wheat	62	49	34	41	47
Rye	254	282	291	276	246
Feed grains	738	816	860	822	862
Total grains	1054	1147	1185	1138	1156
Root crops	52	106	140	308	356
Legumes and pasture	888[a]	959*	994	799	743
Other crops	68[a]	61*	14	12	22
Fallow	248	262	251	225	200
Permanent grass	397	359	322	431	411
Total agricultural land	2707	2893	2906	2912	2888

* Appears not to be fully comparable with data for subsequent years.
Source: For 1876 and 1888: as Table 5.2. For other years, Danish Council of Agriculture (1935).

Exports of livestock products began to expand rapidly as early as the 1870s (Table 5.5); in 1872 the value of these exports first exceeded that of grain; by the 1880s Danish exports of dairy products, meat and livestock were firmly established and Denmark had become an importer of all grain except barley.

Table 5.5: Average annual exports of livestock products

	Fat cattle	Pigs	Beef and mutton*	Pigmeat	Butter	Eggs†
	thousands			*thousand tons*		*million score*
1865/66–1869/70	40	37	5		4	. .
1870/71–1874	48	111	5		9	. .
1875–79	81	186	4		13	. .
1881–85	102	279	1	8	12	3
1886–90	102	131	1	24	24	5
1891–95	103	135	5	41	38	7
1896–1900	57	–	18	65	52	12
1901–05	73	–	24	76	70	18
1906–10	117	–	31	95	83	18

* Including offals.
† Net exports.
Sources: Before 1880, Skrubbeltrang (1953); after 1880 *Statistisk Tabelvaerk, Femte Raekke, Litra D.* (The two sources may not give fully comparable data.)

Agricultural co-operation

The development of co-operation was a distinctive feature of Danish agriculture and played a vital role in its transformation. The principle of co-operation had been established with credit associations founded in the 1850s and with retail co-operatives which first appeared in 1866. The effects of the crisis in the 1880s accelerated the movement.

The growth of livestock production necessitated arrangements for rapid and efficient processing and marketing. This was beyond the means of the individual small producer, but co-operation offered a solution. Co-operative dairies were the first to be formed: previously, butter-making had been left to the farmer's wife, but now butter was processed in bulk, by modern techniques (the centrifugal separator came into use around 1880). The skimmed milk was returned to the farms to be fed to pigs (hence the expression 'the pig hangs on the cow's tail'). Members of the co-operatives received an initial sum for their deliveries, plus a share of the eventual profits in proportion to their deliveries. From 1882, when the first of the co-operative dairies was founded, their number rose to 176 in 1886, 600 in 1890 and 942 in 1900; this represented an average of one for every other parish. This expansion of the dairies was indispensable to the increased output and improved quality of butter, and to the great development of exports.

In 1887 Germany suspended imports of live pigs on the pretext of disease control and in subsequent years this restriction was reinforced, reducing Danish pig exports almost to zero from 1896 onwards (Table 5.5). The export trade in pigs had become very important to the Danish farmer, and this was a severe blow. It was essential to redirect production into bacon for the British market. This was achieved by the creation of co-operative bacon factories: the first was built in 1887, and by 1890 such factories existed in every part of the country.

Associations were formed also for the joint export of eggs. Other co-operatives were set up for the purchase of agricultural requisites, but this only became important after 1900. The importance of the various types of co-operative in 1900 is indicated in Table 5.6.

Table 5.6: Agricultural co-operatives in 1900

	No. of societies *units*	No. of members *thousands*	Annual turnover *million kroner*
Dairies and butter export societies	1035	140	148
Bacon factories	26	60	35
Egg export societies	370	25	2
Purchase of feedingstuffs	110	6	4
Total	1541	231	189

Source: Danish Council of Agriculture (1935).

Danish agriculture after transformation

The adaptation to changed conditions through the shift to livestock production did not prevent the effects of the crisis from being felt. As has already been seen, agricultural prices in general did not revive till around 1900. In the course of the Depression, many farmers found themselves in difficulties: the burden of debt increased and many farms had to be sold. The price of agricultural holdings fell from a peak of about 4000 kroner per hectare in 1880–4 to 2400 kroner in

1895–9. Nevertheless, the population living by agriculture (including horticulture, forestry and fishing) rose from 889 000 in 1880 to 972 000 in 1901, and again to 1 004 000 in 1911. Since the total population rose from 1 969 000 to 2 757 000 over this period, the agricultural population was no longer quite so important as formerly.

The most important result was that Denmark emerged from the crisis with an agriculture adapted to the new situation, profiting from the opportunity of turning cheap grain into livestock products which were increasingly demanded as standards of living rose and which found a ready market in Britain.

Further, Danish farmers had built a highly efficient and integrated structure of production, processing, distribution and marketing, based upon a highly-developed co-operative system. This foundation made possible Denmark's development in the twentieth century as a great agricultural exporter, and enabled Danish agriculture to make an indispensable contribution to the economic growth of the nation as a whole.

Bibliography

Comprehensive accounts in English of the period covered by this chapter have been given by E. Jensen (an official of the US Department of Agriculture) in 1937 and especially by Skrubbeltrang (in an FAO series) in 1953. An interesting study in German is that by von Arnim (1951).

The land reform has been discussed, in Danish, by Falbe-Hansen (1888), Holm (1888) and H. Jensen (1936). (It was unfortunately not possible to take account of a work by Kjaergaard in 1980 on the *Hoveri* question.) It is regrettable that there is no specialised work in other languages on this remarkable period, though Munck (1979) has studied the conditions prior to these reforms and Thorpe (1951) has made an interesting case-study of developments in a single village, while Bjørn (1977) has analysed peasant reactions.

Arnim, W. von (1951) *Die Landwirtschaft Dänemarks als Beispiel intensiver Betriebsgestaltung bei starker weltwirtschaftlicher Verflechtung.* Kieler Studien **17**.

Bjørn, C. (1977) 'The peasantry and agrarian reform in Denmark'. *Scandinavian Economic History Review* **25**, 117–37.

Danish Council of Agriculture (1935) *Danemark: L'Agriculture.* Copenhagen.

Desbons, G. (1917) *La crise agricole et le remède coopératif – l'exemple du Danemark.* Paris.

Faber, H. (1931) *Co-operation in Danish Agriculture.* London.

Falbe Hansen, V. (1888) *Stavnsbaands-Løsningen og Landboreformerne.* Copenhagen.

Holm, E. (1888) *Kampen om Landboreformerne i Danmark.* Copenhagen.

Ilsøe, P. (1940) *Nordens Historie.* 3rd edn. Copenhagen.

Jensen, E. (1937) *Danish Agriculture – Its Economic Development.* Copenhagen.

Jensen, H. (1936) *Dansk Jordpolitik 1757–1919.* Copenhagen.

Koedt, P. (August, 1908) Untitled paper in *Report of the Proceedings of the International Free Trade Congress.* London.

Munck, T. (1979) *The Peasantry and the Early Absolute Monarchy in Denmark 1660–1708.* Copenhagen: Landbohistorisk Selskab.

Skrubbeltrang, F. (1953) *Agricultural Development and Rural Reform in Denmark.* FAO Agricultural Studies No. 22. Rome.

Thorpe, H. (1951) 'The influence of enclosure on the form and pattern of rural settlement in Denmark'. *Transactions of the Institute of British Geography* **17**, 113–129.

Part II:

The Crisis of the 1930s – The Second Wave of Protectionism

Chapter 6

General

The slump

During the First World War, farmers in both European and overseas countries benefited from high prices, which persisted for a few years after the war owing to the shortage of food supplies. But the reduction in soil fertility, the loss of livestock and of capital in the countries of continental Europe were gradually made up, and agricultural production recovered. Overseas production had expanded greatly during the war and afterwards continued to rise. In 1921 the increase in food supplies, combined with the effects of general economic depression, caused a sharp fall in prices. In subsequent years prices remained at a low level, and European farmers found themselves in difficulties. In many cases they had incurred debts during the period of prosperity, and repayments now constituted a heavy burden. Further, the costs of agricultural production – wage costs in particular – were rising, and a price 'scissors' began to operate to the detriment of agriculture. To increase their revenue many farmers sought to expand their output, but this only aggravated the fall in prices.

An improvement in the general economic situation brought a revival in farm prices in 1924, but this was of short duration: in 1926 depression set in again, and the pressure on agricultural prices was accentuated as production continued to rise. The output of grain in Europe rose well above the pre-war figure, and in overseas countries too the expansion continued (Table 6.1). Wheat consumption

Table 6.1: Production of wheat, in annual averages

	1909–13	1921–5	1925–9	1930–34	1935–8
			million tons		
World	106.7	. .	121.3	128.3	142.2
of which:					
USA	18.8	21.9	22.4	19.9	20.8
Canada	5.4	10.2	11.7	9.5	7.1
Argentina	4.0	5.5	6.6	6.6	6.2
Australia	2.5	3.5	3.7	5.1	4.4
Europe (excluding USSR)	26.1	32.3	38.7	43.8	47.3
USSR	20.6	. .	21.5	25.2	37.3

Source: League of Nations, *Statistical Yearbook.*

meanwhile remained almost stationary. In 1928, exceptionally good harvests in Europe and overseas added to wheat stocks which were already heavy: by the middle of 1929, world stocks of wheat stood at 28 million tons, the equivalent of more than a year's exports by all exporting countries. The prices of wheat and other grains fell sharply.

Meanwhile, industrial production in the United States had been expanding rapidly, with great activity in the construction sector. By 1928 there were signs of saturation: wholesale commodity prices began to fall and the volume of new construction declined. Yet on the stock market there was a speculative boom: share prices soared without any relevance to the real earning power of the assets. This could not last while industrial activity was in decline, and the stock market suddenly collapsed in October 1929. The slump in the United States was rapidly communicated to other countries: prices fell, industrial output was drastically cut, unemployment rose.

The industrial countries sought to preserve their market for their own output, and the prohibitive duties of the Hawley-Smoot tariff in the United States (June 1930) set off a chain of tariff increases; 'beggar-my-neighbour' policies became the rule. As the depression deepened and the financial crisis became more acute, the need to protect the balance of payments became the principal motive for increased tariffs and other restrictions.

The fall in the prices of grains was sharply accentuated by the depression: by 1931 the wheat price was barely half the pre-crisis level (figure 6.1). As the diagram also shows, the prices of livestock products at first held up better, but by 1931 they too were drawn into the depression through the reduction in consumer purchasing power. Importing countries reacted to the fall in prices by raising their tariff barriers and by a series of other measures through which their markets became more and more insulated, while exporting countries were in many cases forced to get rid of surplus stocks at almost any cost, and dumped their produce abroad with the help of export subsidies.

The first line of defence – tariff protection

During the First World War and the immediate post-war period of food shortages, agricultural tariffs were generally suspended. In subsequent years they were gradually reintroduced, at levels generally not exceeding those of the pre-war period. Thus France restored a moderate degree of protection in 1919, reinforced it in 1926, and in 1927 revised its tariff in the context of a commercial treaty with Germany. Germany recovered the right to determine its own tariff in 1925, and imposed duties approximately equal to those of pre-war. Belgium revised its tariff in 1924, but its policy remained liberal, especially where foodstuffs were concerned: all grains, except oats, remained free of duty. Norway gave some satisfaction to its farmers with a tariff introduced in 1927. In 1921 Switzerland revised its tariff. Austria, having lost the agricultural regions of its former Empire, sought to develop its own agriculture and reintroduced tariffs on agricultural products in 1924; but the degree of protection remained moderate in deference to Free Trade principles.

The agricultural tariff played a greater role in Italy. Mussolini had taken power in 1922; in 1925 he launched the 'Battle for Wheat'. In the early 1920s, about a third of Italy's wheat requirements were met from imports and these supplies accounted for a large part of the total import bill. Increased wheat production therefore seemed desirable both to achieve greater self-sufficiency in food and to relieve the balance of payments. A big effort was made to raise wheat yields through subsidies, technical advice and propaganda; the major feature in the campaign was the increased wheat price, achieved mainly through increases in the import duty. The duty on wheat, suspended since 1915, was reintroduced in 1925 at 7.50 gold lire per 100 kg, raised after the 1928 harvest

Figure 6.1: Prices of major foodstuffs on world markets
Indices, 1927–9 = 100

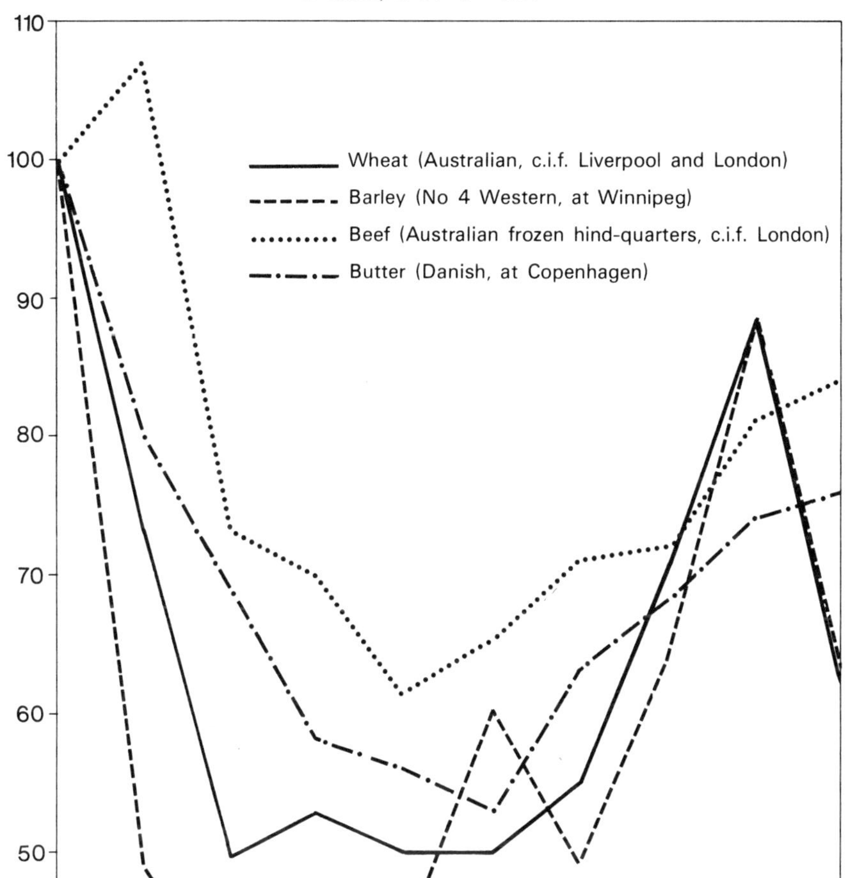

Note: quotations for Australian wheat in Liverpool and London have been used in preference to the price of wheat in the United States. The latter fell more steeply in the early stages of the crisis, but under the influence of the measures of support taken from 1933 on, it recovered more quickly than the world price.

Source: International Institute of Agriculture, *International Yearbook of Agricultural Statistics*. Rome.

to 11 gold lire, and after further increases reached in August 1931 the level of 75 lire in current value, or approximately 19 gold lire; at this rate the duty itself easily exceeded the price of wheat on the world market.

When the crisis broke over European agriculture, agricultural tariffs were raised as a first line of defence in both France and Germany. They reached levels which in normal times would have been prohibitive: with the fall in world prices, they came by 1931 to represent twice or even three times the world market price for some products, in particular grains and sugar. Italy imposed higher duties not only on wheat but also on other grains and on livestock products. Belgium at first resisted the pressure, helped by the fact that the fall in prices at first concerned mainly grains, which were of relatively small importance for Belgian agriculture; but in 1931 the fall in prices of livestock products necessitated increased protection. Several other countries raised their duties on agricultural products.

Table 6.2 shows calculations made by Liepmann of the average level of agricultural duties in 1927 and 1931. It should be recalled (cf. chapter 1) that these figures represent the unweighted averages of duties on thirty-eight important foodstuffs, expressed as a percentage of the export prices of leading European exporting countries. The results cannot be regarded as precise, and it must be remembered that the increased incidence of specific duties reflects the decline in prices as well as increases in the duties themselves. It can however be seen that by 1931 the duties frequently amounted to half or even more of the export price.

Table 6.2: Average 'potential' tariff levels for foodstuffs, as a percentage of export prices in European countries, in 1927 and 1931

	1927	1931
France	19	53
Germany	27	83
Italy	25	66
Belgium	12	24
Switzerland	22	42
Austria	17	60
Sweden*	22	39
Finland	58	102

* Fruit and vegetables not included.

Source: Liepmann (1938).

The most spectacular development in the tariff field was Britain's conversion to protectionism in the autumn of 1931. The measures taken did not at first give any great benefit to British agriculture. In November 1931, fruit and vegetables were made subject to duty, and in February 1932 the Import Duties Act, which replaced the temporary legislation of the previous autumn, imposed duties on some agricultural products but left many of the most important ones free. Later in 1932, the application of new and revised duties in accordance with the Ottawa Agreements brought wheat and other products within the scope of the tariff. But since the produce of the Empire continued to enter free of duty, tariff protection remained of limited importance to British agriculture.

The second line of defence – non-tariff measures

At a time when exporters were prepared to sell at almost any price, tariffs, however high, were an ineffective means of protection. It became necessary to control imports more precisely and directly, and for this purpose a variety of new measures made their appearance.

The first of these, and one of the most effective, was the 'milling ratio' for wheat, sometimes too for rye: millers were legally obliged to use a certain minimum percentage of domestically-produced wheat in their grist. This device seems to have been invented by Norway in 1927; in 1929 both France and Germany adopted it; from 1930 onwards it became widespread in Europe and was applied in some non-European countries as well. It continued to be used up to the Second World War in many cases, and it was reintroduced by several countries after the war.

The ratio fixed between home-produced and imported grain could be made to reflect the market situation and the extent to which the authorities wished to promote domestic production; it could be varied from time to time. In practically all the countries which adopted this device, the proportion of domestic grain was increased as time went on: in some cases it reached 100%, which amounted to prohibiting imports of wheat suitable for milling. Thus in France the proportion of domestic wheat was originally fixed at 97% in December 1929; it was reduced slightly in subsequent years owing to reduced harvests, but was raised to 100% in 1933. In Germany the proportion started at 30% in 1929, but was raised to 97% in 1931. In Italy it began at 75% in 1931 and soon reached 95%. In Sweden the proportion stood at 100% in the 1934/5 crop year. Even in the Netherlands the proportion was raised, from 20% when the system was introduced in 1933 to 35% in 1935. In 1938 Denmark too adopted a milling ratio for wheat of 50%.

Milling ratios by their nature could be used only for bread grains, but they were only one of a family of 'linked-utilisation' measures by which, in one way or another, a domestic product had to be used in connection with a foreign product. Such measures were applied not only to some feed grains and, in Germany, to a few other agricultural products, but also to various non-agricultural products (Italian cinemas were required to show one domestic film for every three foreign ones!). In some countries linked-utilisation regulations were applied on two different products: thus several countries with butter surpluses, including the Netherlands and Sweden, required all margarine to contain a certain percentage of butter; in Denmark, margarine had to incorporate home-produced lard. Farmers producing alcohol from wine, sugarbeet or other products were sometimes helped by the compulsory inclusion of domestic alcohol in petrol.

A similar measure was the 'linked-purchasing' regulation. Latvia in 1931 required ten tons of home-produced sugar to be bought for every ton of foreign sugar imported, and enforced similar measures for wheat, rye and a few other products. The government of Spain in 1935 obtained general authority to impose similar regulations. This device, although not of great importance before the Second World War, became one of the major instruments of Swiss agricultural policy when the *prise-en-charge* system was expressly provided for in the Agricultural Act of 1951 (see chapter 11).

The most important of the new devices was the import quota. Import quotas had been used before the 1930s, but their purpose had been mainly, in connection with a commercial treaty, to guarantee to an exporting country a certain quantity of trade free of duty or at reduced rates. Tariff quotas of this kind were applied by Germany to meat from 1925 to 1930, and to cattle and butter in 1930. The first country to use quotas on a large scale as a means of protection was France: applied first in 1931 as an emergency measure to a few agricultural products, the system was extended in subsequent years to cover practically all agricultural products except wheat, and a number of manufactured goods as well; import quotas gradually became an integral part of French commercial policy. They were taken up to varying extents by other countries: Belgium was one of the first to introduce legislation to this effect, though it did not make full use of the system till 1933. In numerous other countries, import quotas were adopted as an element in measures of market organisation: this was the case in Britain, Germany, the Netherlands and Italy. It was estimated (Gordon, 1941) that in 1939 import quotas were being used on a large scale, mainly for agricultural products, by nineteen European countries and nine non-European ones.

In several countries, import quotas came to be associated with import licence fees. The original object of these was mainly to recover some of the profits made by importers from dealing in a restricted market. Frequently, however, these fees came to have a protective function as a kind of variable import duty and served to eliminate differences between the world price and the domestic price; sometimes the import licence fee was higher than the duty provided for under the tariff. Cases in point were France, Belgium and the Netherlands.

A further step – intervention in agricultural markets

Of all the measures adopted to deal with the crisis of the 1930s, that which had the most significance and the greatest influence on subsequent developments was the attempt to organise domestic agricultural markets. This intervention was usually coupled with regulations concerning imports and, in some cases, exports.

The progression from measures of import control to more far-reaching intervention was particularly clear in France. Output of wheat had been rising fairly steadily and the initial measures of protection prevented the fall in world prices from discouraging this trend. A big harvest in 1932 saturated the market and import controls were no longer adequate to maintain the price. Various measures were taken from then on, involving in particular government purchase of part of the crop at fixed prices, and culminating in 1936 in the *Office National Interprofessionnel du Blé,* with the task of fixing the wheat price and ensuring that it was observed. The *Office* had monopolistic control over all foreign trade in wheat: it could regulate imports and subsidise exports. The other main commodity subjected to market organisation in France was wine: here too the necessity arose in part because a surplus made import control ineffective. Measures of organisation began in 1931 and were reinforced in 1934; they sought in particular to control the amounts marketed by producers and the period of marketing.

In Britain, the wish to avoid restricting imports from the Empire meant that protection for agriculture had to take the form of marketing schemes, designed to strengthen producers' bargaining power, and subsidies. For wheat and sugar-beet, assistance was given by subsidies alone; in the former case, the market was left free and the subsidy was given through 'deficiency payments', reviving a device used in the First World War and destined to play an important role after the Second. Milk was supported by Milk Marketing Boards with the exclusive right to sell milk and fix prices. Imports of beef were regulated by agreement with the main exporting countries, including Empire ones. The markets for bacon, potatoes and hops were supported by marketing schemes together with import controls; that for eggs through voluntary restrictions by exporting countries.

The crisis thus forced Britain to abandon its long-established policy of laissez-faire towards agriculture. Denmark too was compelled, by the difficulties encountered on export markets and the consequent fall in farmers' returns, to abandon liberal practices. The special character of the measures taken reflects Denmark's situation as a large net exporter of the commodities in question. All cattle sold on the home market were subjected to a tax, and the proceeds were used to buy up low-quality cattle in order to raise prices. Taxes were imposed also on sales of butter for domestic consumption, and the proceeds were distributed among producers. A particularly far-reaching scheme, involving what was probably the first agricultural marketing quota in history, was devised for pigs.

The situation of the Netherlands was somewhat similar to that of Denmark. The Dutch government too was forced, partly by falling import prices but even more by the difficulties facing exports, to depart from tradition and to intervene on agricultural markets. Here again the complexity of the measures adopted reflects the problems involved in supporting the prices of commodities which are on an export basis. In a first phase, measures of support were applied to individual products according to needs: thus in 1931 wheat imports were regulated by a milling ratio and a target price on the domestic market, the latter maintained with the help of state subsidies. Subsidies were also paid for potatoes and sugar-beet from 1931. In 1932, as livestock products were drawn into the crisis, these too became the object of intervention: a Central Pig Office was set up to keep the number of pigs in line with market outlets and to operate an export and import monopoly for both live pigs and pigmeat. In the same year, a Central Dairy Office was formed with a monopoly of butter exports, and the domestic milk market was strictly controlled. In 1933 a Central Cattle Office was introduced; it sought to restrict cattle numbers by decreeing the slaughter of part of the herd. Fruit and vegetable producers, from 1931 on, began to hold back part of their supplies from the market in order to maintain a minimum price, helped by a state subsidy for the disposal of unsold stocks; in 1933 imports of fruit and vegetables were made subject to control. A second phase was reached with the Agricultural Crisis Law of 1933, which brought together all these separate regulations and provided a legal basis for subsequent action: the government was given the right to designate 'crisis products' and to regulate production and all aspects of trade in such products. The Netherlands thus became one of the first countries to have an extensively planned organisation of its agricultural markets.

Switzerland was another country in which an attachment to liberal economic principles did not prevent recourse to a significant degree of market organisation in agriculture. Even before the crisis, the Central Union of Swiss Milk Producers had been formed, controlling almost all marketed supplies of milk; also, from 1915 to 1929 the government exercised a monopoly of the wheat trade. In 1929 this arrangement for wheat was replaced by government purchase of domestic wheat and rye at guaranteed prices: millers were bound to take over this grain at prices fixed by the government, but received compensation through the restriction of flour imports. In 1933 a special commission was set up to control imports of feed grains and other feedingstuffs, and import levies were imposed in addition to customs duties, with the object of raising the price of fodder and discouraging the output of milk, already regarded as excessive. Though subsidies had to be given to milk production, an attempt was made to impose production quotas. Output of pigmeat too had increased rapidly and prices were falling: in 1935 limitations were imposed on the number of pigs that could be kept and a scheme for directly restricting production was introduced.

Austria began in 1931 to regulate supplies to the cattle market, and imports were admitted only in so far as they did not endanger sales of home-bred animals. The milk market too was organised from 1931 on, and various measures were adopted for other products. In its trade policy Austria abandoned its liberal principles and began to draw up trade agreements in which the 'most favoured nation' clause was replaced by reciprocal agreements on preferences and quotas: arrangements of this kind were included in the Rome Protocols of 1934 between Austria, Italy and Hungary.

Various other countries adopted measures of a similar nature. Norway began to organise its agricultural market even before the crisis, with a grain monopoly set up under a law of 1926; from 1929 onwards a series of measures were taken, based mainly on sales co-operatives for the major products. Sweden in 1930 reinforced its milling ratio for wheat with government purchases to support prices; organisation of the dairy market and other measures were introduced in 1932.

Germany and Italy – agricultural policy in a totalitarian state

In Germany under National Socialism and in Italy under fascism, agricultural markets were strictly regulated in accordance with general political aims and with an overall economic plan.

The National Socialist government in Germany came to power in 1933 with a clearly-defined philosophy as to the role of agriculture in the nation. This philosophy formed an integral part of National Socialist thinking: its essential features concerned the social and racial importance of the farm population, the need to ensure fair prices to farmers and the importance of national self-sufficiency in food. In a remarkably short time German agriculture was organised in accordance with a prepared plan. The *Reichsnährstand* (State Food Corporation) was set up in September 1933 to organise all aspects of food production and distribution and to regulate markets and prices. Full control of the volume and prices of imports was vested in the *Reichsstellen* (State Boards) which were set up for all important commodities from April 1933 onwards: the *Reichsstellen* could also buy and sell on the domestic market and operate buffer stocks. Under this system imports were reduced and the manipulation of supplies initially enabled prices to be raised above the level to which they had fallen in the crisis: subsequently, however, Hitler had to give priority to keeping down food prices. Food production and self-sufficiency were raised, but by the outbreak of war Germany was still short of livestock fodder and of fats.

In fascist Italy too, economic policy came to involve a high degree of central planning in agriculture as in other sectors. This, however, occurred over a period of years, and was never so thoroughly applied as in Germany. Mussolini did not develop the elaborate racial theories of the Nazis, but like them he regarded agriculture as a basic element in national well-being and saw in the agricultural population a source of strength in wartime. His agricultural policy too was dominated by the desire for self-sufficiency and was intimately linked with his object of raising Italy to the status of a Great Power: this involved promoting an increase in population by all possible means, which made increased food production all the more necessary.

Mussolini's agricultural policy meant an enormous effort to lift Italian agriculture out of technical backwardness and economic depression. Great emphasis was laid on the policy of *bonifica integrale* (integral land reclamation), initiated by a law of 1928 which ordered the improvement of all unused but cultivable land. The early stages of the Battle for Wheat have already been described. After a slow start, considerable success was achieved in raising yields and output; at the same time the increase in price resulting from the import duty reduced wheat consumption, and after a large harvest in 1932 imports were reduced to a low level. This, however, meant that the import duty was no longer so effective in raising the domestic price, and other measures became necessary. A milling ratio was introduced in 1931: at first 75% of domestic wheat had to be incorporated in the flour; soon afterwards the ratio was raised to 95%. Then, however, the situation changed; the 1934 harvest was small and the milling ratio was suspended; moreover, the Abyssinian War of 1935–6 contributed to an increase in prices. In 1935 a target price system was introduced with the object of restraining price increases. This proved ineffective, and was replaced the following year by a system under which wheat could be sold to the mills only by authorisation and at fixed prices. After the end of the Abyssinian War, a full state monopoly was instituted: growers were bound to deliver their wheat at fixed prices, and imports were directly controlled by organisations responsible to the Ministry of Agriculture.

Though the Battle for Wheat largely achieved its immediate object, the excessive emphasis placed by Italian policy on wheat had questionable results. It helped the larger farmers and landowners, but not the peasants. It stimulated wheat at the expense of other products more needed and better suited to Italian conditions. It made only a limited contribution to the balance of payments, for with the low level of world prices wheat could have been imported cheaply; moreover, the rising demand for livestock products had to be met by increased imports.

It is worth noting that the agricultural policy adopted in another totalitarian state, Japan, was closely analogous to those of Germany and Italy. Here too, nationalist aims involved belief in the agricultural population as a vital element in the nation: the Imperial Agricultural Society declared that the farming community was 'the backbone of the nation, the source of its military strength and the guardian of traditional virtues against alien influences'.[1] In Japan too, increased food production with a view to self-sufficiency became a major aim, and an extensive system of control over agricultural production and markets was set up.

Factors contributing to the growth of intervention

The growth of government intervention in agricultural markets during the 1930s took place with remarkably little opposition: there was far less controversy than during the Great Depression of the nineteenth century. Protection was being given to industry in an effort to ward off unemployment and similar assistance could hardly be refused to agriculture. In the face of the crisis, academic theories of Free Trade were irrelevant. Governments did not want to be accused of raising food prices, but it was difficult to criticise measures which sought merely to restrain the fall in prices.

Many of those who accepted the change, however, did so in the expectation that the new measures would be temporary and could be relaxed when economic conditions returned to normal. In fact the opposite happened: measures of support remained and were even reinforced in the course of the 1930s. To some extent, this was necessitated by the fact that world prices continued low till the middle of the decade. However, the various measures of intervention were themselves an important cause of the persisting difficulties on the world market: import restrictions and export subsidies were constantly limiting demand, raising supply and depressing prices, and thus perpetuating the need for their own existence. The cumulative effects of separate national protectionist policies were scarcely understood, and the climate of opinion was not yet ripe for the international action which alone could have reconciled these divergent interests; this point is further discussed in the last section of this chapter.

[1] Quoted by Allen, G. C. (1946) *A Short Economic History of Modern Japan.* London: Allen & Unwin (page 110).

Other factors contributed to the willingness of governments and the public to help agriculture. There was a growing awareness that agriculture was not like other industries: that it was particularly subject to fluctuations in supplies and prices and that a large number of producers acting individually were helpless in the face of market disequilibrium. This realisation contributed to the introduction of marketing schemes, in Britain in particular, the object of which was as much to reinforce the farmers' position on their own market as to protect them against foreign competition.

Added to this problem of market disequilibrium was the old argument about the social value of the farming community. This concept appeared in an extreme form in the racial theories of National Socialism, but it was present to varying extents in most countries. It was particularly important in Switzerland, where the rural population of independent farmers appeared as an essential element in the democracy and a bulwark against socialist tendencies in the urban working population; on several occasions in the 1920s the farmers were instrumental in breaking industrial strikes.

The other old argument, concerning the strategic importance of a large food supply, gained considerable force from the experience of food shortages during the First World War and from the preparations for the Second. Thus in Germany, Italy and Japan, the need for self-sufficiency in food was an integral part of nationalist policies. In Britain, responsible opinion pointed out that a policy of self-sufficiency in peacetime could actually damage the economic strength of the nation in war and there was little deliberate effort to raise food production till close to the outbreak of war.

The increased support for agriculture owed much to the activities of farm organisations. In France there was a big increase in the number of such organisations and in their influence; the initiative passed from the old *Société des Agriculteurs de France* to a right-wing syndicalist movement, which in 1934 and 1935 combined its action with a rural movement dominated by fascist tendencies, forming the *Front Paysan*. Producers' organisations for wine, sugar-beet and wheat appeared as powerful pressure-groups. In Germany, the Prussian landlords continued to be influential in the Weimar Republic; moreover, on a number of points – particularly concerning the importance of agriculture for the nation's military strength – they had views in common with the National Socialists: they concluded an alliance with the Nazis in 1931 and their position was to a large extent respected under the Nazi regime. Farmers' organisations opposed to the regime were forced to close down. In Britain, the old Central Chamber of Agriculture, together with the newer and more vigorous National Farmers' Union, publicised farmers' views, though their influence at this time was somewhat limited.

In spite of all the action that was taken, little thinking was done in most countries into the needs of long-term agricultural policy. In the early stages of the economic crisis, measures of assistance were improvised in response to the emergency; afterwards, these measures were reinforced in various ways. But in most cases, no official policy was evolved as to the place of agriculture in the economy: basic questions as to how much food and what kinds of food should be produced at home, how big the agricultural population should be, what level of farm income was justifiable, were scarcely considered. This lack of a clearly-defined policy was particularly evident in Britain. In France, the instability and short life of the various governments during this period put a consistent agricultural policy out of the question. Even in Italy, where agriculture played a large part in official thinking from 1925 on, the obsession with wheat production blinded the policy-makers to the overall economic problems of agriculture. It was only in Nazi Germany that the role of agriculture was clearly defined in relation to overall political and economic aims, and that the resulting policy was consistently and effectively applied. But their legislation to protect viable family farms was tainted with racial doctrine, while their organisation of production and markets aimed to reinforce national power and the interests of other countries, in agriculture as in other respects, were disregarded.

Some consequences for prices, production and trade

Perhaps the most striking outcome of this period in which the countries of western Europe intervened in a variety of ways to support their agricultural markets was the resulting divorce between the trends of world prices and those of prices received by producers in protected importing countries.

Figure 6.1 has shown the extent of the collapse in the world grain market in 1930 and 1931 and the later but still drastic fall in the world prices of livestock products. It was not till 1933 or 1934 that a recovery took place, and even by 1937 prices were still well below the pre-crisis level.

In most of the European importing countries, wheat producers were to a considerable extent insulated from the fall in prices. Developments are shown in Figure 6.2. It can be seen that the fall in the world price in 1930 and 1931 was not at all reflected in the domestic price in France and Germany, and in Italy was reflected only in part. Subsequently, the wheat price in France fell under the influence of large domestic harvests rather than of the low world price; it recovered in 1936 and by 1937 was slightly above the pre-crisis level. In Germany the price declined from 1931 to 1933; afterwards, the strict controls instituted by the Nazi regime enabled a stable price level to be maintained. In Italy the price recovered by 1937 to nearly the pre-crisis level. When the world price suffered a further fall in 1938, the prices on these protected markets were unaffected.

It is unfortunately difficult to obtain continuous price series for many products throughout the 1930s. Figure 6.3 however shows data for butter. The London wholesale price dipped until 1934, then recovered; the trend in Belgium was similar but the recovery was greater. In Germany, National Socialist market organisation fixed a virtually unchanged wholesale price from 1934 onwards. In France the large fall in 1935 was not due to imports – which had been drastically cut by the import quota system – but to domestic market developments; there

Figure 6.2: Wheat prices
Indices, 1927–9 = 100

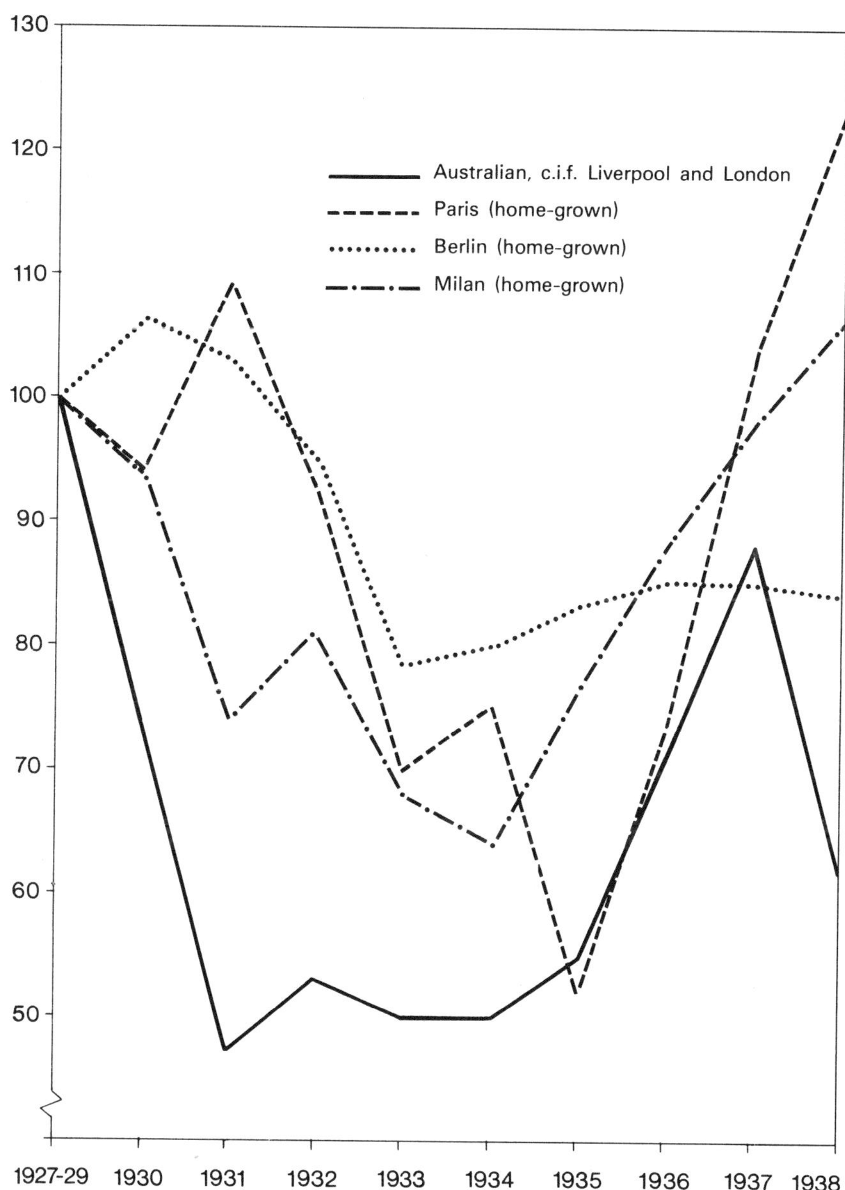

Source: International Institute of Agriculture, *International Yearbook of Agricultural Statistics*. Rome.

was then a large recovery. By the end of the period, price differentials between countries, at current exchange rates, had widened. In 1931 only the French price was above the London price of Danish butter (New Zealand butter was 15% cheaper), but by 1937 the prices on the French, Belgian and especially the German markets were well above those of the UK market. For both butter and wheat, the widening of the differentials was more marked than appears from the diagrams because, over the period, sterling had been devalued in terms of the other currencies.

Figure 6.3: Wholesale prices of butter
Indices, 1931 = 100

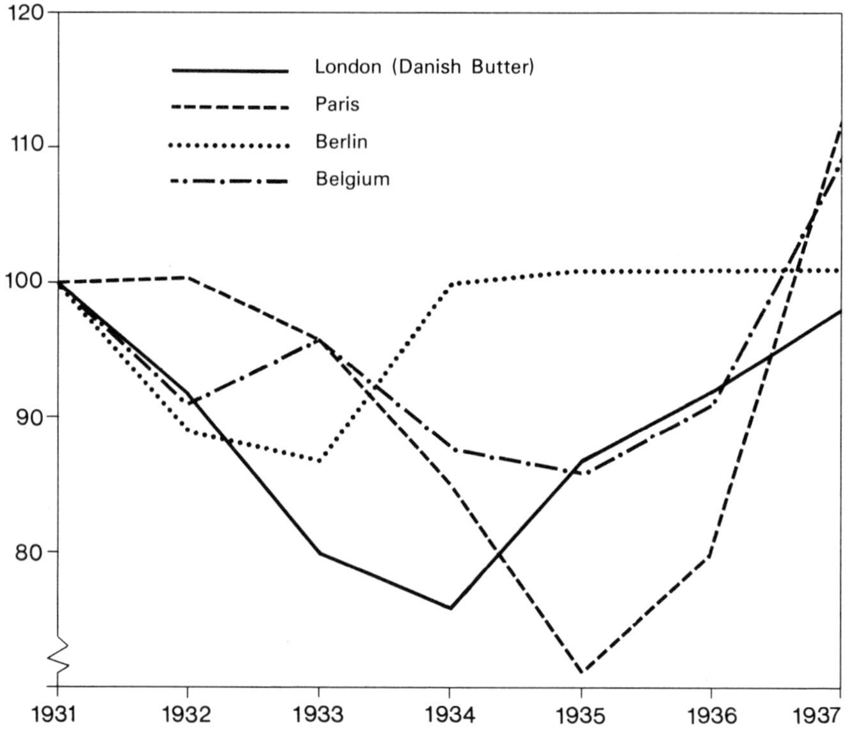

Source: Statistics of the Commonwealth Economic Committee.

As a result of protection and intervention, agricultural producers in most European countries, though they by no means escaped the crisis, were spared its worst effects. Farm output was not discouraged; rather the reverse. As Table 6.3 shows, production of most major commodities in western Europe was higher in the period following 1929 than it had been previously; the expansion in wheat was particularly large. Numbers of livestock also rose.

At a time when the general reduction in purchasing power was depressing

Table 6.3: Crop production and livestock numbers in western Europe

	Unit	1924–8	1929–33	1934–8
Wheat	*million tons*	24	29	30
Barley	” ”	10	11	11
Potatoes	” ”	78	89	94
Sugar	” ”	4.2	4.8	5.0
Wine	*million hectolitres*	131	127	131
		1928	1933	1938
Cattle	*millions*	87	90	92
Pigs	”	52	59	59

Note: It has been necessary to adapt data given by the source for the whole of Europe: as information is not available for all countries, the results include small amounts for certain eastern European countries.

Source: International Institute of Agriculture (1948) *Les Grands Produits Agricoles.* Rome: FAO.

consumption and when prices often were not allowed to fall sufficiently to offset this decline, the increasing trend in home production could only mean a cut in imports. Thus the fall in prices on the world market was accentuated and the volume of trade reduced.

It appears from Table 6.4 that the value of world trade in foodstuffs as a whole was less by about a quarter after the crisis, in 1937, than in 1929 (total world trade fell by just over 20% in value). Both the volume and the unit values of trade had fallen for the products of the temperate regions – grains and livestock products – as well as for sugar, which also competed directly with European agriculture. On the other hand, trade in tropical and semi-tropical products – coffee, tea, oilseeds, certain fruits, etc. – increased in volume, though it suffered a serious fall in prices.

Table 6.5 shows in greater detail the developments in trade for some important 'temperate' foodstuffs, in terms of quantity and indicates the effects on various exporting countries. The decline in the wheat trade mainly affected the United States and Canada; Argentina and Australia did not do so badly. Poland and the countries of south-east Europe managed to increase their exports of wheat and rye during and after the crisis, thanks in part to preferential treatment by France and, more important, to the close commercial links established with Germany after 1933. Trade in barley and oats also declined; the only type of grain for which trade expanded was maize, this expansion being accounted for by imports into Britain (where no maize was grown). Trade in sugar fell, and the pattern shifted: the traditional exporters – Cuba and the Dutch East Indies – lost heavily, while the dependencies of the United States – Philippines, Hawaii and Puerto Rico – together with the countries of the British Empire were enabled by preferential arrangements to increase their trade.

Exports of livestock, of which cattle and pigs were the most important, suffered badly. Britain reduced its imports from the Irish Republic (largely as a result of the trade war between the two countries from 1932 to 1938). Austria, which had been the world's largest importer of pigs, mainly from east and south-east Europe, cut its imports drastically; Germany too reduced its imports from

Table 6.4: World agricultural exports, by commodity

	Current value		Unit values*		Volume index†	
	1929	1937	1929	1937	1929	1937
	$ million		*1913 = 100*		*1913 = 100*	
Cereals	2050	1526	107	88	107	98
Sugar	719	432	84	62	185	152
Livestock products	2132	1557	137	114	135	118
Fruit and vegetables	872	660	105	69	185	214
Coffee, tea, cocoa	946	653	123	79	139	149
Oilseeds and fats	1109	882	92	68	161	171
Other food and drink	626	541	..	..	..	..
Total	8456	6251	113	85	136	133

* Weighted average of unit values of the various items in each group.
† Current value deflated by indices of unit values.
Source: Yates (1959).

Denmark and the Netherlands during the worst of the crisis, but in 1934–8 this trade partly recovered. As for beef and mutton, of which Britain was by far the largest importer, the striking feature was the displacement of South American supplies by Australia and New Zealand as a result of preferential arrangements for the latter. Germany ceased to be a major beef importer. For bacon, import quotas in Britain reduced trade from almost all sources except Canada, which benefited from a particularly large quota. Trade in pigmeat (other than bacon, ham and lard) suffered a decline, with reduced imports by France and Germany in particular: exports by Denmark and the Netherlands, especially the latter, were severely hit. On the other hand, Australia and New Zealand were able to increase their exports of pigmeat to Britain. For butter there was a similar development: Germany reduced its imports in 1934–8, while those of Britain increased, and as a result exports by Denmark and the Netherlands remained at about the same level while those of New Zealand and Australia more than doubled as compared with 1924–8. In cheese too, New Zealand and Australia gained at the expense of Denmark and the Netherlands.

Thus, in this period when trade in 'temperate' foodstuffs in general was declining, the exporting countries which did least badly were those able to get preferential treatment from a major importing country: the American sugar-exporting colonies from the United States, to some extent south-east Europe from Germany and France, and above all the British Empire from Great Britain. France, Belgium and the Netherlands also gave preference to their overseas territories; this however – apart from wine and wheat from French North Africa – mostly concerned tropical and semi-tropical products. On the other hand, the agricultural exports of Denmark and the Netherlands suffered from increased protection in neighbouring countries and from discrimination against them in Britain; those of the United States suffered from the general slump in grains and from import restrictions both on grains and on pigmeat; and those of Argentina and other South American countries suffered from the preferential treatment given to Empire meat on the British market.

Table 6.5: Exports of major agricultural products, by country of origin

	1924–8	1929–33	1934–8
		million tons	
Wheat and rye*			
N. America	14.4	9.5	6.2
Argentina	4.1	4.2	3.5
Australia	2.5	3.4	2.8
E. and S.E. Europe†	1.4	1.8	1.9
Others	3.4	4.9	4.2
World	25.8	23.8	18.6
Sugar			
Cuba	4.4	3.2	2.6
Dutch E. Indies	2.1	1.8	1.0
Philippines, Hawaii, Puerto Rico	1.7	2.4	2.5
British Empire	0.5	0.8	1.2
Others	4.3	4.4	4.1
World	13.0	12.6	11.4
Cattle and pigs		*thousands*	
Ireland	984	1056	757
Denmark and Netherlands	342	218	252
E. and S.E. Europe†	1715	1335	1045
Others††	2302	2026	1986
World	5343	4635	4040
Beef and mutton		*thousand tons*	
Argentina, Uruguay, Paraguay	1042	841	853
Australia and New Zealand	218	196	221
Others	276	254	212
World	1536	1291	1286
Bacon, ham and lard			
Denmark and Netherlands	254	382	225
USA	170	74	32
Canada	41	15	70
Others	87	131	101
World	552	602	428
Other pigmeat			
Denmark and Netherlands	50	29	9
E. and S.E. Europe†	32	11	28
USA	41	32	23
Australia and New Zealand	4	12	37
Others	40	43	35
World	167	127	132
Butter			
Denmark and Netherlands	177	196	199
Australia and New Zealand	113	183	240
Others	169	189	177
World	459	568	616

* Including flour in grain equivalent.
† Poland, Rumania, Hungary, Yugoslavia, Bulgaria.
†† Consisting largely of exports of pigs from China to Hong Kong and of cattle from Mexico and Canada to the USA.

Source: International Institute of Agriculture (1948) *Les Grands Produits Agricoles.* Rome: FAO.

Reactions in agricultural exporting countries

Agricultural exporting countries were forced by their difficulties to take a number of measures to assist their farmers. Those which were adopted in Denmark and the Netherlands have already been described. The remainder do not fall directly within the scope of this study, but may be referred to briefly.

The countries of east and south-east Europe were in a particularly difficult position. These were mostly poor countries where agriculture predominated; the peasant population was often living on a bare subsistence, and there was little possibility for support by the rest of the community. These countries – Bulgaria, Hungary, Poland, Rumania and Yugoslavia – made an attempt to co-ordinate their exports and to obtain preferential treatment from the countries of western Europe; several conferences were held to this end, starting in July 1930. Other exporting countries, however, objected strongly to the idea of preferential treatment for their competitors and not much was obtained. During 1931 some preferential tariff agreements were negotiated, in particular by France with Hungary, Rumania and Yugoslavia: these agreements provided for the reimbursement by France to the exporting countries of a part of the duty on wheat. Germany also offered preferences on grain to Hungary and Rumania, but this idea had to be given up because of the objections raised by countries entitled to 'most favoured nation' treatment by Germany; however, at a later stage, the National Socialist regime entered into close trading links with south-east European countries, from which they derived a limited benefit.

The United States naturally became extremely concerned at the prospects, especially for its wheat exports, as is indicated by the following passage from a study by the Food Research Institute of Stanford University:

> Western Europe aims to raise more wheat. Central Europe seeks preference in the wheat markets of western Europe. The Dominions of the British Commonwealth seek preference in the wheat markets of Great Britain. Great Britain, Holland, Belgium and France extend preferences to their colonies for feedingstuffs. In order to effectuate quotas and preferences, intricate internal regulations and extensive interstate barters become necessary. Russia, Argentina and the United States stand outside the charmed circle. [Food Research Institute (1932) cover note]

In 1933 the US Senate considered an important document on 'World trade barriers in relation to American agriculture', which stressed the extent to which farm prices in the United States had been affected by the fall on the world market.

Unlike most other agricultural exporting countries, the United States had the means to come to the rescue of its farmers. The emphasis was laid at first on government-financed stockpiling operations to relieve the market, but this effort proved largely unsuccessful. In the context of Roosevelt's 'New Deal', the Agricultural Adjustment Act of 1933 provided more effective and far-reaching legislation: it aimed to restore the balance between production and consumption and to give farmers 'parity' prices. Payments were made to farmers for reducing their acreage of wheat, maize, cotton and tobacco; pigs in excess supply were purchased by the authorities and slaughtered. The Commodity Credit Corporation was instituted to buy and sell agricultural commodities, making 'loans' to farmers on the security of the commodities delivered. Surplus stocks were disposed of in various ways at home and abroad.

In Canada too, early efforts to help wheat growers consisted mainly of stockpiling operations. The Canadian Wheat Pool was formed in 1924, and received government support from 1929. As in the United States, the mounting burden of surplus stocks necessitated a change in policy: in 1935 the Wheat Board was established to buy grain from producers and to sell on world markets with the help of export subsidies.

Argentina and Australia lacked the means and the facilities for storing grain. The Argentine Government bought a large part of the exportable surplus from producers and sold it on the world market, usually at a loss. In Australia, the Commonwealth Government made grants to the individual states for distribution to farmers; it did not subsidise wheat exports directly till 1938.

Export subsidies became common in a number of other countries as well. South Africa began in 1931 to subsidise exports of processed foodstuffs and sugar and later extended subsidies to beef and mutton. The Irish Republic tried through export subsidies to counter the increased duties imposed by Britain in the course of the trade war. Even New Zealand began to subsidise its dairy exports in 1936.

An interesting picture of the increase in intervention in both importing and exporting countries was given in a diagram by Mackenroth (figure 6.4). This showed the percentage of agricultural trade which was 'directly or indirectly' affected by 'state measures of control or guidance' other than tariffs, in thirty-two countries accounting for 80–90% of world trade in foodstuffs. It appeared that this percentage grew from about 5% in 1929 to 55% in 1934; moreover, intervention apparently took place in exporting countries to approximately the same extent, and at approximately the same time, as in importing countries.

The failure of international action

With the exception of the Brussels Sugar Convention of 1902, by which sugar exporting countries agreed to refrain from export subsidies, there had been no effective international action during the difficulties of the late nineteenth century. The problems facing agriculture were discussed in 1889 at the International Congress of Agriculture in Paris, and as a result an International Commission for Agriculture was set up in 1891 with the task of studying agricultural problems and organising further international congresses. Agricultural congresses were in fact held every few years up to the First World War, but neither these nor the work of the Commission seem to have had any noticeable impact on official policy. The Commission continued in existence between the wars and called a number of congresses at which economic as well as technical problems were discussed, but the influence on policy remained negligible.

Figure 6.4: Percentage of world trade affected by measures of intervention

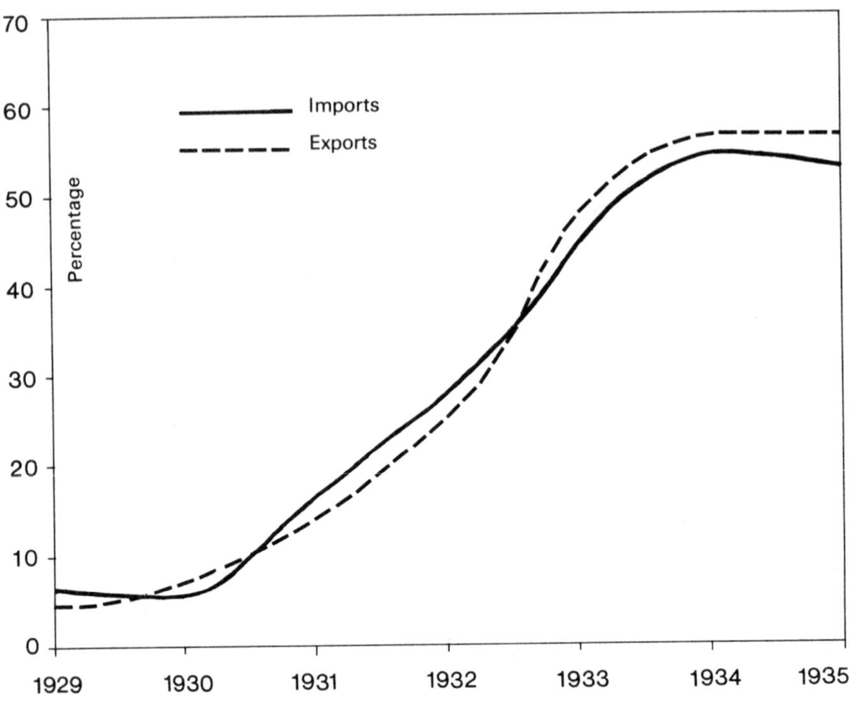

Note: This diagram reflects only the existence or non-existence of a measure – not the extent of its effect upon trade. On the latter score, there is no doubt that import restrictions were by far the greater influence and that they were often the cause of measures applied by exporting countries. It should also be noted that the diagram may be misleading in suggesting that measures were applied simultaneously in importing and exporting countries: a lag of some months between one measure and another might not show up in the calculation.

Source: Mackenroth, G. (1939), untitled contribution to *Proceedings of the Fifth International Conference of Agricultural Economists.* London: Oxford University Press.

Of greater significance was the founding in 1905 of the International Institute of Agriculture. This was an intergovernmental organisation; among its tasks, besides technical action of various kinds, the Institute was to collect and publish information on agricultural production, trade and prices, and to present to governments proposals for furthering the common interests of farmers; it was however specified that all questions relating to the economic interests or legislation of an individual state were outside the competence of the Institute. The Institute had an active existence between the wars; practically all countries in the world became members. It helped to bring about the signature of conventions on a number of minor trade matters. It presented its observations at various international conferences and from 1928 it acted as a consultative body for the League of Nations in agricultural matters (this was a provisional arrangement which was put on a formal basis in 1932). However, on matters of policy, and in particular on problems of international trade in agricultural products, its influence was, at the most, indirect.

The League of Nations' interest in agriculture began with the International Economic Conference of May 1927, which called for a halt to the rising wave of tariffs and discussed the problems of agriculture on the basis of a long report submitted by the International Institute of Agriculture. In January 1930 the League called a meeting of agricultural experts and its Economic Committee subsequently reported that the economic work of the League could not succeed unless it gave due attention to the needs of agriculture. The next year the Economic Committee produced a detailed report on the agricultural crisis, which laid stress on the growing gap between prices paid and prices received by farmers, and declared that the crisis was being aggravated by the numerous restrictions on imports and aids to exports. At the Assembly of the League in 1934 there was a long discussion of the problems of agriculture and as a result, in 1935, the Economic Committee issued a further study on the evolution of agricultural protectionism. This report – a comparatively short one – made a penetrating analysis of the current problem which deserved attention. It began by pointing out that around 1925 the recovery of agricultural production in countries affected by the war had upset the balance of supply and demand and started a decline in agricultural prices, which since 1930 had become acute. Exporting countries had been able to do little to help themselves except build up stocks. Importing countries, however, had resorted to far-reaching measures of protection:

> On the other hand, agricultural producers in the importing countries with a shortage of crops were obliged to appeal to their governments and to national solidarity in order to escape the contagion. This means they have used, and abused: the introduction of duties two or three times higher than world prices, ever stricter rationing, the progressive reduction to close upon vanishing-point of the proportion of foreign products admitted in the various preparations, bounties for production, export bounties, 'schemes', monopolies and various other forms of planned economy. [League of Nations (1935) page 6]

These measures had disrupted trade, and protectionism had proved contagious:

> The consequence of all these measures, applied more particularly by the big industrial countries, was naturally a big-scale reduction, amounting in some cases to the total exclusion, of agricultural imports, and that more particularly in the case of products – such as wheat, meat and butter – playing a fundamental part in the economic life of the producing countries ... This situation, taken as a whole, represents a defensive reaction, often violent and incoherent, but in the main comprehensible, against the dangers of an unprecedented economic depression.
>
> Had it achieved its object, it would be difficult indeed to criticise it; but facts are to hand which prove that this exaggerated policy of protectionism, spreading from one country to another, is tending to prolong the depression which it was designed to combat and to prejudice the interests of the classes

that it aimed at protecting. [Ibid., page 7]

The attempts by the League to combat the rising tide of protectionism met with only limited success. Already in 1929 the Assembly discussed proposals for a 'customs truce', and in February 1930 a conference was held to work out this idea; a Convention was drawn up, but was never put into effect. A further attempt was made at the London Conference in June 1933: all delegations agreed that trade barriers should be reduced, but this agreement was hedged around by reservations and no concrete results could be obtained. In the autumn of 1936, the declaration by the United States, Britain and France expressing the intention of these three countries to safeguard the peace and restore order to international relations was favourably received by the Assembly of the League, and the improvement in the economic situation at that time did permit some reductions in tariffs and an easing of quantitative restrictions. The Economic Committee, in a report to the Assembly the following year, drew up a plan for restoring normal economic relations. But the threat of war then began to dominate the international scene, causing countries to expand production by all possible means and making impossible any further action to liberalise trade.

The effort, already referred to, by the countries of east and south-east Europe to obtain preferential treatment from European importing countries attracted much attention at the time. It was helped by the idea of a European Union put forward by French delegates at the 1929 Assembly of the League and was discussed at a number of conferences held under the auspices of the League. In particular, the Stresa Conference of September 1932 was devoted to this question: it led to a recommendation that limited tariff concessions should be given to the countries concerned and to a financial plan for measures to stabilise the grain market. However, the financial arrangements – on which the other provisions depended – could not be put into force, and the scheme had little practical result.

One of the few arrangements put into force during this period was the first International Wheat Agreement, in April 1933. This was a two-year agreement in the first instance. Exporters undertook to limit their exports to certain quotas and in the second year to reduce wheat production by 15%; however, Russia and the Danubian States did not commit themselves. With regard to importing countries, it was recognised that domestic policies could not easily be changed: still, these countries agreed to take no action which would increase their wheat output, to adopt measures to raise wheat consumption, to reduce tariffs if the price on the world market rose above a certain level for sixteen consecutive weeks, and to accept as desirable a relaxation of import restrictions if prices did in fact improve over a period of a year. An International Wheat Committee was set up in London to administer the Agreement.

Perhaps the most remarkable thing about this Agreement is that it was reached at all: it was the first international agreement of any significance affecting a major foodstuff, with the exception of the Brussels Sugar Convention of 1902. However, it was not a success. Importers did not undertake to import any specified amount, and as prices continued to fall, their obligation to relax import restrictions was not called upon. Further, most exporting countries did not take the necessary steps to reduce production: in the United States, Canada and Australia, the area sown to wheat in 1934 was only slightly less than in the previous year, while Argentina increased her wheat area and shipped more wheat than her quota allowed. It proved impossible to reach an agreement on export quotas for the 1934/5 season, or on acreage controls for the 1935 crop; as a result the Agreement was terminated in 1935, and only a Wheat Advisory Committee continued to function up to the war.

Sugar was also the object of an international agreement, more far-reaching than the 1902 Convention. The first step was the Chadbourne Agreement of May 1931 between seven exporting countries (later joined by two others), which agreed upon export quotas to maintain a stated world price. Importing countries, however, continued to promote their own production, and trade fell even more than had been foreseen; moreover, the United States gave preference to its own colonies, Britain to the British Empire. The Chadbourne Agreement expired in 1935. In 1937 the International Sugar Agreement came into force, for a five-year period: it covered most of the exporting countries as well as some large importers, including the United States and Britain: the United States undertook not to reduce its imports below the 1937 level, while Britain agreed to limit home production to a stated amount and also to limit sugar exports by the British colonies; Australia and South Africa also agreed to limit their exports. This formed the basis of a more effective arrangement, which however did not have long to operate before it was overtaken by the war.

The question arises why international action had so little success in restoring balance to agricultural markets and in curbing the growth of mutually-defeating measures of intervention by individual countries. The secretariat of the League prepared many excellent studies, numerous conferences were held, and many admirable resolutions were passed. Yet the effects on national policies remained marginal. In an interesting report published in 1942, the League itself set out to explain why its efforts to solve the problems of world trade and failed:

> We have noted with what circumspection and with what scant results autonomous tariff reduction was recommended. This was due to the fact that, while each country believed that the tariffs imposed by others were damaging to it, it believed that its own were an asset not readily, certainly not gratuitously, to be sacrificed. There was, that is, no general belief that each extension of the division of labour would bring about an economy in production and hence an increase in welfare, or that each country must gain, even if the degree of gain varied, from a general reduction in trade barriers. [League of Nations (1942) page 131]

As the report of the League observed, the basic difficulty had been that trade barriers could not be reduced while depression and unemployment prevailed. Many statesmen, when in Geneva, approved liberal policies of long-term value, but when they returned to their countries they were compelled to take an opposite course in accordance with immediate needs. The emphasis on the 'customs' truce' was therefore misplaced: the only effective policy would have been to take joint action against depression. But the League of Nations was unable to intervene in matters then regarded as domestic policy.

*

* *

The economic crisis which began in 1929 was thus the cause of a second convulsion in agriculture: following the collapse of agricultural prices on the world market, traditional patterns of agricultural production and trade were transformed and official policies towards agriculture underwent a basic change. In countries where agriculture already received protection, as in France, Germany and Italy, the tariff structure was reinforced by a series of new and far more drastic measures, while countries such as Britain, Denmark, the Netherlands and Belgium abandoned their traditional policies of Free Trade and began to support their farmers in a variety of ways. Among the new measures, import quotas and intervention on domestic markets were destined to play a significant role after the Second World War. Even Britain had introduced Marketing Boards for several commodities; France had organised its markets of wheat and wine in particular; in Germany the National Socialists had taken far-reaching steps to organise all aspects of agricultural production and marketing.

While in all countries the aim had been to protect farmers from the worst of the slump in world prices, in Nazi Germany and in fascist Italy the governments had also sought to promote domestic food production for strategic purposes. Elsewhere the measures taken were generally conceived as temporary: it was widely supposed that once the crisis was over they could be removed. But the effect of measures insulating domestic markets had been to throw world markets into still greater chaos – with serious effects on agricultural exporting countries – and thus to perpetuate both the crisis and the need for protection. In agriculture as in other sectors, 'beggar-my-neighbour' policies were the rule and international action was too weak to modify policies in the direction of the general interest.

The 1930s were characterised also by the growth of political consciousness among farmers and by the development of farm organisations whose demands centered on protectionist measures. Neither they nor their governments showed much interest in basic reforms. In this period, as in the late nineteenth century, the response to challenge was above all defensive and conservative.

Bibliography

For a general survey of protectionist policies during the 1930s, the work by Margaret Gordon (1941) is particularly useful. A penetrating contemporary analysis of the problems involved was made by the League of Nations secretariat in 1935. Lamartine Yates (1940) provided the first comparative review of agricultural conditions and policies in the various European countries.

GENERAL

Ashworth, W. (1952) *A Short History of the International Economy, 1850–1950.* London, New York, Toronto: Longmans.

Bacon, L. B. and Schloemer, F. C. (1940) *World Trade in Agricultural Products.* Rome: International Institute of Agriculture.

Berkelbach, A. and Hutton, D. G. (1932) *The Pinch of Plenty.* London.

Birnie, A. (1961) *An Economic History of Europe, 1760–1939.* London: Methuen.

Clough, S. B. and Cole, C. W. (1952) *Economic History of Europe.* Boston: Heath.

Delle Donne, O. (1928) *European Tariff Policies since the World War.* New York.

Dietze, C. von (1936) *Preispolitik in der Weltagrarkrise.* Berlin: Weidmannsche Buchhandlung.

Food Research Institute, Stanford University (1932) 'Economic nationalism in Europe as applied to wheat'. *Wheat Studies.*

Gordon, M. S. (1941) *Barriers to World Trade.* New York: Macmillan.

Heuser, H. (1939) *Control of International Trade.* London: Routledge.

Liepmann, H. (1938) *Tariff Levels and the Economic Unity of Europe, 1913–1931.* London: Allen & Unwin.

Madison, A. (1962) 'Growth and fluctuations in the world economy'. *Quarterly Review of the Banca Nazionale del Lavoro.* **61**. Rome.

Malenbaum, W. (1953) 'The world wheat economy', *Harvard Economic Studies,* **XCII**. Cambridge, Mass.

Panaitesco, P. N. (1935) *Les contingentements dans les relations commerciales avec les pays agricoles.* Paris.

Royal Institute of International Affairs (1932) *World Agriculture.* London.

US Senate (1933) *World trade barriers in relation to American agriculture* 73rd Congress, 1st Session, Senate Document No. 70.

Warriner, D. (1939) *Economics of Peasant Farming.* London: Oxford University Press.

Yates, P. L. (1940) *Food Production in Western Europe.* London: Longmans.

Yates, P. L. (1959) *Forty Years of Foreign Trade.* London: Allen & Unwin.

INTERNATIONAL ACTION

Guillain, R. (1930) *Les problèmes douaniers internationaux et la Société des Nations.* Paris: Sirey.

Houillier, F. (1935) *L'organisation internationale de l'agriculture.* Paris: Librairie technique et économique.

International Institute of Agriculture (1927) *Agricultural problems in their international aspect.* Documentation for the International Economic Conference, Geneva.

League of Nations (1927) International Economic Conference: report and resolution. Geneva.

League of Nations (Economic Committee) (1931) *The Agricultural Crisis.* Geneva.

League of Nations (1935) *Considerations on the present evolution of agricultural protectionism.* Geneva.

League of Nations (1942) *Commercial Policy in the Interwar Period.* Geneva.

Vitta, C. (1936) *La coopération internationale en matière d'agriculture.* Académie de Droit International, Recueil des Cours, Tome 56. Paris: Sirey.

INDIVIDUAL COUNTRIES

(For the UK, France, Germany and Denmark, see the bibliographies at the end of relevant chapters.)

Austria

Dorfwirth, L. A. (1937/8) *Die österreichische Agrarpolitik seit dem Ende des Weltkrieges.* Wien: Schöler.

Meihsl, P. (1961) 'Die Landwirtschaft im Wandel der politischen und ökonomischen Faktoren'. In Weber, W. *ed. Osterreichs Wirtschaftsstruktur.* Zweiter Band. Berlin: Duncker & Humblot.

Belgium

Baudhuin, F. (1946) *Histoire économique de la Belgique, 1914–1939.* 2e édn. Bruxelles: Bruylant.

Leener, G. de (1934) *La politique commerciale de la Belgique* (publication de l'Institut Universitaire de Hautes Études Internationales, Genève). Paris: Sirey.

Italy

Bandini, M. (1937) *Agricoltura e Crisi.* Firenze: Barbera.

Schmidt, C. T. (1938) *The Plough and the Sword: Labour and Property in Fascist Italy.* New York: Columbia University Press.

Schüttauf, A. W. (Nov. 1936) 'Strukturpolitik und Marktregulierung in der italienischen Weizenwirtschaft'. *Weltwirtschaftliches Archiv.*

Smith, D. M. (1959) *Italy – A Modern History.* University of Michigan.

Ucker, P. (1935) *Die italienische Agrarpolitik seit 1925, unter besonderer Berücksichtigung des 'Kampfes um das Getreide'.* Schweizerische Beiträge zur Wirtschafts- und Sozialwissenschaft. Aarau.

Netherlands

Buning, E. de Cock (1936) *Die Aussenhandelspolitik der Niederlande seit dem Weltkriege.* Berner wirtschaftswissenschaftliche Abhandlungen, 17.

Frost, J. (1932) 'Landwirtschaft und Agrarpolitik in den Niederlanden'. In Friedrich List-Gesellschaft *Deutsche Agrarpolitik: Ergänzungsteil.* Berlin.

Mary, G. (1943) *La politique agricole de l'Etat néerlandais pendant la crise de 1929.* Paris.

Schiller, K. (Sept., 1936) 'Das niederländische Marktregulierungssystem für Weizen und Weizenprodukte'. *Weltwirtschaftliches Archiv.*

Schiller, K. (Sept., 1937) 'Die Regulierung der niederländischen Schweinewirtschaft'. *Weltwirtschaftliches Archiv.*

Scandinavia

International Labour Office (Jan., 1960) 'Agricultural policy in Scandinavian countries'. *International Labour Review.*

Svenska Handelsbanken (March, 1939) 'Government measures for the relief of agriculture in Sweden'. Supplement to *Svenska Handelsbanken Index.*

Switzerland

Bickel, W. (1961) *Landwirtschaft und Landwirtschaftspolitik der Schweiz.* Bern: Unionsdruckerei.

Neuhaus, J. (1948) *Die Entwicklung der bundesstaatlichen Agrarpolitik seit 1948.* Turbenthal: Furrers Erben.

Eastern Europe

Morgan, O. S. *ed.* (1933) *Agricultural Systems of Middle Europe.* New York.

Tibal, A. (1931) 'La crise des États agricoles européens et l'action internationale'. *Conciliation Internationale,* bulletins nos. 2–5, Paris.

Chapter 7

The United Kingdom

The First World War and its aftermath

Britain entered the First World War with no prepared plan for raising food production. In the early stages, farmers were advised to grow certain crops but they were not given any special financial inducement to do so. In July 1915 a Food Production Committee under Viscount Milner reported that if increased wheat production was required, farmers would need to be guaranteed a minimum price over a period of years. (It suggested that this minimum should be implemented by payments to make up the difference between the guaranteed price and the average price in the harvest year – this seems to be the first mention of a system of 'deficiency payments'.) In August, however, the government announced that it did not intend to introduce price guarantees: the harvest in Britain as well as in America promised to be large, and the German submarine campaign seemed to be on the wane.

Farmers therefore made no special effort to raise production. The numbers of livestock were practically unchanged. In 1916 the wheat acreage declined and the total arable area fell below the pre-war level; the harvest was poor. But the submarine campaign intensified. In December 1916 a new government had to take careful stock of the food situation: it decided to stimulate an increase in the home supply of grain and potatoes. A Food Controller was included in the government, and a Food Production Department was formed. County Agricultural Executive Committees were set up, with powers to compel ploughing of grassland and to take possession of badly-cultivated farms. The new President of the Board of Agriculture, R. E. Prothero (afterwards Lord Ernle) had been a member of the Milner Committee, and he now decided to adopt the recommendations of the Committee: a scheme of guaranteed prices for wheat, oats and potatoes was announced in February 1917, followed in August by the Corn Production Act, which guaranteed minimum prices for wheat and oats on a declining scale up to 1922.

These measures encouraged a considerable expansion of the ploughed area in 1917. Market prices were in fact above the guaranteed minima, but the existence of the guarantees gave the farmers confidence. The events of 1917 proved how necessary the change in policy had been. Attacks by submarines reached a peak and shipping problems became acute. In 1918 a crisis in food supplies was avoided only by a further expansion in the crop area and a large harvest. It was estimated that by the end of the war British agriculture was feeding the population for the equivalent of 155 days in the year, as compared with only 125 days at the outbreak of war (Middleton, 1923).

By 1918 extensive control had been established over food supplies: all essential foodstuffs, both imported and home-grown, were bought or requisitioned by the Food Controller at fixed prices; importers, manufacturers and distributors became in various ways the agents of the Controller.

The war had thus necessitated a reversal in the traditional policy of laissez-faire towards agriculture. When it ended, there was much discussion as to the policy which should be followed in peacetime. A Royal Commission was appointed to inquire into the 'economic prospects of the agricultural industry'. It reported in 1919, but its members reached opposite conclusions. A majority of twelve pointed out that farmers were anxious about their future: the costs of arable farming had become very high, with wages in particular rising, and there was a fear that world prices would fall. Farmers generally were prepared to do without guarantees if they were freed from all control, but if the government decided that production should be increased, it was under an obligation to provide a financial guarantee. These twelve members recommended that price guarantees for wheat, barley and oats should be maintained for at least four years. A minority of eleven, however, disagreed: they thought that there was no risk for some years to come of grain prices falling to an unremunerative level. Already the financial situation of agriculture had improved. They were not convinced that it was necessary to maintain agriculture on a wartime basis; the foreign exchange situation could be improved by raising exports rather than by decreasing imports. They concluded:

> We have seen no evidence that will enable us to conclude that the financial position of the country can be improved by embarking on a policy involving the expenditure of public money in diverting agriculture into uneconomic fields. [Royal Commission (1919) page 10]

The government, faced with increasingly insistent demands from farmers for some sort of policy, introduced an Agriculture Act in 1920. This in effect extended and reinforced the Corn Production Act: guaranteed prices were to continue indefinitely, and the level of the guarantees for wheat and oats was raised.

But the government soon had cause to regret the liability it had incurred. The following year the price of grain fell steeply, and the cost of supporting the guarantees amounted to over £18 million. This appeared excessive, and the government abandoned its year-old policy. In August 1921 the price guarantees were repealed.

Farmers now found themselves in serious difficulties, with on the one hand falling grain prices and on the other rising wages and other costs. The government's act of 'betrayal' had far-reaching consequences. In the years to come, more and more land was returned to grass: the arable area fell even below the level of 1914. Many farms were neglected, their capital equipment was allowed to deteriorate and farm labour again began to move into industry. Once again, agriculture was in decline.

The 1920s

Through most of the 1920s, the general price trend was downwards, interrupted only by periods of relative stability from 1923 to 1925 and from 1927 to 1929. Throughout this period, British agriculture was feeling the effects of overseas competition. Wheat-exporting countries were raising their output; the invention of chilling made it possible for the Argentine to send meat which could compete with all but the best British qualities; New Zealand enormously expanded her trade in meat and dairy produce; Denmark was shipping more butter, eggs and bacon; and fruit was being sent in increasing quantities from North America at times when it competed directly with the output of British horticulture. The only course for the British farmer was to concentrate on products, such as milk, eggs and fresh vegetables, for which he enjoyed a naturally sheltered market.

The governments of this period held almost without exception to the policy of laissez-faire. The only direct measure of financial assistance to farmers was the subsidy on beet sugar, introduced in 1924. This subsidy stimulated a rapid expansion in the area under sugar-beet (from 22 000 acres in 1924 to a peak of 396 000 ten years later) and in the number of sugar factories.

There were some departures from strict Free Trade principles, but none that was of any importance for agriculture: the 'McKenna' duties on certain luxury goods, introduced in 1915, were maintained after the war, and the Safeguarding of Industries Act of 1921 introduced duties to protect certain 'key' industries. In the 1919 Finance Act, the principle of Imperial Preference was applied by Britain for the first time, a rebate on the existing revenue duties and on the McKenna duties being granted to Empire goods. At the Imperial Economic Conference of 1923, the Conservative Party announced its support for a much more extensive policy of protection and Imperial Preference, but at the general election of the same year, fought largely on this issue, the electorate showed itself still hostile to the idea. By 1929, Britain's policy was still essentially that of Free Trade, and agricultural products were subject only to certain revenue duties. There was thus little scope for Imperial Preference; sugar was almost the only item where preference had a marked effect in favour of the Empire.

The crisis – Britain abandons Free Trade

In the absence of any measures of protection, the crisis which began in 1929 had an immediate and severe effect on British agriculture. The fall in world grain prices was reflected on the British market and as other European markets were more and more closed by measures of protection, Britain became the object of large-scale dumping. In October 1931 the volume of food imports was 35% above normal: besides grain, imports of dairy produce, meat and other products increased.

General economic depression added to the difficulties of agriculture: Britain's exports fell, industrial unemployment grew and consumer purchasing power was reduced. No effective policy was devised for reviving the economy and the emphasis was laid on measures of protection against imports.

The Labour Government in power at the onset of the crisis resigned in August 1931. The National Government which took its place was a coalition dominated by Conservatives, who urged that a comprehensive tariff system should immediately be adopted. In November the government obtained authority to impose duties up to a maximum of 100% on a large range of wholly or partly manufactured goods; at the same time, protection was given to horticulture. These emergency measures were replaced in 1932 by the Import Duties Act, which remained the basic instrument of the British tariff. It imposed a general tariff of

10% *ad valorem;* however, certain major foodstuffs and raw materials – including wheat, maize, meat, livestock and wool – were exempted from duty. An Import Duties Advisory Committee was set up, which could recommend changes in the duties. Further, the government was authorised to admit goods free or at reduced rates of duty from specified foreign countries, or on the contrary to double the rate on goods from countries which discriminated against British products.

Imperial Preference

Protection in Britain could not take the same course as in other countries. The question of how to treat Empire goods – most of them agricultural – was immediately raised. Negotiations were held at Ottawa in 1932 between Britain and the Dominions and as a result Britain made a series of important concessions which were shortly afterwards put into effect by the Ottawa Agreements Act:
(a) The exemption from duty of Empire goods, already provided for under the Import Duties Act, was confirmed.
(b) New or increased duties were imposed on imports from foreign countries which competed with Empire produce; these included a new duty on wheat and revised duties on butter and cheese, certain fruits and various other products.
(c) A guarantee was given that the existing 10% margin of preference on certain products would not be reduced.
(d) Special preferences on certain other goods from the Empire (including sugar, wine, coffee, tobacco) were guaranteed.
(e) An undertaking was given that imports of meat from foreign countries would be controlled.

In granting these concessions, the British Government made two main reservations: firstly, the new duties on foreign wheat and certain other products could be removed at any time if Empire supplies proved inadequate (this provision was never called into force); and secondly, the Dominions were to be given an expanding share of meat imports only in so far as this was consistent with the development of British production. Also, Britain reserved the right, in the interests of home producers, to review after three years the preferences for dairy products, poultry and eggs. The principle underlying the government's policy was that home producers should have first claim on the market, Empire producers second and foreigners last.

The advantages which Britain obtained from the Ottawa Agreements consisted mainly of increased preferences for British exports and assurances that protection against British goods would not be excessive.

The Irish Free State did not benefit from the Ottawa Agreements. On the contrary, in July 1932, following a financial dispute between the two countries, Britain imposed extra duties on Irish goods, most of which were agricultural products. Britain also prohibited imports of Irish beef and veal and restricted imports of Irish cattle. This discrimination was brought to an end in 1938, and subsequently Irish produce was admitted on the same terms as Empire goods.

Also in 1932, assistance was given to the sugar industry in the West Indies through increased tariff preference for all Colonial sugar (that from the Dominions continued to get the existing rate of preference), together with an extra preference for defined quantities of sugar from specified colonies in the West Indies.

Support for British agriculture

All these measures concerning trade gave little benefit to British agriculture. By the time when farmers in most other European countries were sheltering behind high tariffs and strict import controls, Britain's farmers were still exposed to the full blast of competition from the Empire and even imports of agricultural products from foreign countries were subject only to relatively low duties. Lord Astor and Murray observed:

> The Ottawa Agreements must have made it plain, if it was not so evident beforehand, that British farming cannot expect much relief at the hands of Members of Parliament representing the urban voter. The British delegation set out to seek a freer market for the United Kingdom exports, and these exports are not agricultural products. It has also been made evident that if import concessions are to be made to the Dominions, these concessions must be made in those imports which form the bulk of the Dominions' trade, i.e. in agricultural products. [Astor and Murray (1933) page 136]

Besides these considerations, the traditional policy of cheap food was still an important factor; particularly in a period of acute depression, the government was reluctant to take measures which might be held to raise the price of foodstuffs.

The attempt to help British farmers took the form of a series of measures from 1931 onwards on a commodity-by-commodity basis. These measures were applied piecemeal: they did not correspond to any deliberate policy for agriculture as a whole nor to any agreed priorities for different products. They involved subsidies, marketing schemes, import restrictions and various combinations of these. Thus for wheat and sugar-beet, assistance was given by subsidy alone; for milk, by a marketing scheme and by subsidy; for beef and cattle, by restriction of foreign imports and by subsidy; for bacon, potatoes and hops, by marketing schemes combined with import control; for eggs, by voluntary restrictions applied by exporting countries.

The marketing schemes were based on the Agricultural Marketing Acts of 1931 and 1933; the second of these also provided the authority for import controls. The 1931 Act was introduced by the Labour Government and represented a first attempt to deal with the crisis: the government was not prepared to make a radical breach with Free Trade, and marketing reform seemed to provide an attractive alternative. This Act enabled a 'substantial' majority of producers in any branch of agriculture to adopt a marketing scheme, subject to approval by the Minister of Agriculture and by Parliament; the scheme then became binding upon all producers. The marketing scheme could involve the enforcement of minimum and maximum prices and other regulations, but the 1931 Act made no provision for controlling imports. This loophole was filled by the Act of 1933, which declared that if a satisfactory marketing scheme was evolved for any product, the government would regulate imports; under the marketing scheme itself, the quantities sold by any registered producer could be regulated. With this extension of powers, marketing schemes began to look more attractive and, before long, schemes were in operation for milk, potatoes, hops, pigs and bacon.

The arrangements for the major products are described below.

WHEAT

In 1932 a Wheat Act was passed, which reintroduced guaranteed prices on lines similar to those followed at the end of the First World War: producers were to be ensured an average price of 10s. per cwt. for millable wheat by means of a deficiency payment making up the difference between the average market price and the guaranteed price. To discourage undue expansion, full payment was to be made only for an output up to 27 million cwt.: in excess of this, the rate of subsidy would be proportionately reduced. Finance for the subsidy was obtained from a levy on all flour, whether from home-produced or imported wheat. The scheme was made possible by the fact that home supply formed a small proportion of the whole.

Market prices in subsequent years were generally well below the guaranteed price of 10s. per cwt.: subsidy had to be paid at rates varying, from 1932 to 1936, between 3s. and nearly 5s., and the annual subsidy bill reached £7.2 million in 1933/4.

The result was to give a considerable stimulus to wheat cultivation: area and production rose, largely at the expense of oats; oats and barley did not benefit from a price guarantee until 1937. The system was criticised for encouraging expansion in areas not well suited to wheat. Although in 1932 wheat constituted only about 5% by value of gross agricultural output, it was still regarded by many as a vital crop and the special support it received owed a good deal to this, almost sentimental, attachment. Lord Astor and Rowntree wrote in their important work on British agricultural policy:

> Bread is the staff of life; wheat makes bread; therefore, [it was supposed that] wheat must be the most essential foodstuff and its production should be safeguarded. It was also supposed (quite erroneously) that wheat was the cornerstone of British agriculture, and that any measure which stimulated wheat production would not only feed the people but would stimulate agriculture as a whole. [Astor and Rowntree (1938) pages 82–3]

Astor and Rowntree considered that the support given to wheat, instead of merely countering the instability of world markets, had come to operate as a permanent subsidy. They criticised this development:

> On general economic grounds it would be unwise to give permanent artificial encouragement to wheat production in this country. Wheat can be grown overseas far more cheaply than in Great Britain; it is pre-eminently a crop suited

to large-scale farming methods which we cannot conveniently imitate; it is a commodity which enters largely into inter-Imperial trade. [Ibid., page 88]

SUGAR-BEET

The subsidy granted in 1924 was intended to disappear in ten years: by then, it was hoped, the sugar-beet industry could stand on its own feet. This expectation was not fulfilled. A considerable expansion took place and costs of production were reduced; meanwhile, however, some spectacular improvements were made overseas as a result of which the cost of producing sugar from cane fell to less than two-thirds that of producing from beet. When the ten years were up, the government appointed a committee to inquire whether the industry could be made self-supporting. Two members of the committee recommended that the subsidy should be brought to an end, the third disagreed. The government decided to maintain the subsidy but under new arrangements. The resulting Sugar Industry Reorganization Act of 1935 provided for continued assistance but the quantity of sugar eligible for support was limited; also, the sugar factories, which had been making considerable profits out of the subsidy, were amalgamated into the British Sugar Corporation, and the whole industry was subjected to an independent Sugar Commission which was to determine the rate of subsidy.

MILK

Milk Marketing Boards were set up at the end of 1933 in an attempt to place farmers in a better bargaining position. All milk had to be sold through the Boards, which had powers to negotiate contracts with distributors of liquid milk and to divert milk between alternative manufacturing uses, giving preference to higher-priced outlets. The Board organised a two-tier price system, pushing up the price charged for milk for liquid consumption while fixing relatively low wholesale prices for milk sold for manufacturing into dairy products. The market for the latter was of course influenced by competition from imports: restrictions on imports of dairy produce from the Dominions were precluded under the Ottawa Agreements, and the Dominion governments turned down a suggestion that they should reduce their exports of butter and cheese on the understanding that imports from foreign countries would also be cut. A small temporary subsidy was given in 1934 for home-produced butter and cheese. Irrespective of the use of the milk, all farmers received the same pool price, subject to some differentiation between regions.

BEEF AND CATTLE

Owing to the importance of supplies from abroad, the British meat market was very much exposed and the substantial fall in prices caused a crisis in the British cattle industry. Britain had agreed at Ottawa not to reduce supplies of meat from Australia and New Zealand until the end of June 1934, nor to impose a duty on Empire produce before 1937. Moreover, in an agreement with the Argentine in September 1933, Britain undertook to reduce imports of chilled beef only if this became necessary to ensure remunerative prices, and then subject to limitations.

Within these limits, an attempt was made to restrict imports of chilled beef from foreign countries, starting in 1933. As a result, imports from the Empire increased substantially; a voluntary agreement for restricting them was reached in 1935. Rather more severe restrictions were imposed on frozen beef: supplies from foreign countries were progressively reduced till by the end of 1934 they were limited to 65% of the 1931/2 level. In 1937 the task of regulating imports was passed to an International Beef Conference in which British producers, Empire countries and foreign exporting countries were represented. Quarterly import quotas were determined for chilled, frozen and canned beef. In 1938, mutton and lamb were brought within the scope of the Conference. As has already been mentioned, imports of beef from the Irish Republic were prohibited after 1932, while Irish cattle were subjected to a quota and to a prohibitive duty.

None of these measures succeeded in making prices profitable to British producers, largely because they were inadequate to restrain the growth in competition from the Empire. The government therefore introduced subsidies on cattle in 1934, in the form of a flat rate paid irrespective of market prices. At the same time, duties were imposed on all foreign imports of beef and veal.

BACON

By the beginning of 1932, pig producers were in a critical position. The general fall in prices was superimposed on a cyclical downward movement. A reorganisation committee recommended a system of contracts and quotas with the aim of increasing British production at the expense of foreign supplies. It was decided to stabilise the total supply of bacon and ham at 10 670 000 cwt. annually (this was the estimated average annual supply from 1925 to 1930). Home producers were to have preference in filling this amount: imports would be allowed only so far as necessary to make up the difference. In November 1932, voluntary agreements were reached by which eleven foreign countries undertook to restrict their exports to Britain.

Minimum prices for home supplies of bacon were fixed from time to time. The initial price offered was so attractive in relation to the alternative market for pork that there was a big increase in supply. It was decided to restrict imports still further: as Denmark objected, this had to be done on a compulsory basis by an order of November 1933 which prohibited the import of bacon and ham, except under licence, from any foreign country sending more than 400 cwt. a week to Britain. Imports from foreign countries were cut by about half, and Denmark in particular was severely hit. However, the scheme was administered through export licences issued through the governments of the exporting countries: as a result, the foreign suppliers obtained almost the whole of the increased price resulting from the restriction. The price of British bacon rose relatively little, because of its inferior quality as compared with Danish bacon. No restrictions were imposed on Empire countries, but it was agreed that Canada could supply up to 2.5 million cwt.: this was about ten times Canada's previous exports (it was rumoured that an extra nought had been added by mistake). The result was a rapid increase in Canadian exports.

The import restrictions were relaxed in 1938 under the Bacon Industry Act, and instead a system of subsidies for pig producers was introduced.

Trade agreements

After the introduction of import duties, Britain entered into tariff negotiations with a number of countries and in subsequent years several trade agreements were signed. However, in negotiations with countries whose exports were mainly agricultural, Britain was generally unable to offer concessions without breaking a pledge to one or another of the Dominions as regards the maintenance of a minimum margin of preference. The agreements thus brought about little reduction in import duties in general and in particular had little effect on the agricultural duties.

In the agreement with Denmark in 1933, Britain agreed to import bacon and hams free of duty and guaranteed that the duties on eggs and other products would not be raised above specified levels. However, Britain reserved the right to impose quotas on Danish bacon in connection with domestic marketing schemes; subsequent developments have been described above.

An agreement was also reached with the Argentine in 1933, in which Britain gave the guarantees concerning trade in beef which have already been discussed. Britain also undertook to admit meat and maize free of duty and guaranteed maximum tariff rates for wheat and other products.

The most important agreement reached during this period was with the United States; it was signed in November 1938 and came into force the following January. Britain gave a wide range of concessions on agricultural products, removing entirely the duty on wheat, guaranteeing continued free entry for maize, reducing the duty on rice and removing that on lard, guaranteeing the existing duties on pigmeat and granting an increased quota for hams. Further, the duties on a number of fruits and vegetables were to be reduced during the months when US exports took place, and those on certain canned fruits and fruit juices were removed or reduced. There was to be continued free entry for cotton and no increase in the existing margin of Imperial Preference for tobacco.

The controversy over protection

The reversal of Britain's traditional policy of Free Trade did not take place without much heart-searching. As late as the autumn of 1931, a self-appointed committee of eminent economists, headed by Sir William Beveridge, pronounced themselves uncompromisingly against protection in any form and for whatever purpose, basing themselves squarely on the classical argument for free trade. The work in question included a chapter on agriculture by Professor Lionel Robbins, in which he conceded that a tariff on wheat would for a short time benefit the wheat farmers, who were the most severely hit by the crisis. But before long rents would rise, wages too might absorb part of the increased profit, and competition between the farmers would bring prices down again. Thus Robbins wrote:

> Eventually equilibrium would be established with wheat farmers making no more than was being made elsewhere, a greater proportion of the food supply produced at home, and a national income smaller than it would have been by the extra cost of raising that much more wheat at home rather than procuring it abroad by way of exchange. [Beveridge *et al.* (1932) pages 156–7]

This book was one of the most forthright and closely-reasoned statements of the Free Trade position ever produced in Great Britain. Yet within a few months all its recommendations had become a dead letter as the government resorted to protection and Imperial Preference. It would be hard to demonstrate more conclusively the futility of abstract economic reasoning when the livelihood of much of the population is at stake and strong political forces are at work. Ten years later, Benham, who had been one of the co-authors of Beveridge's book, lamented in the following terms:

> Free trade would almost certainly have come in Great Britain in the nineteenth century if no abstract reasoning had ever been advanced in its favour, and all the economists – who, after all, are the specialists in this subject – were quite powerless to stem the tide of protection during and after 1931. It is a sad thought. [Benham (1941) page 24]

Indeed, no government in 1931 and 1932 could ignore the problems of depression and unemployment or turn a deaf ear to the many groups of producers clamouring for assistance. Measures other than tariff protection could no doubt have been taken, but Keynes had not yet written his General Theory and there were sufficient voices laying blame on imports and dumping for the government to direct its attention that way. Moreover, the Empire enthusiasts were out in force and now enjoyed much greater influence within the government than they had done at the turn of the century. As for agriculture, the Central Chamber of Agriculture and the National Farmers' Union were agitating strongly for tariffs. The Farmers' Unions (for England and Wales, Scotland and Ulster), however, were not yet very influential, though their membership was expanding. The measures taken during the 1930s were simply notified to the Unions as government decisions, except for the Agricultural Marketing Schemes where some Union participation was required.

The unemployment problem influenced the action taken with regard to agriculture. The idea was going round that a large-scale programme for settling workers in smallholdings could do much to relieve unemployment, and in 1930 the Labour government introduced a Land Utilisation Bill with this in mind. The old argument about keeping work at home was also much in evidence: if food were produced at home instead of being imported, it was thought that there would be more work for farmers and at the same time a larger market for home industry. Highly optimistic estimates were put forward as to the proportion of food that could be produced at home. On the other hand, Professor Robbins, in the work already referred to, pointed out tersely that if the object was to raise employment, it would be preferable to choose an industry where wages were relatively high and not one such as agriculture where they were exceptionally low. Several authorities, in particular Astor and Rowntree (1935), stressed the danger that by cutting down food imports Britain might injure the capacity of other countries to buy British manufactures.

Much was heard also about the need for a 'balanced economy', by which was meant a higher proportion of people on the land. There was talk of the advantages of rural life, of better health and morality and so forth. Opponents of this view were quick to point out that those who talked this way usually did not live on the land themselves.

Emphasis was laid too on the need to secure food supplies: the dangerous experience of the First World War was often recalled. The counter-argument was that strength in war depended on many factors besides food supply, which could be harmed by a policy of agricultural self-sufficiency in peacetime: such factors included the nation's overall economic and financial situation, the size of its shipping fleet and the extent of activity in shipbuilding. Moreover, excessive cultivation of the soil over a long period could exhaust its fertility. To the exponents of these views, the best preparation for war seemed to lie in building up reserves of those foodstuffs that could be stored.

A basic difficulty for the advocates of protection was to reconcile the interests of British agriculture with those of Empire producers, while at the same time paying at least lip-service to the ideal of cheap food. A pamphlet written early in the crisis by Lord Beaverbrook and circulated by the 'Empire Crusade' was entitled, rather hopefully, *The Farmers' Crusade: How Empire Free Trade will help British Agriculture*. This was an unsuccessful attempt to have the best of both worlds. British farmers were promised Protection against foreign imports while consumers were promised Free Trade in Empire products. The basic contradiction was scarcely concealed in the following passage:

> The policy of Empire Free Trade is in effect a new form of the Protection for British Agriculture which farmers have so long demanded.... The new policy of Empire Free Trade or Empire Protection [sic] is designed to give the same assistance to the farmer as the old, but at the same time to obviate the difficulty of dear food. [Beaverbrook (undated) page 2]

Another extreme statement of the case for agricultural protection, by Fordham (1932), argued in favour of import controls to give the British farmer the first claim on the home market, with the ultimate objective of reaching as near as possible to self-sufficiency in food. Fordham completely ignored the issue of Empire produce. He thought moreover that his proposals need not lead to increased food prices, mainly on the unlikely supposition that domestic production could be greatly increased at little or no extra cost per unit.

Thus, while it was pointless and even cynical to reiterate laissez-faire principles in the face of obvious distress, the arguments of the Imperialists and Protectionists were often specious, sometimes absurd. The most important work on the problems of British agriculture at this time was that by Astor and Rowntree in 1938. This was based on sound economic reasoning, but recognised the difficulties with which agriculture was faced and the need for assistance in some form. The views of the authors with regard to wheat policy have already been mentioned. In general, they thought that protective action was justified to avoid the extremes of price fluctuations; thus they were in favour of a minimum price for wheat at a level approximating to 'normal' world prices. In their conclusions they expressed the view that the introduction of protection for industry gave agriculture a claim to a similar degree of state assistance, not merely as a matter of equity but because farming costs were raised by tariffs on manufactures. Nevertheless, they were firmly opposed to any policy of protection designed to promote the expansion of domestic agricultural output, and wrote:

> We are convinced ... that those are profoundly mistaken who aim at effecting an expansion of British agriculture by further measures of protection against overseas competition, or by further subsidies for commodities which can be produced considerably more cheaply or more efficiently abroad. Our national interests in the maintenance of a large-scale international trade, the interests of the consuming population in the provision of cheap food, the interests of the British Dominions and Colonies as agricultural producers, the growing budgetary difficulties of the British Treasury, the complications of the structure of British agriculture itself, combine to render any such policy the height of unwisdom. [Astor and Rowntree (1938) pages 441–2]

Consequences for agriculture and trade

Not surprisingly, in view of the nature of the measures taken, the market prices of most important foodstuffs remained low. In 1933 the overall agricultural price index (Table 7.1) reached its lowest point at 76% of the pre-crisis level; it then recovered slowly, but even by 1938 it was only at 88%. The inclusion of the subsidies on various products raises the overall index by only a couple of points. The subsidies, however, were important for wheat, where the market price (except in 1937) remained very low but the subsidised price approached the pre-crisis level. Livestock prices fell later than those of crops, but from 1931 onwards they did little or no better. The subsidised price of fat cattle remained well above the market price; on the other hand, the subsidy on milk for manufacturing made little difference to the overall milk price.

The effect of intervention was felt most markedly in the cultivation of wheat. Between 1932 and 1938, the wheat area in the United Kingdom as a whole expanded from 1.3 million acres to 1.9 million, while the area of most other crops declined and the total arable area fell from 13.6 to 13.0 million acres (Table 7.2). There was an increase in the number of dairy cows and other cattle, as well as in the number of pigs; poultry and sheep declined in numbers.

The gradual recovery of prices from 1933, and the help given through subsidies, provided some relief to farmers' incomes. In the northern and western parts of the country, the livestock farms regained reasonable prosperity, often with increased emphasis on dairying; the arable areas of the east and south were

Table 7.1: Agricultural price indices for England and Wales: 1927–9 average = 100

	1930	1931	1932	1933	1934	1935	1936	1937	1938
Crops	74	88	89	71	72	76	90	98	80
			(93)	(78)	(79)	(82)	(94)	(99)	(86)
of which:									
Wheat	80	56	56	49	47	51	71	90	59
			(74)	(94)	(89)	(85)	(91)	(96)	(94)
Barley	80	80	71	83	86	80	86	109	84
Oats	68	68	74	61	69	71	70	92	79
Livestock	97	81	75	75	78	75	78	87	86
					(78)	(77)	(80)	(88)	(88)
of which:									
Fat cattle	100	91	86	76	75	69	73	82	84
					(78)	(80)	(83)	(92)	(95)
Bacon pigs	104	74	63	70	77	70	78	83	84
Milk	95	82	84	84	88	85	85	94	100
					(89)	(88)	(87)	(95)	(102)
Eggs	91	77	72	71	68	72	78	83	85
All agricultural products	91	84	81	76	77	79	81	89	88
			(81)	(77)	(79)	(81)	(83)	(91)	(90)

Note: Data in parentheses take account of subsidies for wheat, cattle and milk.
Source: Ministry of Agriculture and Fisheries, *Agricultural Statistics,* 1939.

generally in less good condition. Agricultural wages were fairly well maintained even during the crisis, but the number of agricultural workers in Great Britain fell from 857 000 in 1930 to only 697 000 in 1938.

Perhaps the most striking result of the various measures of intervention was a shift in the pattern of food imports away from foreign countries in favour of the Empire. By 1938, as compared with 1927–9, total food imports from foreign countries had fallen by 17% in volume, while those coming from the Empire had

Table 7.2: Crop area and livestock numbers in the United Kingdom

	1929	1932	1935	1938
		million acres		
Area				
Wheat	1.4	1.3	1.9	1.9
Oats	3.1	2.7	2.5	2.4
Barley	1.2	1.0	0.9	1.0
Roots, other crops and fallow	4.0	4.0	3.8	3.7
Temporary grass	4.6	4.6	4.4	4.0
Total arable land	14.3	13.6	13.5	13.0
Permanent grass	18.4	18.7	18.5	18.8
Total crops and grass	32.8	32.3	32.0	31.8
Livestock		*millions*		
Dairy cows	3.4	3.6	3.8	3.8
Other cattle	4.5	4.7	4.8	5.0
Sheep	24.3	27.2	25.1	26.8
Pigs	2.7	3.6	4.5	4.4
Fowls	55.5	73.5	75.1	69.1

Source: Annual Abstract of Statistics.

Table 7.3: Index numbers of the volume of food imports into the United Kingdom: 1927–9 average = 100

	Wheat and flour	Meat, incl. bacon	Dairy products	Eggs	Fruit	Vegetables	All food
1930	99	106	109	107	99	101	105
1931	110	117	123	104	120	156	117
1932	96	112	127	84	118	142	111
1933	103	101	133	76	107	87	107
1934	95	94	142	81	88	81	104
1935	92	92	137	84	111	88	103
1936	92	92	138	102	88	100	105
1937	89	95	136	103	85	86	103
1938:							
Total	92	95	138	114	107	79	106
Empire	129	142	158	63	217	130	143
Foreign	59	71	119	125	61	66	83

Source: Murray, K. A. H. and Cohen, R. (1934) *The Planning of Britain's Food Imports.* Agricultural Economics Research Institute, Oxford.

risen by no less than 43% (Table 7.3). This shift in the sources of supply was evident for all major foods except eggs, for which in any case trade consisted mainly of small amounts from Ireland. Overall, there was relatively little change in the volume of imports – an increase of 6%. The only commodity for which total imports rose substantially was dairy products, where there was an increase of 38%; total imports of wheat and meat were reduced. It thus seems that British trade policy, helped by other measures taken during this period (in particular the formation of the sterling area), succeeded in ensuring a larger market for the Empire while bringing total imports down to nearly the level which prevailed before the crisis.

Bibliography

The titles listed reflect extensive contemporary controversy over the issue of introducing protection. Two works stand out: that by Richardson (1936), which despite its title is an excellent account of domestic market organisation as well as of protective measures and Imperial Preference; and that by Astor and Rowntree (1938).

Some of the works listed deal only with the organisation of food production during the First World War and its immediate sequels – in particular Middleton (1923) and Beveridge (1928); Ernle's work in its later editions just covers this period.

Addison, Lord (1939) *A Policy for British Agriculture.* London: Left Book Club.
Astor, Viscount and Murray, K. A. H. (1933) *The Planning of Agriculture.* London: Milford.
Astor, Viscount and Rowntree, B. S. (1935) *The Agricultural Dilemma.* London: King.
Astor, Viscount and Rowntree, B. S. (1938) *British Agriculture – The Principles of Future Policy.* London: Longmans.
Beaverbrook, Lord (undated but probably about 1930) *The Farmers' Crusade: How Empire Free Trade will help British Agriculture.* London.
Benham, F. (1941) *Great Britain under Protection.* New York: Macmillan.
Beveridge, Sir William (1928) *British Food Control.* London.
Beveridge, Sir William, *et al.* (1932) *Tariffs: The Case Examined.* 2nd edn. London.
Edminster, L. R. (February, 1939) 'Agriculture's Stake in the British Agreement and Trade Agreements Program', *International Conciliation.*
Ernle, Lord (1962) *English Farming, Past and Present.* 6th edn. London: Heinemann.
Fordham, M. (1932) *Britain's Trade and Agriculture.* London.
Hall, A. D. (1917) *Agriculture after the War.* London.
Harkness, D. A. E. (1941) *War and British Agriculture.* London: King & Staples.
Hobson, J. A. (1916) *The New Protectionism.* London.
Jeffcock, W. P. (1937) *Agricultural Politics 1915–1935, being a History of the Central Chamber of Agriculture during that Period.* Ipswich.
Kirk, J. H. (1979) 'UK Agricultural Policy 1870–1970'. In *The Development of Agriculture in Germany and the UK.* Wye College, Ashford, Kent.
McGuire, E. B. (1939) *The British Tariff System.* London: Methuen.
Middleton, T. H. (1923) *Food Production in War.* Oxford.
Murray, K. A. H. and Cohen, R. (1934) *The Planning of Britain's Food Imports.* With supplements. Oxford: Agricultural Economics Research Institute.
National Institute of Economic and Social Research (1943) *Trade Regulations and Commercial Policy of the United Kingdom.* Cambridge University Press.
Plummer, A. (November, 1932) 'The British Wheat Act, 1932', *Quarterly Journal of Economics.*
Richardson, J. H. (1936) *British Economic Foreign Policy.* London.
Robbins, L. C. (1935) 'L'agriculture dirigée'. Dans *L'Economie dirigée:* publications à part de la *Revue d'Economie Politique.* Paris.
Royal Commission on Agriculture (1919) Interim Report (Cmd. 473).
Venn, J. A. (1933) *The Foundations of Agricultural Economics, together with an Economic History of British Agriculture during and after the Great War.* Cambridge.

Chapter 8

France

The First World War and the 1920s

The war brought severe losses for French agriculture, both directly through the depletion of its livestock and the destruction of the battle areas (which included some of the most productive land in France) and indirectly through difficulties in preserving the fertility of the soil and in maintaining farm equipment. The greater part of the farm labour force was called into military service, fertiliser and fuel were in short supply, transport was lacking. Production of wheat in 1917 was less than half what it had been at the outbreak of war, and it was necessary to import, on the average, three times more wheat and flour than before the war.

'Protectionism is not a policy for wartime', wrote Augé-Laribé (1950). The existing duties on major foodstuffs were suspended early in the war to facilitate imports and to alleviate the rise in the cost of living. The State took over the purchase and distribution of foodstuffs; from 1916 to 1921 imports of grain and flour were carried out by an official body. The export of certain foodstuffs was prohibited. The preoccupation with keeping food prices down, combined with the mounting costs of agricultural requisites, tended to discourage output; incentives to production were granted only after some delay, and then not very systematically.

In the period of reconstruction which followed the war, food was still short and there was fear of a continued rise in the cost of living. Protection was restored to agriculture only by degrees. In 1919 the pre-war tariff was reintroduced, with a system of coefficients to take account of the increased price level; the coefficients applied to foodstuffs were relatively low. Various modifications were made to the tariff in subsequent years, particularly in 1926 when overall increases of 30% were applied to the coefficients twice in the year. An attempt was made in 1927 to carry out a general revision of the tariff, but before this could be completed the government revised almost half the items of the tariff in the context of a commercial treaty with Germany; the agricultural duties however were not affected.

By the mid-1920s agriculture was suffering from a cost-price squeeze and the shift of population to the cities was resumed. Agricultural associations demanded greater equality of treatment with industry. The *Confédération Nationale des Associations Agricoles* complained in a pamphlet in February 1927 that 'since the war, our entire customs and economic policy seem to have been directed to one sole aim: that of developing France into a great industrial power, regardless of the interests of agriculture'; it maintained that the current tariff proposals would raise duties on industrial products above the pre-war level to a far greater extent than those on agricultural products. In April 1927 a conference of agricultural associations demanded that agriculture and industry should be given equal tariff treatment. Industrial interests, on the other hand, opposed any increase in the agricultural duties and a bitter struggle ensued. In March 1928 the agricultural duties were raised, but the representatives of agriculture were not satisfied. In November 1929 the presidents of the *Chambres d'Agriculture* opposed the League of Nations project for a tariff standstill, declaring that this would merely confirm the inadequate treatment of agriculture. This view seems to have carried weight, since in February 1930 the French delegation to the League of Nations declared that France could not accept the project.

Apart from tariff policy, little was done in this period to improve the situation of French agriculture. Perhaps the most significant development was the setting-up of specialised producer organisations. Twice the beet growers and wine producers joined forces to persuade the government to enact arrangements for distilling their respective products (the government agreeing to buy surplus alcohol and to mix it with petrol to make a *carburant national*).

The crisis – tariffs and import quotas

In 1925 the French grain crop for the first time reached the level of the pre-war harvests. From about 1927 the pressure of supplies on the world market made itself increasingly felt. In the autumn of 1929 the crisis broke and the world grain market collapsed. The French harvest of 1929 was large, contributing to the fall in prices.

At the onset of the crisis, France was in a strong economic and financial position, and the general price level was at first relatively well maintained; the French market thus provided an attractive outlet for foreign supplies, while French exports encountered increasing difficulties.

The first reaction was to raise tariffs wherever possible. Most agricultural products were subject to the *loi de cadenas* of 1897, under which the government was entitled to raise the duties on certain products without waiting for the approval of Parliament. The scope of this law was extended in 1929 and again in 1931, so that instead of covering only forty-six items it became applicable to almost all products competitive with French agriculture. This authority was used to raise the duties on wheat and most other grains, as well as those on some livestock products, wine, sugar and other foodstuffs. In many cases, successive increases in 1930 and 1931 resulted in the duty being at least doubled from its previous level.

However, tariffs were found to be an inadequate method of protection. In the first place, it was feared that even bigger increases in the duties would be necessary to compensate for the fall in world prices; elections were due in 1932, and the government did not want to be accused of raising the price of food. Moreover, some of the duties on agricultural products (including barley and most dairy products), and most of those on manufactured goods, had been consolidated in treaties since 1927 and could not easily be raised. The device of import quotas offered a solution. Import quotas had been used before by various countries, generally in connection with commercial agreements when tariff concessions were given for specified quantities of imports. France, however, was the first country to make systematic use of them as a means of protection.

The first import quotas, for timber and wine, were introduced by a decree of

27 August 1931. Others followed in rapid succession: cattle, pigs, beef, pigmeat, butter and cheese on 30 September; mutton, poultry and eggs on 10 November; sugar on 5 December. In the course of 1932 sheep, potatoes, onions, fruit and barley were added to the list; in 1933 maize, oats, rye and margarine. Practically all agricultural products, with the important exception of wheat, thus became subject to quota. In a valuable contemporary study, Moroni commented:

> What in 1931 was only an emergency measure had gradually changed its character and become a method of protecting and organising national production. Quotas were applied without limit, not only to products which constituted a direct threat to domestic production, but also to substitutes and derivations of these products. If the market for feed grains is in difficulty, imports of maize are restricted in order to divert demand to barley, rye and oats. Imports of bananas are restricted, because bananas may be a dangerous competitor for home-grown fruit. Once it had embarked on this course, it was difficult for the French Government to stop half-way, and at the present time [May 1934] all or nearly all agricultural products are subject to quota. [Moroni (1934) page 73]

Import quotas were at first regarded as a temporary expedient, but with the continuing instability of the world market they remained in force. The originally deficient techniques were gradually perfected (this point is further discussed in a later section) and quotas became a major instrument of economic control, extended from 1932 onwards to a considerable number of industrial goods. From about 1934 they were used as a weapon in commercial negotiations, concessions in the size of quotas being offered in return for advantages to French exports. The quotas, however, did not apply to imports from French North Africa or the colonies.

Special measures of protection were devised for wheat. The large harvest of 1929 caused serious difficulties: export possibilities were severely limited and substantial imports had been contracted for early in the year. The average price in 1929 was 135 francs per 100 kg as compared with 152 francs the previous year. The import duty was raised from 35 francs to 50 francs in May 1929, and again to 80 francs a year later, but this was not enough to maintain prices. The main instrument of support was the law of 1 December 1929, which authorised the Minister of Agriculture to prescribe the minimum percentage of domestic wheat which millers must use in their flour. This was fixed at 97% in the first instance, but the proportion was changed from time to time in accordance with the market situation. Wheat from French North Africa was treated as domestic produce in determining the milling ratio, but special annual quotas were later applied to this trade. In addition to these measures, a law of 30 April 1930 authorised action to build up wheat stocks and to spread the marketing of the crop over a longer period. As it turned out, the harvests of 1930 and 1931 were mediocre and the price revived to around 150 francs; manipulation of the milling ratio proved sufficient to balance supply and demand fairly satisfactorily.

State intervention in agricultural markets, 1933–9

Even these perfected instruments for regulating imports were inadequate to maintain prices once domestic supplies themselves increased to the point of depressing the market. A growing need was felt for intervention on the domestic market as well as for import control. In a study of the agricultural crisis in France, Nogaro commented:

> Once the French home market had been isolated, through the policy of import quotas, from the world market, it was found that, for the majority of products, domestic production was fully sufficient to meet demand. As a result, the home market underwent a crisis similar to that of the world market, and from 1933 in particular French policy entered a new phase, that of 'organising' the principal markets. [Nogaro (1936) pages 23–4]

Wine and wheat were the commodities most concerned in the new measures of market organisation.

WINE

In 1931 wine began to be in excess supply, at least so far as the ordinary varieties were concerned, mainly as a result of rising imports from French North Africa. The average price per hectolitre fell from 183 francs the preceding year to 121 francs. A series of measures were taken from 1931 onwards. Originally, if total supplies exceeded 72 million hectolitres, producers could be compelled to deliver part of their output to the State Alcohol Office for distillation; also, part of the harvest could be blocked on the vineyards. The larger producers were forbidden to plant new vineyards and high-yielding vineyards were taxed. In August 1931 import quotas were imposed, but imports from North Africa could not be restricted and continued to rise. After a bumper wine harvest in 1934, prices fell disastrously, reaching 64 francs per hectolitre in 1935 (Table 8.1). A complex

Table 8.1: Wine: production, trade and average prices

	Production	Imports*	Exports	Prices
		million hectolitres		*fr. per hectolitre*
1925–8†	54.8	10.5	1.5	163
1929	65.0	12.0	1.4	154
1930	45.6	13.4	1.1	183
1931	59.3	15.9	0.8	121
1932	49.6	14.1	0.7	128
1933	51.8	17.5	0.7	117
1934	78.1	12.8	0.7	78
1935	76.1	12.6	0.7	64
1936	43.7	12.9	0.8	138
1937	54.3	12.5	0.9	180
1938	60.3	16.3	1.0	169

* Mainly from Algeria.
† Annual average.
Source: Statistique Agricole Annuelle.

marketing scheme was introduced, under which producers had to obtain permits to sell their wine, these permits being issued only in the amounts necessary to maintain a minimum price. Again, part of the output of the larger producers was blocked and they could be obliged to sell part of their produce at very low prices for distillation. A decree of 30 July 1935, provided subsidies for uprooting vines. In 1936 production fell, and by 1937 the average price was back to 180 francs.

WHEAT

The pressure on the wheat market was renewed in 1932 with a large harvest in both France and French North Africa; prices fell to an average of 117 francs per 100 kg for the year (Table 8.2). In December 1932 the milling ratio was raised to 99%, in April 1933 to 100%, so that foreign imports were virtually excluded. In October 1932 the government drew up a plan for subsidising farmers willing to store part of their crop until the following harvest year, but most farmers

Table 8.2: Wheat: production, trade* and average prices

	Production	Imports† *million tons*	Exports	Prices *fr. per 100 kg*
1925–8**	7.6	1.3	0.02	153
1929	9.2	1.4	0.01	135
1930	6.2	1.1	0.9	152
1931	7.2	2.4	0.6	153
1932	9.1	1.8	0.2	117
1933	9.9	0.5	0.2	106
1934	9.2	0.5	0.5	118
1935	7.8	0.5	0.9	75
1936	6.9	0.4	0.4	156
1937	7.0	0.3	0.1	189
1938	9.8	0.4	0.1	208

* Including flour in grain equivalent.
† Partly from Algeria.
** Annual average.
Source: Statistique Agricole Annuelle.

preferred to sell even on a falling market. In January 1933 a law was passed authorising the government to purchase wheat at 109 francs, later at 115 francs, but the funds available were too limited for this intervention to make much impact. With the expectation of a further large harvest in 1933 (in fact it turned out to be a record), further measures appeared necessary. A law of 10 July 1933, established a price of 115 francs as the legal minimum for all sales of wheat (the price on the world market by this time was only about 25–30 francs); this minimum was subsequently raised to reach 131.50 francs in July 1934. Those who did not observe the minimum were liable to heavy penalties and at first the regulation was respected. Soon, however, the market became disorganised, as small mills bought wheat illegally below the minimum price while the larger ones, whose operations were more easily supervised, refused to purchase. The government had to relieve the market by exports, with subsidies to reduce prices to the world level, and by 'denaturation' (rendering wheat unfit for human consumption and feeding it to livestock). To defray part of the heavy cost of these measures, levies were imposed on producers and millers. In December 1934 the legal minimum price was abolished (though temporarily retained for purchases by the government): instead, the government was authorised to block part of the harvest on the farms, to constitute stocks and to intensify export aids and 'denaturation'. Combined with the import duty, the milling ratio, area restrictions and other measures, this action constituted a very far-reaching intervention on the market. From 1933 onwards, imports were reduced to a low level and with large exports in 1935 France for the first time in many years had net exports of wheat.

So far the various measures taken had mainly been improvised to meet immediate needs. The Popular Front Government which took office in June 1936, with Léon Blum as Prime Minister and Georges Monnet as Minister of Agriculture, aimed at more systematic legislation with the basic object of insulating the market from the forces of supply and demand. A law passed (with difficulty) on 15 August 1936 set up the *Office National Interprofessionnel du Blé,* with the task of fixing the wheat price and ensuring its observance. The *Office* had monopolistic control over all foreign trade in wheat and could decide on the measures necessary to absorb surpluses. A contemporary study by Maspétiol commented:

> By the novelty of the solution which it brings to the French wheat problem, the recently-established *Office* has a significance which is not limited to this country nor to this commodity. The experience which is gained will be closely followed both by those who place their hopes in an extension of state activity in agriculture and by those who, on the contrary, would prefer an organisation giving a flexible type of co-ordination and leaving the activities of trading and cultivation free.
>
> In any case it would be a mistake not to appreciate the scope and importance of the initiative, which constitutes, among all the measures taken by the government of M. Léon Blum, the greatest departure from the classical liberal economy. Moreover, in the mind of M. Monnet, the Minister of Agriculture who initiated the reform, this is only a first step, the principles of which are to be extended to the entire agricultural market. [Maspétiol (1937) pages 545–6]

In fact the Popular Front Government remained in power only until 1938, and the *Office du Blé* was to remain its only major contribution to a more systematic organisation of agricultural markets. The 1930s were indeed marked by acute political instability: the regime was menaced by the rise of fascism on the right and the counter-activity of the communists on the left, and one government succeeded another. In these circumstances a consistent agricultural policy was hardly possible.

The import quota system

In view of the important role which import quotas have since come to play, it is of interest to examine some of the problems which France encountered in the early stages. The following account is largely indebted to the excellent study of this question made by Moroni in 1934.

The original measures rested on a rather doubtful legal basis, but in the atmosphere of the crisis no one was inclined to quarrel over niceties of interpretation. Other countries too were more inclined to take similar steps themselves than to question the legality of the French move.

The techniques used were originally simple. The quotas were announced on a global basis, without any specification as to where the imports should come from or who should carry out the imports; the frontier was closed to imports from all sources as soon as the quota was declared filled. A difficulty arose over the treatment of goods in transit at the time the frontier was closed, but this problem was dealt with in December 1931 by an announcement that goods would be admitted if proof were available that they had been consigned before the quota had been declared complete.[1] Problems arose also from the system whereby goods could be temporarily admitted into France free of duty: mills owning large stocks of grain imported in this way were liable to declare them officially for customs purposes just before the frontier was closed, and it proved necessary to restrict the use of temporary admissions.

A serious difficulty was the tendency for imports to flood in as soon as the quota was opened: frequently, as a result of delays in closing the frontier, the quota was exceeded in the first few weeks. An attempt to solve this problem was made in November 1931, when import licences were distributed to importers in relation to their past imports. This enabled a precise control to be exerted over the volume of imports.

However, there were difficulties in deciding who should get the licences and it was found necessary to set up special committees of government representatives, traders and producers to allocate them. Another solution tried was to hand over the administration of the quotas to the exporting countries. Besides relieving the French administration of the embarrassing task of allocating the licences, this served as a gesture to placate countries which were annoyed at the French restrictions. The export country was thus responsible for distributing licences between its exporters, and French customs offices would admit goods only if they were accompanied by the necessary authorisation. Agreements to this effect were reached with a dozen countries early in 1932, concerning in particular cattle, meat, dairy products, fruit and vegetables. The foreign administration was expected to space out the exports over the period to which the quota applied and to ensure that traditional trade channels were used. The latter undertaking was not always respected: French importers were by-passed in a variety of ways.

[1] The story was told of a consignment of bulls from Czechoslovakia which arrived at the French frontier just after the quota for bulls was declared filled, but which, with the assistance of a veterinary surgeon, were admitted on the quota for bullocks!

What was more serious was that the foreign firms possessing export licences were in a strong bargaining position and could extract a high price from French buyers. Another problem was that some foreign countries, when they could not fill their quota, obtained supplies from elsewhere and sent them to France, concealing their real origin. French traders complained at these practices, and the government was forced gradually to abandon this system and revert to the allocation of import licences.

French importers fortunate enough to obtain licences were generally able to make considerable profits. On the other hand, the restricted volume of trade meant that customs receipts were diminished. Consequently, in February 1933 taxes on import licences were introduced. These soon came to have an additional function: that of an extra – and variable – protective duty. The amount of the tax was often calculated so as to compensate for the difference between foreign and domestic prices, and in some cases the tax on a product was considerably higher than the import duty (Table 8.3). The taxes, moreover, were included in trade negotiations.

Table 8.3: Import licence taxes and import duties on major agricultural products, 1934

Product	Tax	Import duty minimum	Import duty general
		francs per 100 kg	
Cattle (live weight)	50	100	200
Pigs (live weight)	75	150	300
Beef and veal – fresh or chilled	100	175	350
– frozen	100	90	180
Pigmeat – fresh or chilled	100	250	500
– frozen	100	130	260
Eggs in shell	150	24	72
Butter	50	700	1400
Cheese	300	100	200
Barley	25	22.50	45
Apples	125	7.50	15
Pears	175	20	40

Note: All licence taxes were established by a decree of 12 May 1933, except apples and pears (28 December 1933) and cheese (30 March 1934).
Source: Moroni (1934).

Perhaps the most serious problem was the distribution of quotas between exporting countries. The global quotas which were used at first put the more distant countries at a disadvantage. In January 1932 quotas were allotted to each country on the basis of its past exports. The difficulty here lay in deciding what years to take as the basis, especially in cases where trade had already been restricted in one way or another. The original basis was 1927 –31, but within this period greater weight was given to the years 1927–9, which were regarded as more normal. For products to which quotas were applied after July 1932, the base period was the three years preceding the introduction of the quota.

This arrangement seemed fairly equitable, but it had the serious disadvantage of crystallising the previous pattern of trade. It also hampered commercial negotiations, since countries whose quotas were fixed had no further restrictions to fear and nothing better to expect: they could even discriminate against French exports with little danger to themselves. On the other hand, good customers of France could not be given advantages on the French market. As a result, an important change of policy took place in September 1933: quotas became a subject for negotiation, and only a quarter of the total quota was now allocated on the basis of past trade. Agreements involving quotas were reached with a number of countries in the course of 1934. Other countries, however, objected to this procedure: Germany and Great Britain both retaliated against French exports and the existing trade agreements were denounced. Nevertheless, with the use of quotas as a bargaining-counter, the first phase in which they were simply a means of protection passed into a second phase in which they became an instrument of economic control, intimately linked with overall trade policy.

The justification of agricultural support

The crisis which began in 1929 was so acute and the effects of rising imports in the early stages were so obvious that there was little opposition to the measures of protection adopted. Free Trade arguments would have appeared irrelevant in the circumstances, and in any case there was no strong body of Free Trade opinion, as there still was in Britain. Though there was some concern that the price of foodstuffs should not be raised, the question in the early stages at least was rather one of preventing a collapse in the market.

Responsible opinion at first stressed the exceptional nature of the situation and accepted protection through import quotas and other means as inevitable in the circumstances; the corollary of this view was that import restrictions should be relaxed once conditions returned to normal. This for example was the attitude taken in the first report of the *Conseil National Économique* on the trade problem in 1932: the *Conseil* criticised the quota system for its rigidity, but recognised that it was the only effective measure available. Augé-Laribé echoed widely-held views when he wrote in 1933 that 'removal of protection would make it practically impossible to practise agriculture in Europe, to the benefit of the new countries where the land is worked either by machines or by cheap labour' and concluded that France's economy had to be isolated until conditions on the world market improved.

In fact, world markets were slow to recover, and in 1935 the *Conseil National Économique* observed that protection by tariffs and quotas was still necessary; however, efforts should be made to resume normal trade and quotas should now be used not to restrict imports but to permit a gradual increase.

However, the tendency – as has been seen above – was to a greater degree of state control both of imports and of domestic markets. This development was far from being unanimously accepted. Maspétiol, whose study of the *Office du Blé* has already been mentioned, regarded this innovation with disquiet. In an analysis of the import quota system, Long pointed out that quotas, once installed, tended to perpetuate themselves:

> Protection prevents adaptation, and the lack of adaptation calls for protection. Consequently, unless the steps are retraced – and this appears less and less likely because it becomes increasingly painful – protective measures follow one another. This explains why, some seven years after the first quotas were introduced, we now find the quota system firmly and widely embedded in the French economy. [Long (1938) page 175]

Such criticism, however, was swamped in the rising tide of interventionism. The evolution that had taken place in general attitudes was reflected in a further report by the *Conseil National Économique:*

> Agriculture in France plays an important role. It is an essential element in the economic and social balance. It is necessary that the nation as a whole should, even at the cost of some sacrifice, ensure to farmers a satisfactory income which will keep them on the land and enable them to raise their standard of living on an equitable basis. [*Conseil National Économique* (1939) page 17]

This was a significant passage: it was probably the first authoritative statement in France (the *Conseil* was responsible for advising the government) of an income policy for agriculture. The *Conseil,* moreover, departed from its earlier, comparatively liberal views and advocated a comprehensive policy in which France would aim to obtain its supplies from its overseas territories, while the latter would as far as possible buy from France foodstuffs such as flour, butter and cheese. This would be achieved by mutual preferences, with imports from foreign countries restricted by tariffs and quotas except in cases where France and its territories could not attain self-sufficiency.

The experience of the 1930s left a permanent impression on French thinking about agricultural matters. After the war, Augé-Laribé wrote the following passage, which gives food for thought:

> On account of the structure of her agriculture, of the devastation suffered by her best regions (which were repaired but slowly) and of the imperfect ways in which production and distribution are organised, France is a country of high costs: her market attracts competition from producers throughout the world who have an advantage in one respect or another. If France had not maintained her customs barriers (which she was not the first to use), if towards the end of the inter-war period, she had not resorted to the method of import quotas (which she did not invent), if in other words France had not taken the precautions necessary to attenuate – but not entirely to suppress – the import of goods which can be produced on her own soil, then French agriculture would have ceased to exist. [Augé-Laribé (1961) page 29]

Augé-Laribé went on to point out that after the First World War France attached great importance to recovering a balance between industry and agriculture: the war had demonstrated the danger of dependence on imported supplies.

The fact that home produce cost more than imported was a secondary consideration. Social reasons also were important, and the following passage provides an insight into a widespread attitude to agriculture in France:

> Agriculture occupies too large a place in the French economy for it to be quickly suppressed and replaced. Neither the factories nor the accommodation available in the towns would be able to absorb all those of the farm population who would be obliged by Free Trade to give up the cultivation of the soil, even if a large number remained to breed livestock or engage in specialised activities. France, though she is a country of revolutions, has a profound desire for stability; she likes to stick to her old habits. She lets some young enthusiasts embark on adventures (where they have had some success) but she wants a large number of her sons to cultivate the land on which their ancestors, painfully but not unhappily, gained their daily bread. [Ibid., pages 29–30]

The growth of agricultural organisations

The French peasants, through being mobilised in the First World War, acquired a different outlook. They obtained new points of comparison, they realised how unsatisfactory was their situation, and they learned how to complain. Improved education, communications and transport had their inevitable effects on the mentality of the peasants and on their ability to organise themselves.

Chapter 3 has described the beginnings of agricultural organisation in the late nineteenth century. The right-wing, aristocratic *Société des Agriculteurs de France* in the rue d'Athènes[1] had founded in 1886 the *Union Centrale des Syndicats des Agriculteurs de France* (UCSAF). Its Republican and anti-clerical rival, the *Société Nationale d'Encouragement ā l'Agriculture* (in the boulevard Saint-Germain) had promoted the growth of co-operatives and *'mutuelles'*: in 1910 these too acquired a national organisation in the *Fédération Nationale de la Mutualité et de la Coopération Agricoles.*

In 1919 there was an attempt to unify the different organisations in the *Confédération Nationale des Associations Agricoles,* but this was not successful. Augé-Laribé, who was Secretary-General of the *Confédération* for ten out of the twenty years of its existence, observed that it was handicapped by the unwillingness of the various associations to obey a central body and to give it adequate funds.

The creation of *Chambres d'Agriculture* was provided for in a law passed in 1924, but they came into existence only in 1927. These were public bodies, financed by an additional levy on the land tax. There was one Chamber for each Department, with its members elected by farmers, landowners and farm workers and one member in five nominated by the agricultural organisations. The task of the Chambers was essentially to provide a link between the central authority and

[1] As the organisations multiplied and their titles grew increasingly complex, it proved easier to refer to them by their addresses.

the agricultural population. The Presidents of the various Chambers were eventually grouped in a central body, the *Assemblée Permanente des Présidents des Chambres d'Agriculture* (rue Scribe).

Following the crisis of 1929 and 1930, several of the syndicates affiliated to the rue d'Athènes found themselves in financial difficulties, and the associated *Caisse Centrale de Crédit Agricole* collapsed in 1931; the rue d'Athènes was thereby seriously weakened. However a new type of right-wing syndicalism issued from the crisis: advocates of corporatist doctrine gained control of the UCSAF in the rue d'Athènes, renaming it the *Union Nationale des Syndicats Agricoles* (UNSA): from new headquarters in the rue des Pyramides, this proved to be a powerful body.

Reference has already been made to the specialist producer organisations: the most successful were those grouping producers of wine, sugar-beet and wheat. These organisations had the advantage of being able to concentrate their demands on precise action to assist individual commodities, and were effective agricultural pressure-groups. The *Confédération Générale des Betteraviers* and the *Association Générale des Producteurs de Blé* were controlled by big farmers, but they sought to enlist as many small producers as possible and claimed to speak for their interests.

General political movements also found expression in the countryside. On the left, the Communist Party was the first in the field with its *Confédération Générale des Paysans Travailleurs,* instituted in 1929: its impact was limited. The socialists followed two years later with the *Confédération Nationale Paysanne* (CNP): this had some over-idealistic schemes for 'peasant councils', but also stressed the need for technical progress, pointing out that tariff protection was not an all-time cure. The CNP however slipped back into obscurity after the Popular Front disintegrated in 1938.

On the extreme right, a more spectacular development was the action of Henri Dorgères, who had avowedly fascist views and advocated an illegal seizure of power: from 1931 onwards he campaigned on the theme of peasant misery, stressing that the peasantry should defend itself against the State, which was rotten and incapable, and should unite in a compact bloc to impose a strong government. Dorgères aimed to create this bloc in a *Front Paysan,* in which he combined his action with the UNSA and another organisation, the *Parti Agraire et Paysan Français,* which had been founded in 1927. There were violent scenes at meetings organised by the Front in 1934 and 1935. But the following year a scission between fascist and democratic elements caused the *Parti Agraire* to break up, and with it the *Front Paysan.* Dorgères however continued his action with his 'green-shirts', a semi-military organisation on fascist lines, and he subsequently played a significant role in the agricultural policy of the Vichy Government.

A fair sample of current right-wing views can be found in the following passage from a book by one of the officers of the *Parti Agraire:*

Will the tragic decline of French agriculture continue to the point of ruin and

> until our villages are finally abandoned? ... The future of our distressed agriculture depends on the rank given to it in the national economy and in public opinion. If society as a whole considers that the peasantry's task is to serve other sectors ... if the policy of the treaties of 1860, which still dominates [sic], is pursued, agriculture will never recover; it will moreover be condemned to death if the proletariat of the towns does not recognise that the price of agricultural products, in a country like France where small peasants predominate, is a wage which the consumer must pay, at a fair rate, in order to enable those who work the land to live with dignity. [Braibant (1936) pages 164–5]

The proliferation of agricultural organisations, and the violence with which many of them expressed their views, ensured that the complaints of agriculture were kept well in the foreground. The political parties and the governments of the time were bound to pay attention to the agricultural problem. However, the nature of the demands expressed and the conflicts between the different agricultural organisations, did not favour the formulation of a coherent, long-term agricultural policy. It was easier to give satisfaction to agricultural producers by measures for supporting prices than to insist on the need for painful measures of adjustment. Few people were bold enough to suggest that if agriculture suffered from low incomes, the cause lay at least in part in the inefficiency of its structure and methods. Certainly none of the precarious and short-lived governments of the time was in a position to carry through far-sighted reforms; the structure of holdings remained essentially what it had been and the problem of fragmentation was tackled only in a few regions.

It might have been hoped that the growth of farm organisations would at least lead to an increase in constructive co-operative action between agricultural producers themselves. In fact, little progress was made; by 1939 only about one producer in four belonged to a syndicate or co-operative and the French peasant remained essentially an individualist. In particular, there was no significant extension of co-operative marketing: the marketing system for most products remained unwieldy and unprofitable for the peasant, whose produce passed through a long chain of middlemen before it reached the consumer. A bill which Georges Monnet introduced in 1937 to legalise collective marketing agreements by producers of specific crops was hotly debated but finally buried by the Senate (a similar scheme was however revived by the Fifth Republic).

Consequences for agriculture and trade

The level of agricultural prices in France was initially affected to some extent by the crisis on world markets and the consequent increase in supplies entering the French market. Before long, however, the various measures of intervention insulated the French market for major agricultural products to such an extent that it was no longer subject to fluctuations in world prices. French market prices then became dependent primarily on domestic supplies and, in some cases, on the volume of imports from French North Africa.

The developments for wheat and wine have already been described. Unfortunately there seems to be no reliable price index for agricultural products as a whole. The available estimates indicate that there was only a moderate fall in prices in 1930 and 1931, and that a more serious decline did not occur till 1933, following the increases in domestic supply. On the whole, livestock products seem to have done better than crops, but even among the former there were cases where domestic production was in surplus or near-surplus and where import quotas failed to prevent a fall in prices: thus the quotas for beef and beef cattle were progressively reduced after their introduction and became nil in the first quarter of 1933, but beef prices continued to fall during this period. In other cases there were temporary increases in the price level, resulting from unduly restrictive import quotas.

At any rate the crisis was not allowed to cause any drastic change in the pattern of French agricultural production. Table 8.4 shows that the distribution of the agricultural area as between arable land, grassland, etc., remained practically unchanged: the slight fall in the arable area does not appear significant. The numbers of cattle and pigs increased at a moderate rate, roughly in line with past trends; the number of sheep, already reduced to a low level, remained fairly stable.

Table 8.4: Agricultural area and numbers of livestock

	1920–24	1925–9	1930–34	1935–9
		million hectares		
Area				
Arable land	22.6	22.2	21.4	20.6
(of which wheat)	(5.4)	(5.3)	(5.4)	(5.2)
Grassland	11.0	11.2	11.3	11.6
Vineyards	1.6	1.6	1.6	1.6
Other cultivations	1.1	1.1	.9	.9
Total	36.3	36.1	35.2	34.7
Livestock		*millions*		
Cattle	13.6	14.9	15.6	15.7
Pigs	5.3	5.9	6.6	7.1
Sheep	9.8	10.6	9.8	9.8
Goats	1.4	1.5	1.5	1.3

Source: Statistique Agricole Annuelle.

On the other hand, agricultural imports were substantially reduced. The drastic fall in wheat imports has already been shown in Table 8.2: a more general picture is given by Table 8.5. The sharp increase in imports of most foodstuffs at the beginning of the crisis is clearly shown; but from the time when import quotas were introduced (late 1931 for livestock, meat, butter and cheese, and 1933 for feed grains), imports of almost all products were reduced to levels well below those prevailing before the crisis. The major exception was wine, as a result mainly of increased imports from Algeria.

Table 8.5: Imports of major foodstuffs

	Grain and flour	Sugar, raw and refined	Live animals	Meat, fresh and frozen	Butter and cheese	Wine
	volume indices: 1925–8 average = 100					
1929	120	129	61	37	128	114
1930	101	104	205	84	165	128
1931	193	86	325	117	264	151
1932	190	104	134	52	166	134
1933	89	100	117	43	140	167
1934	76	108	85	31	95	122
1935	75	92	79	23	78	120
1936	85	82	108	22	78	123
1937	69	98	82	30	65	119
1938	62	80	72	26	69	155

Source: Statistique Agricole Annuelle.

As Table 8.6 shows, the cut in imports took place at the expense of foreign countries. Imports of foodstuffs from the French overseas territories were maintained and even increased in value: wine and other foodstuffs from Algeria formed a large part of this trade.

The reduction in imports was not an unmixed blessing, even for French agriculture: in several cases it led to retaliation by foreign countries against French exports, many of which were agricultural. Thus Switzerland retaliated against French restrictions on dairy products by prohibiting imports of French cheeses, and Denmark, badly hit by the measures taken in France, imposed quotas on almost all French products, wine and spirits in particular.

Table 8.6: Total imports of foodstuffs, by origin

	All sources	Foreign countries	French territories*	of which Algeria
	billion francs, in current prices			
1925–8†	11.8	8.0	3.8	2.0
1929	13.2	8.3	4.9	2.6
1930	11.8	7.1	4.7	2.9
1931	14.0	9.0	5.0	3.2
1932	11.0	5.6	5.4	3.1
1933	9.6	3.9	5.7	3.6
1934	7.5	2.8	4.7	2.6
1935	6.3	2.2	4.1	2.1
1936	7.8	2.3	5.5	2.6
1937	10.7	3.4	7.4	3.4
1938	12.5	3.2	9.3	4.4

* French colonies, protectorates and mandated territories, including Algeria.
† Annual average.
Source: Tableaux Généraux du Commerce de la France.

Bibliography

Augé-Laribé's history (1950) is the obvious starting-point for reading on this period, as for the late nineteenth century; his article of 1933 suggests that even when close to events he was able to take a reasonably detached attitude. Many of the other references reflect positions taken in the contemporary controversies: some, such as Braibant (1936), were frankly propaganda. The best and most detailed accounts of the new protectionist measures were given at the time by Lautman (1933), who was an official involved in the administration of quotas but nevertheless provided an excellent and objective account, and by Moroni (1934).

The political aspects of the crisis were described by an English author (Hunter, 1938) and were thoroughly studied in the work by Wright (1964).

Augé-Laribé, M. (1950) *La politique agricole de la France de 1880 à 1940.* Paris: Presses Universitaires Françaises.

Augé-Laribé, M. (mars–avril, 1933) 'Nouveaux fondements du protectionnisme agricole'. *Revue d'économie politique.*

Augé-Laribé, M. (1961) Introductory chapter to Cépède, M. *Agriculture et alimentation en France durant la IIème guerre mondiale.* Paris: Genin.

Augé-Laribé, M. and Pinot, P. (1927) *Agriculture and Food Supply in France during the War.* Yale.

Barral, P. (1968) *Les agrariens français de Méline à Pisani.* Paris: Colin.

Braibant, M. (1936) *L'agriculture française – son tragique déclin, son avenir.* Paris.

Confédération Nationale des Associations Agricoles (1927) *La nouvelle loi douanière va-t-elle consacrer la déchéance de notre agriculture?* Paris.

Conseil National Économique (1932) *Les échanges internationaux et la politique douanière française au cours de la crise économique.* Paris.

Conseil National Économique (1935) *La politique agricole de la France.* Paris.

Conseil National Économique (1936) *La protection et les encouragements à donner par les pouvoirs publics aux diverses branches de l'économie nationale.* Paris.

Conseil National Économique (1939) *La politique agricole destinée à réduire le déficit de la balance commerciale et à coordonner les productions métropolitaines et coloniales.* Melun.

Fouchet, J. (1938) *La politique commerciale de la France depuis 1930.* Paris.

Haight, F. (1941) *A History of French Commercial Policies.* New York.

Hallé, P. and others (1930–31) *Conférences sur la crise agricole.* Paris: Institut National Agronomique.

Hauser, H. (avril, 1937) 'La concurrence internationale et le problème de l'économie dirigée. *Revue économique internationale.*

Hunter, N. (1938) *Peasantry and Crisis in France.* London.

Lautman, J. (1933) *Les aspects nouveaux du protectionnisme.* Paris.

Long, O. (1938) *Le contingentement en France.* Paris.

Maspétiol, R. (juin, 1937) 'L'Office français du blé'. *Revue économique internationale.*

Moroni, P. (1934) *L'agriculture française et le contingentement des importations.* Paris.

Nogaro, B. (janvier, 1936) 'La crise de l'agriculture et la politique agricole en France'. *Revue économique internationale.*

Nogaro, B. et Moye, M. (1931) *Le régime douanier de la France.* Paris.

Perthuis de la Salle, J. (1935) *La politique française de contingentement. Macon.*

Prault, L. (1935) *Protectionnisme douanier et commerce extérieur français: Agriculture – Industrie.* Amiens.

Proix, J. (1931) *La politique douanière de la France.* Paris.

Queuille, H. (1932) *Le drame agricole.* Paris.

Tugwell, R. G. (June, Sept., Dec., 1930) 'The agricultural policy of France'. *Political Science Quarterly.*

Vignaud, M. du (1931) *La réforme de la Loi de Cadenas en France.* Paris.

Wright, G. (1964) *Rural Revolution in France – The Peasantry in the Twentieth Century.* Stanford University Press.

Chapter 9

Germany

A The Weimar Republic

The First World War and the 1920s

At the outbreak of the First World War, Germany was dependent on imports not only for large quantities of foodstuffs, in particular bread grains from North America, but also for very considerable amounts of animal feedingstuffs and fertilisers. Moreover, foreign labour was important to German agriculture at seasons of peak employment.

During the war, Germany's imports were cut off to a far greater extent than those of Great Britain and as the war progressed the blockade of Germany was intensified. The food shortage was all the more serious because inadequate plans had been made beforehand and hardly any stocks had been built up, largely because the war was expected to be short. Further, in the first few months, the soldiers at the front were sent more food than was necessary and there was considerable wastage. It was only when it began to be realised that a long struggle lay ahead that measures to deal with the food shortage were introduced: by mid-1916, food supplies in general were subject to control and rationing was in force.

But there were serious obstacles to an increase in production. The shortages of feedingstuffs and fertilisers, and the severe lack of farm labour, led to a sharp fall in the yields of crops and livestock. Price controls aimed mainly at preventing the cost of food from rising and they were not counterbalanced by adequate measures to encourage production. By the end of the war the level of nutrition had fallen drastically.

For some time after the war, food remained short and price controls were maintained; they were gradually relaxed from 1920 on and then food prices shot up, giving large temporary gains to farmers. However, after the currency reform in November 1923, the prices of farm products rose more slowly than costs of production – wages in particular. Demand for foodstuffs was restricted by growing industrial unemployment, and German agriculture was subject to the full force of international competition since tariffs on foodstuffs had been suspended at the outbreak of war and had not yet been restored. Farmers thus found themselves in serious difficulties. Indebtedness was widespread and many farms had to be sold.

In 1925, however, Germany recovered its right (suspended by the Treaty of Versailles) to determine its own tariff. There was a vigorous controversy over the question whether protection should be restored to agriculture; arguments similar to those used before the war were revived. Thus some economists argued for Free Trade in grains, wanting Germany to develop a livestock industry in the way that Denmark and the Netherlands had done. In the *Reichstag* the left-wing groups, as well as the parties representing the small peasants, were against protection. Nevertheless, the conservative party and the big landowners obtained a majority for the new tariff, which came into force on 1 September 1925. In general it re-established, at least nominally, the pre-war Bülow duties, but the rates were somewhat higher for livestock products and, initially, slightly lower for grains.

At the same time exports of agricultural produce were again permitted (they had been forbidden since the war), and the system of import certificates was revived in the same form as before the war: exports of grain gave entitlement to certificates which could be used to pay the duty on the import of a corresponding quantity of grain, which could be of a different type.

The re-introduction of tariffs helped to restrain the fall in prices; the import certificate system in particular, by encouraging exports of grain, relieved the pressure on the market. The prices of grains, which till then had generally been below the world price, now rose slightly above it. Livestock prices fluctuated, but these on the whole were not so much affected by foreign competition at this period. However, agricultural prosperity was by no means restored. An overriding preoccupation of the government was to promote exports of manufactured goods, especially because of the burden of war reparations: protection for agriculture remained moderate, and the duties of the 1925 tariff were breached in several bilateral trade treaties. Farmers remained dissatisfied, and political agitation intensified in the countryside.

The economic crisis – tariffs and other measures

The agricultural situation had seemed bad enough already by 1925 to be referred to as a 'crisis'. But the fall in prices which occurred after 1929, in Germany as elsewhere, was much more critical. It was particularly serious for the many German farmers who still bore a heavy burden of debt, resulting from expenditure during the relatively prosperous years of the early 1920s.

The agricultural organisations, united in the 'Green Front', demanded action to stabilise grain prices, reviving the Kanitz plan for a grain import monopoly.[1] The government rejected this idea as both impracticable and contrary to existing commercial treaties. However, by a law of 4 July 1929, it introduced a compulsory milling ratio, with the minimum proportion of domestic wheat in the flour fixed at 30%; in July 1931 this was raised to 60% and shortly afterwards to 97%.

Further, in 1929 the government received certain powers to adjust tariffs by decree, without passing through Parliament, and as the crisis deepened these powers were extended; by March 1931 the government had acquired responsibility for the entire range of duties on agricultural products. The existing commercial treaties with Sweden and Finland were denounced because they involved commitments affecting duties on livestock products. The agricultural duties were then raised to high, often prohibitive levels, greatly exceeding the current world price: the duty on wheat rose to 25 marks in October 1930, by which time the

[1] After the war, the *Bund der Landwirte* was replaced by the *Reichslandbund*, which was similar in character and ideology to its predecessor; it too was led mainly by the Prussian landlords. The Green Front was a union of the *Reichslandbund* with the Catholic peasants and some other peasant associations which had grown up; the latter however played a subordinate role.

German import price was only about 10 marks; for other grains the situation was similar. The duties on livestock products were also raised, though they generally remained at about half the world price.

Though the government had rejected the demand for an import monopoly for wheat, in April 1930 it introduced a state monopoly for all trade in maize; since domestic production was small, this was in practice an import monopoly only. The main reason for this step was that the import duty on maize was bound at a low level in a trade agreement with Yugoslavia, which the government apparently did not want to denounce. This however did not prevent the new body from imposing a levy which was several times as high as the tariff.

Special measures had to be taken also for rye (a bigger item of production in Germany than wheat). The difficulties of 1925 had already caused the government to support the price through purchases on the market; these were carried out by private bodies with government assistance. In 1928 the government bought up two trading businesses and in 1929, after a big harvest of rye, used them for large-scale market operations. This action proved expensive and not very effective. It was difficult to dispose of the government-owned stocks and it proved necessary to render part of the supplies unfit for human consumption and sell them at a low price for animal feeding. At first this action was financed by direct government subsidies; later, purchasers of the 'denatured' rye were induced to pay a higher price by being allowed to import feeding barley at substantially reduced rates of duty.

The difficulties on the rye market led in July 1930 to the abolition of the import certificate system. As the German import duty – and hence the value of the import certificates – now exceeded world prices, there was a danger that German exporters would *pay* customers abroad to take their rye.

In contrast to France, import quotas did not play an important role. On a few occasions Germany granted certain countries an import quota at a reduced rate of duty: this was the case in 1930 for cattle from Sweden and butter and cheese from Finland. On these occasions Denmark received the same quotas as the other countries, though its previous trade with Germany had been much greater: though Denmark protested and demanded a distribution of these quotas in relation to past trade, Germany took no action till 1932, when it introduced a global quota for butter allocated in proportion to imports in 1929–31 (see also chapter 10).

An interesting feature of many of the measures taken at this time was the inclusion of clauses designed to protect consumers; this was essentially a concession to the Social Democrats and other left-wing groups. Some of the provisions of the 1925 tariff reflected a concern with the interests of consumers. The sugar duty, introduced in 1928, was to be lowered if the price rose above a certain level. When the duties on wheat, rye and pigs were raised at the end of 1929 and beginning of 1930, it was laid down that the government should attempt to maintain prices on the domestic market at certain levels by varying the duty within prescribed limits; in fact it proved impossible to keep prices up to the targets without raising the duties still further, and later the limits on the duties were abolished. The concern with the consumer interest found its most complete expression in the 'Index Law' of 28 March 1931, which instructed the government to eliminate the divergence between the price index of agricultural products and the other price indices, but also required it to keep the bread price below a specified level and to take appropriate action if the food price index should rise beyond a certain point. These provisions reflected a belief that conditions would soon return to normal; in fact, for several years to come the problem continued to be one of falling rather than rising prices.

Opinions on the agricultural crisis

The extent of concern with the agricultural crisis was reflected in a surge of literature on the subject from 1930 to 1933. In particular, two mammoth symposiums appeared in 1932, one edited by the eminent economist Max Sering, the other published by the *Friedrich List-Gesellschaft,* both dealing with the relation of German agriculture to the world economic crisis. Sering, who by nature was a liberal economist, wrote in his foreword to the first of these works:

> Is it possible for the food necessary for 65 million persons, tightly packed in a restricted area, to be obtained from our own soil? Is it possible for German farmers to withstand the competition of countries with extensive land resources without a degree of protection such that this nation, which since the war is poorer in raw materials than any other industrial country, becomes unable to compete on world markets? Is it possible for German farmers to keep pace even in livestock production with neighbouring countries, when capital is available in the latter at half its cost to the German farmer? [Sering *et al.* (1932) page III]

There was no real conflict over points of principle between the various writers in these two works: it was generally accepted that the circumstances of the time made protection inevitable. Thus Baade, in the second work, laid stress on the instability of the world market and the increasing degree of intervention by both exporting and importing countries. A similar point of view was expressed in a book by Salin, who referred to Britain's departure from Free Trade and observed:

> The logical correctness of the Free Trade doctrine is in no way upset by this development. But the conditions necessary for its application in practice have been removed to such an extent that it would be quite inappropriate and perhaps pointless to try to apply this historical ideal to the present situation. [Salin (1932) page 83]

There was general agreement too that the orthodox processes of commercial treaties, 'most favoured nation' clauses and binding duties might have suited a situation where world trade was stable and relatively free, but that they had now become inappropriate. New methods were necessary, particularly since the increased degree of self-sufficiency in Germany was making tariffs less effective.

This question of self-sufficiency in food received an increasingly prominent role in the discussions. There was a widespread desire to raise the degree of self-sufficiency, but opinions varied as to what this degree should be. Responsible opinion pointed out that full autarky, in Germany's situation, was out of the question. Also, the more imports of food were reduced, the greater would be the difficulty in selling German manufactures abroad. In the symposium published by the *Friedrich List-Gesellschaft*, Landmann attempted to prove that restrictions on agricultural imports had not significantly affected industrial exports, because the latter consisted mainly of capital goods and high-value consumer goods, demand for which had fallen relatively little; he also claimed that increased food prices resulting from agricultural protection did not necessarily raise the level of wages and hence of industrial costs of production. These views were criticised by other writers in the same volume on the grounds that they did not take account of secondary effects – thus if Germany imported less wheat from Canada, Canada might import less manufactures from Britain, and Britain in turn might import less from Germany.

The debate among academic economists was thus rather inconclusive. In the field of practical politics, however, the issue was soon to be decided by the advent of National Socialism, with an aggressively autarkic policy.

German agriculture around 1933

The increased duties and the other measures taken from 1929 to 1933 succeeded in bringing about a drastic reduction in imports of foodstuffs, which by 1933 were only 62% of their level before the crisis (Table 9.1).

As a result, prices on German markets were prevented from falling to anything like the same extent as prices on world markets. The fall in prices was nevertheless severe, as Table 9.2 indicates. The protective measures seem to have been relatively successful for wheat, but the rye market – as has been seen – came under severe pressure (and the output of rye was twice as big as that of wheat). The relatively large fall in the prices of livestock products from 1930/31 onwards suggests that the fall in consumer purchasing power resulting from the crisis was a factor perhaps as important as foreign competition – and one against which import restrictions were useless.

The result was a serious fall in farm incomes: it was calculated that the net income of agriculture before the crisis amounted to about 2.2 billion marks, but by 1932/3 it had dropped to under a billion marks (Table 9.3). The fall would have been even greater but for the reduced cost of feedingstuffs.

Sering (1934) pointed out that the prices of manufactured consumer goods fell by about as much as those of agricultural products; at the same time, the output of agriculture rose while that of industry fell, so that agriculture had a larger share of a reduced national income. This however was not much compensation for the many farmers who were ruined, and the distress in the countryside made it easy for Nazi propaganda to take root there as well as in other depressed sectors.

Table 9.1: Imports of foodstuffs: volume indices at 1928 prices: 1928 = 100

	1928	1929	1930	1931	1932	1933
Crop products	100	92	80	69	71	61
Live animals	100	104	85	49	49	47
Livestock products	100	104	100	83	79	62
Other foodstuffs*	100	100	100	87	77	80
All foodstuffs	100	96	87	74	74	62

* Including beverages and tobacco.

Source: Statistisches Jahrbuch für das Deutsche Reich.

Table 9.2: Agricultural price indices: 1928/9 = 100

	1928/9	1929/30	1930/31	1931/2	1932/3
All crop products	100	87	86	85	71
of which: Rye	100	82	77	92	74
Wheat	100	115	120	108	91
All livestock products	100	89	76	65	59
of which: Cattle	100	105	96	62	53
Pigs	100	102	72	57	52
Butter	100	89	76	66	58
All agricultural products	100	95	81	67	58

Source: As Table 9.1.

Table 9.3: Receipts and expenses of agriculture

	1928/9	1929/30	1930/31	1931/2	1932/3
			billion RM		
Receipts from farm sales	10.2	9.8	8.6	7.3	6.4
Crop products	3.8	3.6	3.2	3.0	2.6
Livestock products	6.5	6.2	5.5	4.4	3.8
Farm expenses	8.0	7.9	6.9	6.1	5.5
of which: Feedingstuffs	1.5	1.3	0.8	0.8	0.7
Wages	1.7	1.8	1.8	1.5	1.3
Net farm income	2.2	1.9	1.7	1.2	0.9

Source: As Table 9.1.

Structural policy etc.

The Weimar Republic, faced throughout its lifetime with a series of crises requiring immediate attention, was hardly in a position to develop long-term policies. Moreover, it was only after the war that a Ministry of Food and Agriculture for Germany as a whole had come into existence and it initially had some difficulty in asserting its authority over the seventeen *Länder*, especially Prussia: some matters – such as farm consolidation, research and technological development – remained in the hands of the *Länder.* Thus there was no national law to promote the consolidation of fragmented farms: most of the *Länder* pursued modest efforts to this end, but the amount of land affected remained limited.

A serious problem was farm debt: post-war inflation had at least helped farmers by wiping out old debts (including even some still outstanding from land purchases under the nineteenth-century land reforms), but after currency stabilisation in 1923, indebtedness again became a major issue and various debt relief schemes were introduced, particularly in the context of aid for the eastern regions (*Osthilfe*).

Many *Raiffeisen* credit co-operatives, however, foundered in the inflation and the movement recovered only slowly; purchase and sales co-operatives made some progress.

The pre-war policy of 'internal colonisation' was pursued under the resettlement law (*Reichssiedlungsgesetz*) of 1919. The programme aimed to provide smallholdings not only for the younger sons of farmers and for landworkers, but also for the numerous people uprooted by the war, including refugees from the province of Posen and from the parts of west Prussia which had been lost to Poland. Under the 1919 law, land was to be provided from public land, from the reclamation of moors and wasteland, by a right of pre-emption on holdings in excess of twenty-five hectares that came on the market, and – the most controversial item – by the provision that in all districts where estates of more than 100 hectares formed more than 10% of the farmland, up to one-third of the area should be made available. This so-called 'Baltic Third' (*baltische Drittel*) originated with an undertaking by the Baltic nobility in 1915 to surrender to the government, for modest compensation, one-third of their land to permit resettlement, and was supposed to correspond to the restitution of land usurped by the Prussian landlords after the land 'reform' of the early nineteenth century (see chapter 4). The Junkers however were still well entrenched in positions of power and were able largely to fend off these attacks: only about a quarter of the area theoretically available was in fact given up. In various parts of Germany, settlement companies (*Siedlungsgesellschaften*) pursued their task with varying success. During the economic crisis from 1929 on, they were often the only buyer when bankrupted farms came on the market, but at the same time there were not so many candidates to be placed, and sometimes the settlement companies had to run the farms themselves until a settler could be installed. Altogether, during the Weimar Republic, a modest total of some 57 000 new farmers were installed on an area just over 600 000 hectares.

A decree of 1920 provided some protection to tenant farmers: tenancy was by then reduced to less than 15% of the land area, but was recognised by some at least to be potentially an efficient form of tenure.

As with tariff policy, these rather hesitant policies of the Weimar Republic were soon absorbed in the much more determined action of the Third Reich.

B The Third Reich

The agricultural philosophy of National Socialism

The governments of the Weimar Republic reacted to the problems of agriculture by a series of largely disconnected expedients. The National Socialists came to power with a clearly-defined policy for agriculture, forming an integral part of their overall aims.

This policy was formulated in particular by Walther Darré, whom Hitler in July 1930 put in charge of a new agrarian cadre within the National Socialist party. The basis of Darré's thinking was set out in his work entitled *The Yeomanry as the Life Source of the Nordic Race,* which was first published in 1928 and ran into six editions.[1] The Germanic tribes, according to Darré, belonged to the settler as opposed to the nomadic types and were thus closely rooted to the soil. Moreover, the inheritance of farms and farming traditions from one generation to another ensured that the nobility of the blood and moral integrity were preserved. The Nazi slogan *'Blut und Boden'* (Blood and Soil) neatly expressed these ideas.

It followed that the farmer was not just an economic factor and should not be subjected to market forces: on the contrary, he should receive special care from the State and be assured of fair prices for his produce. The clearest expression of this idea appears in a speech by Darré on 19 September 1933:

> We must be quite clear about this: the farmer is not an entrepreneur in the usual sense. The production of food cannot be subjected to the free play of market forces and exposed to the risks which that entails, for agriculture's duty to the nation is immensely important. We need the farmer as the blood source of the German people; we need him too as the source of our food supply. This is not so much a question of ensuring that the farmer gets as high a price as possible for his produce so that his holding produces the maximum rent, as of making certain that the farmer is firmly rooted to his land through a German law of land tenure and that he gets for his work a fair wage – in other words, adequate, equitable prices. The farmer must always see his activity as a duty towards his race and his people, never as a mere economic, money-spinning operation. [Darré (1940) page 359]

Further implications for agriculture arose from the nationalistic urge of National Socialism: the nation's prestige and power could be ensured only if it was the master of its food supply. In the following passage, written in 1932, Darré clinches the long debate in Germany as to the role of agriculture in the economy and the desirable degree of self-sufficiency:

> If we except abnormal circumstances – such as military occupation by an enemy and commitments under treaty – the basis of foreign policy remains the following: the efficacy of every action undertaken by a nation in foreign affairs depends directly on the extent to which it is able to be and remain independent of other nations in its food supply. [Ibid., page 332]

Darré considered, moreover, that if food had to be obtained from abroad, the least objectionable course was to get it from German colonies, but in this case the supply lines had to be safeguarded.

The point was made even clearer by another National Socialist writer:

[1] *Das Bauerntum als Lebensquell der nordischen Rasse.* Darré deliberately revived the archaic term *'Bauerntum'* to describe the farming community. The translation 'yeomanry' was suggested by Holt (1936) and it does seem to have roughly the right historical connotations.

> The National Socialist leadership was in no doubt that the battle for honour and equality among nations could be won only on the firm basis of security in food supplies. So long as the food of the people was not assured against all circumstances, the policy of national liberation could not attempt any serious trial of strength. Together with the construction of a powerful navy and a highly productive industry, it was therefore essential to develop food production to the point where every individual in the nation could be sure of his daily bread. [Schürmann (1940) pages 330–31]

Darré swept aside the old objection that restrictions on imports of food would lead to reduced demand for German industrial exports. By controlling the import trade, it would be possible to ensure that purchases were made only from countries willing to take German goods in exchange. In 1935, in a speech to the Hamburg Senate (a body likely to be sensitive on this issue), Darré referred to past controversies and declared:

> Today, however, the new German agricultural policy has shown the way out of this labyrinth and has built a bridge between the opposing interests.... The German farmer now has no advantage in campaigning for tariff protection and thus obstructing German trade policy. [Darré (1940) page 447]

In a work published in 1958, Haushofer drew a distinction between the policy of Darré and that of other Nazi leaders. He considered that Darré, though he believed in the 'yeomanry as the life source of the Nordic race', did not subscribe to Hitler's *'Herrenvolk'* theories. Haushofer pointed out that the fulfilment of Darré's agricultural policy required peace, and referred to Darré's speech at the International Congress of Agriculture held in Dresden in 1939, in which he wished for a removal of misunderstandings and the avoidance of war. Indeed, Darré offered his resignation early in the war and played no part in events from 1942 onwards: Hitler by then needed ruthless efficiency untrammelled by romantic ideals. The fact remains that Darré's ideas on the racial importance of the yeomanry fitted conveniently into general Nazi thinking, and that the policy put into effect during his period as Minister of Agriculture served Hitler's aggressive purposes.

What were Hitler's own views on agricultural matters? In an important study entitled *The Plough and the Swastika* (1976), Farquharson pointed out that for Hitler the peasantry was important for practical reasons: it held back Marxism, it yielded food and it provided recruits for military service. From 1924 onwards Hitler stressed the need to obtain increased food supplies for a rising population and to reduce dependence on imports. In power, however, Hitler had to arbitrate between conflicting objectives: he was prepared initially to restore prosperity to the farming community, but subsequently – as will be further indicated below – the government had to give priority to price stabilisation in the interests of the urban population and of the rearmament programme. It is probable that early difficulties in the campaign to expand food production convinced Hitler that self-sufficiency within the nation's frontiers was impossible, and contributed to his determination to seek *Lebensraum* to the east.

The philosophy in action

The tremendous expansion of the National Socialist Party, from twelve seats in the *Reichstag* in 1928 to 288 in 1933, was due at least in part to support from much of the agricultural population during the years of crisis. From 1927 onwards the party had realised the potential voting strength of a discontented peasantry and it set out to take advantage of agitation in the countryside. In the elections of September 1930 the party gained a substantial rural vote: its programme offered relief from the burdens of debt, taxation and falling prices. (See Tilton (1975) for an analysis of the factors leading to Nazi successes in the rural areas of Schleswig-Holstein.) The party workers stressed the failure of the government to deal with the problems of agriculture. They took care not to alienate the farmers, suspicious of the 'socialistic' aspects of the new party: the programme was to strengthen the private ownership of the land, not to abolish it. Speculation in land would be ended and inheritance laws would be reformed.

The Nazis also succeeded in joining forces with the Prussian Junkers: in October 1931 the 'Harzburg Front' was formed, including the Nazis and two groups dominated by the Junkers: the German Nationalists (*Deutschnationale Volkspartei*) and the *Reichslandbund.* It has indeed been pointed out (McBride, 1945) that on a number of points the views of the Junkers and those of the Nazis were similar, despite the gulf in class: the stress on raising domestic food production and on the importance of agriculture for military strength, the belief in an autocratic society (though they differed as to who should be the autocrats) and the opposition to Jewish traders and speculators, were features in common. The grain-importing monopoly proposed by the Junker Count Kanitz in 1894 was the forerunner of arrangements introduced by the Nazis, described later in this chapter. The Nazi programme, moreover, laid down no systematic rule about the size of holdings and it was implied that, though the main aim was to have many viable small-to-medium holdings, large estates also had their place.

Following the 1930 elections, the party's agrarian cadre, under Darré's leadership, not only gained influence within the *Reichslandbund* (which was the largest farmers' union) but also infiltrated the Chambers of Agriculture and successfully attacked the hostile Catholic farm union. By early 1933, a powerful party machine had been built up in the countryside.

In January 1933, Hitler was appointed Chancellor. In his election address he promised to relieve the peasants' misery; in his May Day speech he declared that national recovery had to start with the peasants.

Initially, Hitler's government was a coalition, in which some of the ministerial posts were occupied by German Nationalists, including the Minister for Economics and Agriculture, Hugenberg. Hugenberg pursued the previous policy of restricting agricultural imports, partly by raising duties still further but also by a significant new step: two state import boards (*Reichsstellen*) were set up, one for grains, feedingstuffs and various agricultural products, the other for dairy products, oils and fats. Fats were a special problem and Hugenberg introduced a plan giving support to farmers by raising the prices of milk and butter, restricting production of margarine and taxing it – a scheme which Hitler was initially unwilling to accept because of its adverse effects on consumers (though the revenue from the margarine tax was used to subsidise a coupon system for the poor).

In June 1933, Hugenberg was forced to resign and the German Nationalist party collapsed. Hitler appointed National Socialists to take his post: Schmitt as Minister of Economics and Darré as Minister of Agriculture. Darré was now in a position to put all his ideas into practice. He moved swiftly and within a few months important new legislation had been introduced. This included reorganisation of the whole system of food production and marketing, and provisions designed to preserve viable family farms.

(A) CORPORATE ORGANISATION OF AGRICULTURE AND AGRICULTURAL MARKETS

A far-reaching new organisation of German agricultural production and markets was quickly set up. First, a law of 15 July 1933 declared that the organisation of agriculture, hitherto within the competence of the *Länder,* was now a matter for the Reich. Then a basic law of 13 September empowered the Minister to set up the *Reichsnährstand* (State Food Corporation).[1]

The *Reichsnährstand* (RNS) was a comprehensive organisation of all aspects of food production and distribution. All farmers and their families, all agricultural workers, all processors and traders dealing in agricultural produce and agricultural requisites, were legally obliged to belong to the RNS. Existing agricultural organisations of a public character (such as the Chambers of Agriculture) were absorbed into it; farmers' unions were either associated with it (the case of the *Landbund*) or were wound up (the case of opposing unions such as the Catholic one). Agricultural co-operatives were affiliated to the RNS.

Its internal administration consisted of a chain of responsibility based on the Führer principle; at the head was Darré himself as *Reichsbauernführer* and below were offices responsible, firstly, for each region, then for areas within each region and finally for localities within each area. At each level the leader was appointed by the head of the group immediately above. This ensured firm party control of the apparatus. Instructions were passed down the chain and recommendations passed up. At each level there were three *Verwaltungsämte* (administrative offices) for each of the main aspects of policy, described as *'Mensch, Hof, Markt':* i.e. the agricultural population, production and market regulation. For each major commodity group (ten in all) producers, co-operatives, traders, processors and distributors were grouped in unions (*Verbände*) at the regional level and federated in *Vereinigungen* at the national level. These latter, without actually trading themselves, regulated conditions of trade, prices, deliveries, etc. Thus farmers could be required to deliver specified percentages of their crops by certain dates and they had to sell to authorised intermediaries. In the milk sector, supplies to dairies and the output of the various dairy products were strictly regulated. For most of the major products – in particular bread grains,

[1] 'Corporation' is the usual translation. The word *'Stand',* however, means rather more than this, something like 'estate of the realm'.

feed grains, flour, livestock, meat, butter, eggs, wool and potatoes – limits were prescribed between which market prices could vary, but in fact these limits were so close together that the price was practically fixed. Wider price margins were laid down for sugar-beet, hops and some other foodstuffs. In a few other cases – wine in particular – only minimum prices were prescribed. There was provision for variations in price according to quality and to region.

The regulations were intricate and much supervision was needed; marketing efficiency seems to have suffered from too much organisation. The bureaucracy and expense of the RNS came under criticism, particularly from the Ministry of Finance. By June 1938 it had about 73 000 employees, although, of these, some 55 000 were the local leaders (*Ortsbauernführer*) who were unpaid; it must also be remembered that the RNS replaced numerous former organisations. However, substantial contributions were demanded of the farmers, co-operatives and processing or marketing firms.

(B) IMPORT CONTROL AND TRADE POLICY

Control over imports was vested in the *Reichsstellen,* or State Import Boards. These Boards were directly responsible to the Minister of Agriculture: they were not part of the RNS, but they formed an indispensable complement to its action. Boards were set up for the following products:

(a) Dairy products, oils and fats (April 1933).
(b) Grain, feedingstuffs and various agricultural products (May 1933).
(c) Livestock and livestock products (April 1934).
(d) Eggs (early 1934).
(e) Fruit, vegetables and wine (November 1936).

All imports of these products had to be offered by the importer to the appropriate Board, which, if it decided that the necessary demand existed, authorised the importer to sell the produce on the domestic market after payment of a levy corresponding to the difference between the import price and the domestic price level which it was desired to maintain. The Boards could also take over the produce from the importer and resell it at a higher price, and they could import on their own account (this applied mainly to grains and feedingstuffs). They could also buy and sell on the domestic market and keep stocks. This role gained increasing importance and the action of the Boards made it possible to even out fluctuations in supplies and prices, thus making an essential contribution to the work of the market regulation bodies of the RNS.

The Boards not only enabled complete control to be exerted over the volume and prices of imports; they also made it possible to decide from which sources imports would be accepted, and thus served the aims of overall trade policy. The *Neue Plan* of 1934 stated that Germany should import only from countries prepared to buy German goods. The traditional system of commercial treaties with 'most favoured nation' clauses was abandoned and replaced by bilateral arrangements involving reciprocal guarantees of import quotas and preferential tariffs. Special governmental committees were set up to supervise the development of trade between Germany and the countries concerned. Under the *Neue Plan,* Germany's trade was directed away from the democracies of western Europe and North America, and towards Italy, south-east Europe and Latin America.

(C) PROTECTION OF THE FAMILY FARM

The *Blut und Boden* philosophy was embodied above all in the legislation aiming to ensure the preservation of viable farms in the current and succeeding generations, and in the occupation of families of approved German stock. This was the basic aim of the *Erbhofgesetz* of September 1933. This law applied to farms qualifying at *Erbhöfe* and to farmers qualifying as *Bauer* according to the conditions laid down.[1] The *Erbhof* had to be of a minimum size sufficient to support a family and not above a certain maximum. Local courts decided in each case if these conditions were met and the range generally applied was from 7.5 to 125 hectares. The farm had to be owner-occupied; the law did not apply to tenanted holdings. The *Bauer* had to be of Germanic or similar stock, and the sole possessor of the *Erbhof.* Special courts were set up as arbiters of these matters and, inevitably, political reliability was investigated.

Recognised *Erbhöfe* could not be sold without permission of the special courts, could not be mortgaged, and on the death or retirement of the owner had to be passed intact to a single heir (the law also prescribed the order of succession, putting females last). These provisions represented a determined effort to end speculation in farmland, in effect removing *Erbhöfe* from the land market, and also to deal with the problem of fragmentation at each generation. This latter feature was potentially the most significant, especially in regions where farms were customarily split up between heirs (*Realteilung*). Moreover, because monetary compensation to heirs who did not inherit the farm (*Abfindung*) had been a major cause of indebtedness, restrictions were placed on this practice. These provisions naturally caused much concern in farming circles and made many farmers reluctant to register their farms as *Erbhöfe.* Another problem was the greater difficulty in obtaining credit since the farm could not be used as security. Further, the power given to RNS officials to dispossess inefficient farmers, and the restrictions on the owner's liberty to dispose of the *Erbhof,* appeared to reduce the *Bauer* to the status of manager on his own holding, recreating in effect a kind of feudal tenure with the State as landlord.

In fact many farms, even within the prescribed size limits, did not qualify as *Erbhöfe* because they were too fragmented to be viable, others because they were in joint ownership, and some farmers did not meet the requirements. Altogether, by mid-1938, 684 997 *Erbhöfe* had been registered, covering about half of the agricultural area. Perhaps the main immediate benefit to the farmers

[1] *Erbhof* may be translated literally as an 'entailed farm', i.e. one which may not be disposed of by the current owner and which must be passed intact to succeeding generations of heirs, but this is rather too legalistic a term. For *Bauer,* as has been suggested earlier, the term 'yeoman' has approximately the right connotations.

concerned was a scheme whereby the State took over farm debts in return for smallish annual repayments. The practical impact of the provisions concerning inheritance was limited because the collapse of the Nazi regime put an early end to the legislation.

Darré also aimed to intensify the resettlement programme and there was debate as to the right size of farm to aim at. The farms created under the Weimar regime had mostly been under ten hectares which seemed too small – Nazi policy envisaged about fifteen hectares. Darré would have liked to take land from the big estates, but even under the Nazi regime the Junkers were still influential. In 1933–8, only 20 408 new farmers were settled on 328 500 hectares. As Farquharson (1976) pointed out, it is questionable whether Hitler was very interested in 'internal colonisation' once his mind was set on external expansion.

The problem of fragmentation was tackled more seriously by the Nazi regime than by previous governments. Laws were introduced making consolidation a *national* responsibility and laying down appropriate procedures. It was currently estimated that about 22% of the area was in need of consolidation. The rate of progress was approximately doubled under the National Socialists.

Tenancy did not fit into the ruling philosophy and was not encouraged; some steps were however taken in 1937 to extend the period of leases where this would promote investment on the farm.

(D) PRICE AND PRODUCTION POLICY

The aim of the market organisation described above was to control supplies and to ensure producers the 'fair prices' to which, under National Socialist philosophy, they were entitled. A contemporary writer put it as follows:

> The main object of the market regulation is to bring about fair prices for the main products of agricultural labour. Sufficient and stable prices, bearing a correct relationship one to another, are to assure the farmer of an appropriate return for his crop and to eliminate the disturbing influences of the market. The law of supply and demand, which regulates prices and hence production in the capitalist economy, becomes inoperative: automatic price formation is replaced by the will of authority. [Mehrens (1938) page 2]

But conflicts of priorities soon arose. Bad harvests in 1934 and 1935 led to food shortages and increases in consumer prices: public complaints were heard about the prices of bread, milk, butter, etc. Already in 1934 Hitler was demanding explanations from Darré and towards the end of that year appointed a Price Commissioner. The latter issued instructions for meat prices to be kept down, recommended a cut in the margarine price, demanded an investigation into distribution costs and accused the guaranteed price system of simply helping inefficient farmers. Conflict with Darré and the RNS was inevitable and was never resolved. Hitler's course of action was clear: having made initial concessions to restore profitability to farming, he had now to consider wider interests. Unemployment was high; job-creation schemes with public funds required a stable currency; and avoidance of inflation was an overriding concern.

At the same time, increased food production was essential. Germany's foreign exchange situation in 1933 was disastrous and imports of food were a large element in the adverse balance of payments. The Battle of Production (*Erzeugungsschlacht*) was launched with maximum publicity in November 1934. The aim was to increase output in all lines of production while economising on inputs. As feedingstuffs for livestock were a major import item, more home-grown fodder was essential; this in turn should permit higher production of pigs, both providing meat and helping to meet the deficit in fats. Higher output of fibre-producing plants was required to reduce the need for imported fibres.

Guaranteed prices, debt redemption, exhortation by farm leaders, all played their part in the campaign. Annual production targets were laid down, and the RNS issued many and detailed instructions as to how the land should be used. In 1936 a further step was taken in the context of the Four-Year Plan, under which the whole economy was to be prepared for war. Increased funds were allocated for investment in agriculture, particularly for soil improvement and land reclamation; the fertiliser price was cut; new settlers in particular got machinery subsidies.

Hitler's refusal to grant further increases in farm prices, however, meant that by about 1938 farmers were again in a difficult financial situation and generally unable to undertake new investments. There was, moreover, a continued drain of labour from the land, intensifying as urban employment opportunities improved through work-creation programmes and through the economic recovery spurred on by rearmament. In rural areas, working and living conditions were poor. For example, nearly two-thirds of farms still had no running water and villages had few amenities. As mechanisation was not far advanced, farm work was endangered by labour shortages. Various schemes, voluntary or compulsory, were introduced to provide farm labour, but it still proved necessary – though much against party principles – to recruit foreign labour.

Results, 1933–9

It is not easy to assess the results of Nazi agricultural policy because of the lack of objective analysis within the regime. There is no doubt that agriculture gradually recovered from the crisis, but this was a feature common to all countries and cannot be entirely attributed to Nazi policy. Also, the recovery was no doubt due as much to the revival in purchasing power in the economy as a whole as to measures taken in the agricultural field.

The regulation of markets certainly stabilised prices and contributed to a gradual increase in the overall price level from 1932/3 to 1935/6, after which prices of almost all products were kept practically unchanged and the overall agricultural price index stood at 76–8% of the pre-crisis level (Table 9.4).

The role played by import control in stabilising prices is reflected in Table 9.5. The overall volume of imports was kept well below the pre-crisis level. The effects of intervention made themselves felt in the imports of individual commodities which varied quite widely from year to year; imports of barley were drastically cut, but replaced by imports of maize from south-east Europe. Total imports of bread grains were substantially reduced, likewise imports of sugar and of livestock products.

Agricultural net income was enabled to recover from a very low point of under a billion marks in 1932/3 to about 2.6 billion by 1934/5; after this, however, there was no further improvement in view of the price stabilisation.

Table 9.4: Agricultural price indices: 1928/9 = 100

	1932/3	1933/4	1934/5	1935/6	1936/7	1937/8	1938/9
All crop products	71	71	83	85	85	88	89
of which: Rye	74	72	75	77	77	87	86
Wheat	91	85	93	93	93	93	88
All livestock products	59	69	70	72	73	74	76
of which: Cattle	53	59	72	90	87	87	90
Pigs	52	56	62	67	66	67	69
Butter	58	70	71	72	73	73	75
All agricultural products	58	64	71	77	76	77	78

Sources: Statistisches Jahrbuch für das Deutsche Reich and *Statistisches Handbuch von Deutschland 1928–1944.*

Table 9.5: Imports of major agricultural products

	1929	1930	1933	1934	1937	1938
			thousand tons			
Rye	144	59	238	53	181	..
Wheat	2141	1197	770	647	1219	1268
Barley	1766	1523	235	552	242	456
Maize	669	651	254	338	2159	1895
Sugar	58	34	29	21	10	..
Butter	135	133	59	62	87	..
Meat, bacon, sausages	343	112	49	54	111*	..
Eggs	168	160	84	76	100†	..
		volume index at 1928 prices, 1928 = 100				
All foodstuffs	96	87	62	64	71	..

* All meat and meat products.
† Including egg whites and yolks.
Source: Statistisches Jahrbuch für das Deutsche Reich.

The Battle of Production seems to have given mixed results. Harvest fluctuations make assessment difficult (Table 9.6 shows two-year averages which even out the fluctuations to some extent). The grain crops of 1932–3 were good, those of the four following years relatively poor, but the 1938 harvest was a record and that of 1939 was also above the average of the previous years. Production of sugar and of potatoes also increased. Fodder supplies however remained short and limited the growth in the numbers of livestock. There was increased output of textile crops (hemp and flax) which had been an item weighing on the balance of payments. Overall production in 1938–9 was reckoned to be 20% above the level of ten years previously.

Consumption of most foodstuffs increased: total meat consumption, for example, rose from 43.5 kg per head in 1930 to 48.6 kg in 1938, butter consumption from 8.1 kg to 8.8 kg. The main problem remained that of fats: per capita consumption of margarine and of other fats and oils was reduced and Germany remained largely dependent on imports.

While the degree of self-sufficiency had been raised for bread grains, pulses, eggs, and to some extent for fats – so that the overall degree of self-sufficiency rose from 75% in 1932 to 83% in 1937 – dependence on foreign sources for fats and for livestock feedingstuffs remained the main problem and the main factor making Germany's food supply vulnerable during the year. In a study of the Third Reich's policy for attaining general self-sufficiency, Petzina (1968) concluded as regards agriculture that neither the strategic aim of the Four-Year Plan, i.e. that of achieving independence in wartime, nor the commercial aim, that of strengthening the balance of payments, had been achieved.

Table 9.6: Production of major agricultural products and numbers of livestock

	1930–31	1932–3	1934–5	1936–7	1938–9
			million tons		
Rye	7.2	8.5	7.5	7.1	8.5
Wheat	4.1	5.5	4.7	4.5	5.3
Barley	2.9	3.3	3.3	3.5	4.0
Oats	5.4	6.2	5.1	5.8	6.2
Other grain	0.6	0.7	0.8	1.1	1.4
Total grain	20.2	24.2	21.4	22.0	25.4
Potatoes	42.7	42.9	42.6	50.8	51.2
Sugar-beet	13.0	8.2	10.5	12.9	16.1
			million head		
Cattle	18.9	19.5	19.1	20.3	19.7
Pigs	23.7	23.5	23.1	24.9	24.4
Sheep	3.5	3.4	3.7	4.5	4.8
Hens	88.7	86.3	86.2	86.9	89.0

Source: Länderrat des Amerikanischen Besatzungsgebiets (1949) *Statistisches Handbuch von Deutschland 1928–1944.*

Wartime economy

At the outbreak of war, all the necessary machinery existed for controlling food supplies and prices. There was some reorganisation of the RNS to provide even tighter control; by a decree of 7 September 1939, all important foodstuffs were requisitioned and were to be sold only by order of the authorities; rationing was imposed; reserve stocks were built up.

Until 1944, food production was kept up reasonably well. The major problem of labour shortage (nearly two million men were conscripted from the farm sector) was partly dealt with by importing foreign workers and by the use of prisoners-of-war. Fodder shortages however could not be overcome and the number of pigs had to be reduced; this in turn meant less manure and a drop in fertility and crop yields. In 1944, the position became serious, the more so as Germany lost the territories it had occupied. In 1945, the daily calorie ration was down to about 2000 and meat and fats, in particular, were very scarce.

In the countries which they occupied, the National Socialists imposed their corporative organisation of agriculture and sought to make Continental Europe into a single, self-sufficing unit. Herbert Backe, who after being Secretary of State under Darré replaced him as Minister of Agriculture in 1942, wrote in that year that liberalism had led most of the European countries into a misguided dependence on colonial territories for their food supplies: the result had been to weaken their own agriculture. National Socialism was the first movement to restore strength to agriculture. He declared:

> The war of 1939–42 will demonstrate to the other nations of Europe the correctness of the German method, and will lead them to recognise that Europe's food supplies can only be assured by a community of Continental Europe within which the products of labour are freely shared. [Backe (1942) page 261]

Some observations on Nazi agricultural policy

Many of the basic ideas of Nazi agricultural policy were already implicit in measures taken previously in Germany and in the action of some other countries during the 1930s. The belief in the social importance of the farm population, the desire to shield them from market forces and give them fair prices, the aim for national self-sufficiency in food – all these preoccupations were present to varying degrees in the majority of countries. But only the National Socialists in Germany combined these elements into a coherent philosophy, forming an essential part of their overall policy; and they alone systematically carried these ideas into practice.

Their agricultural policy was thus an important event in the history of European agriculture, and if Germany had not lost the war this policy would have been extended to the greater part of Continental Europe. Moreover, although the career of the Nazis came to an abrupt end, much of their thinking on agricultural matters has persisted, both in Germany and elsewhere, and the methods of market organisation they devised to carry out their policy were the predecessors of those which were widely adopted after the Second World War. In structural policy they were the first to try to concentrate aid on a defined category of potentially viable farms, though the *Blut und Boden* doctrine led them into the unrealistic aim of wanting apparently to preserve these farms for all future generations. Their policy in this respect would not have survived in an age of general economic and technical progress.

The internal logic of National Socialist thinking cannot be denied. Unlike previous political parties (including the German Nationalists), their agricultural policy was not merely based on a desire to pander to the agricultural interest and gain its support. Their policy for agriculture was part of their overall aim, that of ensuring, by force, if necessary, the political and economic strength of the German nation. Given that aim, it was rational to seek to safeguard the nation's food supplies by encouraging domestic production through controlled markets and price supports and to restrict imports by all possible means. These considerations were all the more compelling for a densely populated country dependent on imported food and liable in war to be cut off from overseas supplies. The argument was also forceful at a time of severe agricultural depression, particularly when reinforced by an eloquent philosophy as to the role of the 'yeoman' in the nation. It must be granted that Walther Darré's writings and speeches made a remarkable impression: the ideas are clear and forceful, and are in sharp contrast with the anxious deliberations of government and economist alike that preceded the advent of National Socialism.

Moreover, the ideas were put into practice rapidly and uncompromisingly. The Nazis were certain of what they wanted to do and they lost no time; from the moment they took office their policy was consistently applied. In the *Reichsnährstand* and the *Reichsstellen* a remarkably efficient instrument was forged.

This policy could be objected to on grounds of economic liberalism: extensive protection was likely to maintain inefficient branches of production and to lead to a maldistribution of resources. But it might be answered that the exercise of control according to a constructive plan permitted rapid economic growth and that in a time of acute depression it was essential to preserve the nation's productive apparatus from destruction.

There was also the objection that drastic protection must reduce the earnings of other countries and thus endanger exports. As has been seen above, Darré contested this on the grounds that the planned direction of trade resolved the old conflict between agriculture and industry on this score, enabling export markets to be maintained. It was even suggested by Nazi writers that other countries benefited from the new policy because the contracts they obtained gave them an assurance that did not exist under the free market. But on these points the argument seems vulnerable: the fact remained that the more food was produced at home, the less was imported from abroad, and hence the less other countries could afford to buy from Germany. The effects might not be significant so long as only one or a few industrial nations restricted their imports of food, but clearly all countries could not follow the same policy and still hope to export their manufactures.

Thus the National Socialist theory was weakest at its foundations. It made sense only in the context of a single nation aiming to reinforce its power; it was geared to the expectation of war if not the desire for it. Further, the pursuit of such autarkic aims was bound to contribute to international tension and to make the outbreak of war more likely. The National Socialist system was by its nature incompatible with long-term economic and political stability. As a French observer pointed out:

> If all countries imitated Germany and sought a solution to their agricultural problem in complete self-sufficiency in food supplies, there is no doubt that their standard of living would fall to a frightening extent. From an agricultural point of view, all countries have interests in common because their products are complementary. In the capitalist, liberal economy, their food supplies are only assured by international trade. If they set about organising their own agriculture, that is all the more reason for needing additional supplies. Autarky therefore does not make sense except for those who subordinate everything to military requirements. The real solution to the problem of market organisation can only be an international one [Bertrand (1937) page 340]

Bibliography

The standard history of the inter-war period as a whole is provided by Haushofer (1958); in English, policies under the Weimar Republic and the beginnings of the National Socialist regime were surveyed by Holt (1936); also in English, Gerschenkron (1943) carried the story a little further.

The sequels of the food shortages of the First World War were dealt with in particular by Aereboe (1927). The works listed for the Weimar Republic mostly reflect contemporary preoccupations with the already critical situation.

As for the National Socialist period, the 1940 collection of Darré's speeches and papers is particularly revealing as regards the official policy; Schürmann's work (1940) was advertised as a textbook of Nazi agricultural policy; and the book by Backe (1942) – who in that year replaced Darré as Minister of Agriculture – is a fascinating insight into how Germany would have organised Europe's agriculture and food economy if Germany had won the war.

Inevitably, there was no objective assessment from German writers during the Nazi regime: the most thorough contemporary account was by a Frenchman (Bertrand, 1937). Even since the war, the subject has not been studied in depth by German economists or historians, though Petzina (1968) included agriculture in a general assessment of the Third Reich's drive for self-sufficiency. Tornow (1972) provided a useful but uncritical chronology of legislation; Schreiner (1975) described various aspects of structural policy. The comprehensive study in English by Farquharson (1976) is thus particularly valuable.

THE INTER-WAR PERIOD AS A WHOLE

Gerschenkron, A. (1943) *Bread and Democracy in Germany.* California University Press.

Haushofer, H. (1958) *Ideengeschichte der Agrarwirtschaft und Agrarpolitik im deutschen Sprachgebiet. Band II: Vom ersten Weltkrieg bis zur Gegenwart.* München: Bayerischer Landwirtschaftsverlag.

Holt, J. (1936) *German Agricultural Policy 1918–1934.* University of North Carolina.

Macbride, H. (1945) 'Note on the economic basis of the Junker class'. In Viner, J. *et al. The United States in a Multi-National Economy.* New York: Council on Foreign Relations.

Schreiner, G. (1975) 'Die Entwicklung der deutschen Agrarstrukturpolitik von der Reichsgründung im Jahre 1871 bis zum Ende des zweiten Weltkrieges'. *Berichte über Landwirtschaft* **53** (2), 219–315.

THE WEIMAR REPUBLIC

Aereboe, F. (1927) *Der Einfluss des Krieges auf die landwirtschaftliche Produktion in Deutschland.* Stuttgart.

Beckmann, F. (1926) *Die weltwirtschaftlichen Beziehungen der deutschen Landwirtschaft und ihre wirtschaftliche Lage, 1919–1926.* Berlin.

Brentano, L. (1925) *Agrarpolitik.* Berlin.

Dietze, C. von (1925) *Die deutsche Landwirtschaft und die neue Handelspolitik.* Schriften des Vereins für Sozialpolitik, 171.

Friedrich List-Gesellschaft (1932) *Deutsche Agrarpolitik im Rahmen der inneren und äusseren Wirtschaftspolitik.* Berlin.

Fuchs, C. (1927) *Deutsche Agrarpolitik vor und nach dem Kriege.* 3. Auflage. Stuttgart.

Häfner, K. (Juli, 1934) 'Die Politik der mengenmässigen Einfuhrregulierung'. *Weltwirtschaftliches Archiv.*

Ritter, K. (1925) *Die deutschen Agrarzölle.* Schriften des Vereins für Sozialpolitik, 171.

Röpke, W. (1934) *German Commercial Policy.* London.

Salin, E. (1932) *Wirtschaft und Staat – Drei Schriften zur deutschen Weltlage.* Berlin.

Sering, M. (1925) *Agrarkrisen und Agrarzölle.* Berlin.

Sering, M. (1925) *Schutzzoll oder Freihandel.* Schriften des Vereins für Sozialpolitik, 170.

Sering, M. (1934) *Deutsche Agrarpolitik auf geschichtlicher und landeskundlicher Grundlage.* Leipzig.

Sering, M. *et al.* (1932) *Die deutsche Landwirtschaft unter volks- und weltwirtschaftlichen Gesichtspunkten.* Berlin.

Weber, W. (1967) 'Reichsregierung und Agrarpolitik in der Republik von Weimar 1920–1932'. *Berichte über Landwirtschaft* **XLV**, (1), 31–50.

THE AGRICULTURAL POLICY OF NATIONAL SOCIALISM

Backe, H. (1942) *Um die Nahrungsfreiheit Europas: Weltwirtschaft oder Grossraum.* Leipzig.

Bertrand, R. (1937) *Le corporatisme agricole et l'organisation des marchés en Allemagne.* Paris.

Brandt, K. (May, 1934) 'Farm relief in Germany', *Social Research.*

Brandt, K. (February, 1937) 'German agricultural policy', *Journal of Farm Economics.*

Darré, R. Walther (1937) *Das Bauerntum als Lebensquell der Nordischen Rasse.* 6. Auflage. München.

Darré, R. (1940) *Um Blut und Boden* (Reden und Aufsätze). München.

Dietze, C. von (1935) 'La lutte contre la crise agraire'. Dans *L'Economie dirigée:* publication à part de la *Revue d'Economie Politique.* Paris.

Farquharson, J. (1976) *The Plough and the Swastika – the NSDAP and Agriculture in Germany 1928–45.* London and Beverley Hills: Sage.

Gatheron, J.-M. (undated) *Hitler et la Paysannerie.* Paris.

Higgins, B. (May, 1939) 'Germany's bid for agricultural self-sufficiency'. *Journal of Farm Economics* **XXI.**

Lovin, C. R. (1969) 'Agricultural reorganisation in the Third Reich: the Reich Food Corporation (*Reichsnährstand*), 1933–36'. *Agricultural History.* **XLIII** (4), 447–61. University of California Press.

Luxenberg, B. (1932) *Der Niedergang der deutschen Landwirtschaft: ein Rückblick auf den landwirtschaftlichen Markt der Jahre 1930 und 1931.* München.

Mehrens, B. (1938) *Die Marktordnung des Reichsnährstandes.* Berlin.

Meyer, K. (ed.) (1939) *Gefüge und Ordnung der deutschen Landwirtschaft.* Berlin.

Petzina, D. (1968) *Autarkiepolitik im Dritten Reich.* Stuttgart: Deutsche Verlagsanstalt.

Schürmann, A. (1940) *Deutsche Agrarpolitik.* Neudamm.

Tilton, T. A. (1975) *Nazism, Neo-Nazism and the Peasantry.* Indiana University Press.

Tornow, W. (1972) 'Chronik der Agrarpolitik und Agrarwirtschaft des deutschen Reiches von 1933–45'. *Berichte über Landwirtschaft.* Sonderheft 188.

Chapter 10

Denmark

The crisis – the loss of export markets

During the First World War, Denmark remained neutral and was able to maintain her exports to both Germany and Great Britain. Her trade continued to expand after the war and was not seriously affected by the restoration of tariffs in Germany in 1925. In Britain, Denmark gained a larger share of the markets for eggs and bacon, but lost part of the butter market to increased supplies from New Zealand. The years 1928 and 1929 were reasonably prosperous, and even after the world grain price had begun to fall Danish agriculture at first did not feel the effects, since the prices of livestock products fell relatively little and this fall was offset by increased production, making use of cheap imported grain and other feedingstuffs. The maintenance of purchasing power in Danish agriculture sustained other sectors, and as a result unemployment did not increase significantly till the first half of 1931. In 1931 Denmark followed Great Britain off the gold standard, and devalued again in the autumn of 1932.

In 1931 Danish exports of livestock products were severely affected by the fall in purchasing power abroad and also by the increased tariffs and import restrictions applied in several countries, Germany in particular. On the British market, Danish butter continued to suffer from the competition of New Zealand and Australia and in 1932 all Danish exports to Britain were subjected to import duties, while Empire produce continued to enter free. As a result, in 1931 and 1932, the prices of livestock products in Denmark fell sharply (Table 10.1). The fall in agricultural receipts was not offset by a corresponding reduction in farm expenses, except for feedingstuffs, and as a result farmers made little or no profit in 1930/31 and 1931/2 (Table 10.2). The distress of farmers was particularly severe since they had incurred a heavy burden of debt during the period of rising prices and land values early in the 1920s.

Table 10.1: Agricultural price indices: 1928 = 100

	1928	1929	1930	1931	1932	1933	1934
Crop products	100	93	68	62	66	69	94
Livestock products	100	105	87	65	56	68	79
All products	100	104	85	64	57	69	79

Source: Danish Council of Agriculture (1935).

Table 10.2: Receipts and expenses on sample farms

	1924/5 –1928/9	1929/30	1930/31	1931/2	1932/3	1933/4
			kroner per hectare			
Gross receipts	902	817	651	560	562	530
Production costs*	798	682	637	571	502	460
Net income	104	135	14	–11	60	70

* Including allowance for the labour of the farmer and his family.

Source: As Table 10.1.

As in the depression of the 1880s, but now to an even greater extent, Danish agriculture's dependence on trade made import restrictions practically useless. Since Denmark's difficulties arose primarily from protectionist measures in other countries, her first reaction to the crisis was to seek better conditions for exports to the countries with the largest markets, Britain and Germany. This attempt was only partly successful. In 1932 exports of the main agricultural products were brought under the control of ministerial Export Boards, with the task of licensing exports and regulating prices by imposing and distributing export levies. Further, Denmark was gradually forced to adopt a series of measures affecting her domestic market and also to control some imports, in particular grain. This action represented a departure from the principles of laissez-faire to which the farmers themselves adhered and was adopted only reluctantly.

Efforts to improve trade relations

(A) WITH BRITAIN

As other markets became increasingly restricted, Denmark sought to expand her sales to Britain. This gave rise to a vigorous campaign in some sections of the British press, where it was pointed out that the large British purchases from Denmark were not reflected in equally large Danish imports from Britain: on the contrary, Denmark's imports of manufactured goods still came mainly from Germany. The Danish agricultural organisations had some sympathy with this point of view and already at the end of 1929 had exhorted Danish traders and consumers to give preference to British goods. At first there were no great results, but in January 1932, following a further large increase in the German butter duty, the Danish Exchange Control Office was established: all imports had to be authorised by this Office, and it sought to restrict imports from Germany in favour of Britain and other countries.

However, the British Import Duties Act of February 1932 imposed duties of 10% *ad valorem* on most Danish agricultural exports, and subsequently the Ottawa Agreements Act replaced most of these by specific duties which generally had a higher incidence. Free entry was maintained for bacon and hams, but – as has been seen in chapter 7 – an import quota was imposed which halved Danish exports as compared with the peak figures of 1932; Canada, on the other hand, was granted a quota far in excess of her previous exports to Britain.

Negotiations for an Anglo-Danish commercial treaty began in December 1932 and an agreement was signed which came into force in June 1933; it was valid for three years. Britain undertook not to impose duties on bacon and ham, but reserved the right to limit the quantity of imports in connection with domestic marketing schemes, and Denmark was unable to obtain a guarantee for any specific quantity: the only concession was that Denmark could supply at least 62% of all bacon imports from non-Commonwealth countries. For butter and eggs, Britain agreed to bind the duties at the existing levels; again, she reserved the right to impose import quotas, but in this case restrictions were not in fact applied.

On the whole, the treaty offered Denmark only the possibility of maintaining its exports at about the same level as before, with the disadvantage of a tariff giving preference to the British Empire. In the case of bacon, the import quota, though it reduced the volume of trade to about half that of 1932 (Table 10.3), led to prices more than double the previous level because of the inelastic demand by British consumers for Danish bacon. The outcome for butter was much less satisfactory, for New Zealand was helped by the preference to raise its exports considerably, capturing from Denmark the first place on the British market. In the case of eggs, however, Britain's trade war with the Irish Republic helped to increase the volume of imports from Denmark.

In negotiating with Britain, Denmark was in a weak bargaining position because of her dependence on the British market and in the 1933 treaty she was forced to grant Britain substantial concessions, including the removal of certain import duties on manufactures and the reduction or binding of many others, together with an undertaking to purchase certain quantities of British goods such as coal, iron and steel. The treaty was renewed in 1936 and remained in force until the outbreak of war. The advantages it gave to Britain, together with the activities of the Danish Exchange Control Office, caused Britain's exports to Denmark to double in value from 1932 to 1938, while the value of Danish exports to Britain recovered only slightly and failed to reach the level of 1929 (Table 10.4).

Table 10.3: Major agricultural exports

	1928	1929	1930	1931	1932	1933	1934	1935	1936	1937	1938
						thousands					
Cattle	255	270	169	124	116	48	71	97	166	172	134
of which to Germany	255	261	153	91	75	17	50	81	149	141	121
Pigs	45	51	62	34	24	73	57	53	184	167	114
of which to Germany	31	45	52	30	2	32	31	44	162	157	114
						thousand tons					
Bacon	272	249	306	376	390	294	223	200	176	182	179
of which to UK	271	248	306	372	384	284	219	197	174	178	174
Butter	148	159	169	172	158	151	150	138	146	153	158
of which to UK	101	108	116	123	129	126	124	109	110	116	119
of which to Germany	40	43	42	30	13	16	20	25	34	35	36
						million score					
Eggs	40	39	43	49	55	54	56	59	70	81	78
of which to UK	29	31	37	42	38	38	38	39	48	60	57
of which to Germany	11	9	6	7	17	12	13	13	17	18	19

Source: Statistisk Aarbog.

Table 10.4: Value of trade with the United Kingdom

	1929	1932	1935	1937	1938
			million kroner		
Danish imports	263	255	479	638	567
Danish exports	963	728	731	824	861

Source: Gøtrik (1939).

(B) WITH GERMANY

Until the crisis, Danish exports of livestock products to Germany benefited from relatively low rates of duty, largely as a result of concessions made by Germany to other countries and extended to Denmark through the 'most favoured nation' clause. At the end of 1929, however, Germany terminated its agreement with Sweden and in February 1930 raised the duty on cattle from 16 marks to 24.50 marks per 100 kg. Sweden however was allowed to export to Germany, at the old rate of duty, a certain number of cattle (5000 for one year, 6000 for a second year and 7000 for a third): this amount was roughly equivalent to Sweden's previous exports to Germany. But whereas Denmark's exports of cattle to Germany were far greater (255 000 head in 1928), Germany interpreted the 'most favoured nation' clause in such a way as to allow Denmark only a tariff quota of the same absolute amount as Sweden. Denmark attempted to obtain a quota in proportion to past supplies, but without success, so from the beginning of 1930 most Danish cattle exports were subject to the higher rate of duty.

Similarly for butter, Denmark enjoyed the benefit of a reduction in the German duty granted in a treaty with Finland in 1926. In November 1930 Germany raised the duty from 27.50 marks to 50 marks per 100 kg. Finland was granted a quota of 5000 tons at the previous, lower rate: this was a quantity nearly as large as her past exports. Denmark, whose exports had exceeded 40 000 tons, was given only the same absolute amount. In 1932 Germany raised the duty still further, to 100 marks and introduced an 'exchange premium' of 36 marks on imports from countries which had devalued their currency (this included Denmark). The duty applied to Danish exports was then 86 marks per 100 kg for the quota of 5000 tons and 136 marks per 100 kg for the remainder.

The Netherlands suffered in much the same way as Denmark from these measures by Germany and both countries complained in the League of Nations against the German interpretation of the 'most favoured nation' clause. In June 1931 the League's Economic Committee produced a report that substantially agreed with the Danish case, to the effect that tariff quotas should be related to past trade. Germany however took no action till October 1932, when she introduced a complete quota system for butter and distributed the global quota of 55 000 tons among exporting countries in proportion to their share of exports in 1929–31.

When the National Socialists came to power in Germany, they denounced existing treaty commitments and – as has been seen in chapter 9 – subjected almost all imports of agricultural products to the control of state import boards. A bilateral agreement of 1 March 1934 between Denmark and Germany stated that the two countries would endeavour to promote trade between them and to settle difficulties by negotiation. Trade between Denmark and Germany then became subject to a considerable degree of intervention by both governments. When German food requirements rose in 1935 and 1936, additional purchases were made in Denmark, but only on condition that Denmark should increase her imports from Germany by a corresponding amount. As there was difficulty in meeting this requirement the Danish Government was obliged, at the end of 1936, to take action to raise purchases from Germany. Starting in 1937, Germany subjected her imports from Denmark in each quarter to a maximum, determined by the total payments made by Denmark for German goods in the previous quarter.

The demands made by both Britain and Germany for a larger share of the Danish market aggravated Denmark's difficulties in finding enough foreign exchange to pay other foreign countries for the feed grains and other feeding-stuffs needed by Danish livestock farmers.

Intervention on the domestic market

The first intervention occurred in 1930 and consisted of subsidies to growers of sugar-beet to compensate them for low prices and to encourage production. In 1932 this scheme was replaced by arrangements under which licences for producing and refining sugar had to be obtained from the Minister of Commerce, and prices were fixed for both sugar-beet and sugar. Imports were prohibited except under licence. In 1933, arrangements for potatoes for industrial purposes were introduced along similar lines.

The import restrictions by Germany led to a sharp decline in meat prices and measures of support were introduced in 1933. The details of the system underwent several changes: the final arrangement was that a slaughterhouse tax was levied on all cattle sold for the home market, the revenue being used to buy up lower-quality cattle in order to raise market prices. From 1933 to 1938 some 600 000 cattle were bought in this way and slaughtered, many of them cattle infected by tuberculosis, and the market price was doubled. The effectiveness of the scheme was due to the relatively large proportion of the total output of beef and veal that was consumed in Denmark (about 75%).

The low level to which butter prices fell also necessitated measures of support. In 1933 variable levies were imposed on butter sold for domestic consumption, in order to maintain a minimum price of 2.15 kroner per kg. The receipts from these levies were distributed to milk producers in proportion to their deliveries. A tax was also imposed on margarine to discourage any shift of consumption. In 1934, prices improved and the scheme was allowed to lapse, but it was reintroduced in 1937, levies being imposed on all sales of liquid milk as well as butter; the minimum butter price was fixed at 2.60 kroner and was raised to 3 kroner in 1939. The benefits of the scheme were rather limited since, in contrast to beef, only about a quarter of total milk output was consumed on the home market in the form of milk or butter: the amount which could be collected through the levies was thus limited.

In the case of bacon, the British import quota made it necessary to limit Danish production. A scheme was introduced in 1933 which constituted the most far-reaching innovation of the period. General authority was given to the Minister of Agriculture to regulate the production of pigs and the number slaughtered and to order that each producer should be paid a preferential price for a certain number of pigs delivered, with a lower price for the remainder. 'Pig cards' or licences were distributed to producers, and pigs supplied to bacon factories received the preferential price only if they were accompanied by a card. The preferential price corresponded to the price that could be got on the British market; the price for excess deliveries was that available on other markets. The result of this scheme was a reduction in the number of pigs from 5.4 million in 1932 to 3 million in 1934; the bacon price rose from 0.75 kroner to 1.55 kroner per kg.

This scheme, probably the first marketing quota system to be introduced for an agricultural product in any country, led to the inevitable difficulties in distributing licences. A compromise was finally reached with a complicated system in which the number of 'pig cards' allotted to each producer was calculated according to a number of criteria. To begin with, a small number of cards were allocated in an equal number to each farm regardless of its size (except for a reduction in the case of very small holdings); then a certain number were distributed on the basis of the land value of the farm, up to a specified maximum; a further quantity was distributed in proportion to the amount of skimmed milk which the farm used for feeding to pigs; and finally an allocation was made on the basis of past deliveries of pigs. This arrangement deliberately gave an advantage to the smaller producers, through the initial distribution of a fixed amount and through the maximum placed on the land value criterion.

The larger farmers, for whom grain was a more important product, were compensated by a system of guaranteed minimum prices for the various grains, implemented by means of a variable levy on imports. As a further step the proceeds of the levy were distributed to the small farmers who needed to buy feed grain for their livestock. In spite of this arrangement, and in spite of the moderate price level that was guaranteed, many farmers felt this levy on grain imports to be a serious departure from Free Trade principles, and the scheme was decided upon only after long deliberations among the agricultural organisations. The government, however, was anxious to prevent an excessive increase in grain imports which would have further endangered the balance of payments.

In 1938, owing to a renewed fall in world grain prices, imports of wheat and rye were suspended entirely. As an additional measure to limit imports and maintain the price level, compulsory milling ratios were introduced, the minimum quantity of domestic grain in the flour being fixed at 50% for wheat and 30% for rye.

Bibliography

The works by Gøtrik (1939) and Skrubbeltrang (1953), in English, and that by von Arnim (1951), in German, adequately cover the field.

Arnim, W. von (1951) *Die Landwirtschaft Dänemarks als Beispiel intensiver Betriebsgestaltung bei starker weltwirtschaftlicher Verflechtung.* Kieler Studien, **17**.

Danish Council of Agriculture (1935) *Danemark: L'Agriculture.* Copenhagen.

Gøtrik, H. P. (1939) *Danish Economic Policy, 1931–1938.* Report to 12th International Studies Conference at Bergen. Copenhagen: Institute of Economics and History.

Nash, E. F., and Attwood, E. A (1961) *The Agricultural Policies of Britain and Denmark: A Study in Reciprocal Trade.* London: Land Books.

Pedersen, E. H. (1973) *The Danish Agricultural Industry 1910–39.* University of Copenhagen: Institute of Economic History.

Schürmann-Mack, F. (Sept., 1937) 'Die Marktregulierungen in der dänischen Vieh und Fleischwirtschaft'. *Weltwirtschaftliches Archiv.*

Skrubbeltrang, F. (1953) *Agricultural Development and Rural Reform in Denmark.* FAO Agricultural Studies No.22. Rome.

Part III:

The Challenges of Economic Change and European Integration

Chapter 11

Post-war recovery and government intervention

The Second World War

From the outbreak of war, the overriding preoccupation throughout Europe became the maximisation of food production in the face of numerous difficulties: traditional sources of supply were cut off while labour, machinery, fuel, feeding-stuffs and fertilisers were short.

Germany, in the countries of the Continent which it occupied, aimed to redirect agricultural production in ways which would meet its industrial and military needs. In France, this aim coincided with a *'retour à la terre'* philosophy in the Vichy government. Pétain himself came to be known as *'Le Maréchal paysan'*. Shortly after taking office in 1940 he invited the country to follow a new path, calling a halt to excessive industrialisation and declaring that 'France shall become once more what she should never have ceased to be, an essentially agricultural nation' (see Braibant, 1943). In December 1940 a *Corporation Paysanne* was instituted. This had similarities to the German *Reichsnährstand,* but also fulfilled aspirations of a substantial body of right-wing agrarian opinion in France. All existing farm organisations were forcibly united in the *Corporation* which had the object of defending 'the moral, social and economic interests of the great peasant family'. Various measures were taken to reform aspects of legislation affecting the peasantry, to improve the standard of rural education, and so on. The prices of agricultural products were officially fixed and these fixed prices were progressively raised in the course of the war. The movement of labour out of agriculture practically ceased. The peasantry, however, subjected to increasingly burdensome exactions for the benefit of the occupying power, were hardly in a mood to appreciate Vichy's action in their favour. Even the prosperity which they gained from high official prices and from still higher earnings on the black market was largely illusory: they were running down their capital for lack of replacements. Agricultural production by 1945 had fallen to two-thirds of its pre-war level.

The ravages wrought throughout most of Continental Europe by the campaigns in 1944 and 1945 disrupted food production and all other economic activity. When hostilities finally ceased, food supplies were woefully inadequate, especially in Germany itself, and agriculture's productive capacity was seriously depleted.

The food situation in the United Kingdom was better: although food imports had been drastically cut, encouragements given to farmers had borne fruit in an increased contribution by British agriculture to the food supply (in terms of calories, the proportion coming from home production rose from 30% at the beginning of the war to 40% at the highest point in 1943/4). In contrast to what had happened in 1914, Britain entered the Second World War with a prepared plan for maintaining food supplies. The Ministry of Food was set up and became the sole buyer and importer of all major foodstuffs; existing stocks of the main foods were requisitioned; price control was imposed at one or more stages of distribution; rationing was introduced. Purchase at fixed prices assured farmers of a market and of satisfactory profits; special price incentives were given for certain products, in particular wheat, milk and potatoes. In November

1940, the government undertook to maintain fixed prices and guaranteed markets for the duration of the war and at least one year thereafter.

At first the government was reluctant to encourage a large expansion in home food production, fearing difficult readaptations after the war: Churchill instructed his Minister of Agriculture to aim at a 'large but not excessive' increase in food production. However, the steady fall in imports made a greater effort unavoidable. The Ministry of Agriculture argued that if farmers were to raise their output still further they needed more security: the memory of the slump in prices and the 'betrayal' by the government after the First World War was still in farmers' minds. As a result, the government agreed in 1944 to give producers a guarantee, lasting up to 1948, of prices not lower than those then prevailing for milk, cattle and sheep; however, it would give no similar assurances for other products and warned farmers that the prices of grains and potatoes would probably have to be reduced. After discussions with farmers' representatives, a further agreement was reached in December 1944: in future years, the government would review the level of farm prices in consultation with the National Farmers' Union; at this annual review, to take place in February each year, all factors relevant to price determination would be taken into account, including certain basic data derived from farm accounts.

From shortages to surplus

After the war, the immediate concern throughout western Europe was to raise agricultural production as rapidly as possible. Besides the problem of food shortages, there was a general need to save foreign exchange by keeping imports as low as possible. The recovery was rapid, and was greatly helped from 1948 on by American aid under the Marshall Plan, administered through the Organisation for European Economic Co-operation (OEEC). In the crop year 1949/50, agricultural production in the OEEC area exceeded the pre-war level (Figure 11.1); even in western Germany production was back to the pre-war level by 1950/51. Thereafter, the increase continued at a slower but still considerable rate, till by the end of the 1950s agricultural production in the OEEC area was some 50% higher than before the war, for a total population which had increased by about 20%.

This growth in production took place in spite of a large and steady fall in the numbers employed in agriculture. Output per man in agriculture rose rapidly: in most countries labour productivity increased faster in agriculture than in other sectors (Table 11.1). The outflow of farm labour was offset by greatly increased efficiency in the use of the remaining labour force. Machinery of all kinds multiplied on the farms: from 1947 to 1960, the number of tractors in OEEC countries rose from half a million to three million and the amount of motor fuel used on farms almost doubled between 1952 and 1960. From 1947 to 1960, the consumption of fertilisers of all kinds more than doubled. Combined with rapidly spreading knowledge of new techniques, fostered by the activity of farm advisory services, all this progress was reflected in greatly increased yields from crops and livestock.

Table 11.1: Output per man in agriculture and in all sectors, in 1959*
Indices: 1949 = 100

	Agriculture	All sectors
Denmark	177	130
Germany	172	153
Italy	169	159
France	161	154
Austria	156	151
Belgium	156	132
United Kingdom	146	120
Netherlands	142	139
Norway	137	131
Ireland	127	129

* Based on changes in gross product (in constant prices) and in numbers employed.

Source: OEEC and OECD statistics.

Food consumption, on the other hand, grew relatively slowly once the shortages had been overcome. Gradually therefore the problem became that of overproduction of several commodities. By around 1960, the markets of western European countries were frequently oversupplied with dairy products, mainly in the form of butter: most countries had closed their markets to imports and the only large remaining market, that of London, was liable to be swamped with excess supplies. There were periodic surpluses of pigmeat, steadily growing more acute; and in various countries the markets for individual products such as eggs, beef and wheat, were giving rise to concern. Since the middle of the 1950s, the United States had been grappling with surpluses of grains, dairy products and cotton, and other overseas exporting countries were faced with similar difficulties.

The progress accomplished by western European agriculture in raising its efficiency and output was not matched by a comparable advance in farm incomes. Though the product of agriculture rose in value, it rose considerably less fast than the national income. The 1950s were, for most western European countries, a period of fast economic growth and rapidly rising wages, in the industrial sector in particular. However, food consumption increased much less quickly than demand for other goods, and as the output of other sectors rose, the share of agriculture in the national income diminished. The outflow of labour (between 1950 and 1960 the agricultural labour force decreased by about 15% in the OEEC countries) made it possible to increase incomes per head in agriculture but still farm incomes remained in most countries well below incomes in other sectors. Figure 11.2 gives some idea of the situation: between 1955 and 1960, agriculture's share in both gross national product and in total employment fell in all countries. In most countries, however, the percentage of national product derived from agriculture was less than the percentage of active population employed by agriculture – in other words, average product per head from agricultural employment was relatively low. The difference was particularly large

Figure 11.1: Agricultural output and employment in OEEC countries
Indices, pre-war = 100

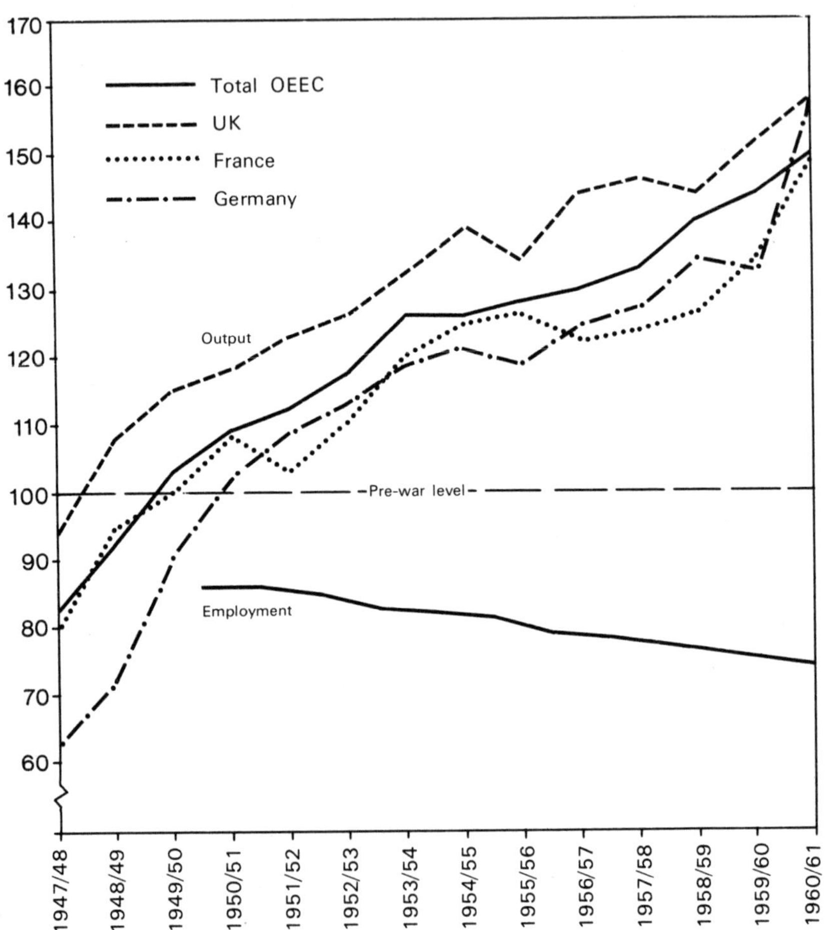

Note: The output series relate to *total* agricultural output. An alternative method of calculation, deducting imported feedingstuffs and store cattle, gives significantly higher results for the UK (from index 118 in 1947/8 to 190 in 1960/61): in other words, UK agriculture was producing more from its own resources. For most other countries the difference between the two series is not significant.

Source: OEEC and OECD statistics.

in Italy, Austria, Norway, France, Sweden and Germany (and probably in Switzerland though data are lacking). A situation approaching 'parity' obtained only in Belgium, the Netherlands and the United Kingdom.[1]

The growth of state involvement

In the immediate post-war period of food shortages, the aim was to expand agricultural production by all possible means, both to raise food supplies and to relieve balances of payment. For this purpose, income guarantees were given to farmers, price supports were introduced or reinforced, farm investments and improved farming methods were encouraged by credits and subsidies.

From about 1953 there was a change in emphasis as agricultural production caught up with demand. The aim was no longer to raise production at all costs, but to achieve selective expansion and to raise agricultural efficiency. At the same time, concern with the relatively low level of farm incomes was increasingly felt, and governments were placed in a quandary as the price guarantees they offered farmers tended to stimulate excess production. Policies tended to become increasingly complicated, and increasingly costly; attempts were made to limit expenditure and to restrain the growth of output of commodities in surplus.

This development was evident in the United Kingdom. The Agriculture Act of 1947 established the objective of policy as being that of:

> promoting and maintaining, by the provision of guaranteed markets and assured prices ... a stable and efficient agricultural industry capable of producing such part of the nation's food and other agricultural produce as in the national interest it is desirable to produce in the United Kingdom, and of producing it at minimum prices consistent with proper remuneration and living conditions for farmers and workers in agriculture and an adequate return on capital invested in the industry.

This text left room for differences of interpretation, in particular as to how much food it would be desirable to produce at home and as to what would constitute 'proper' remuneration. Nevertheless, the Act of 1947 constituted a clear commitment by the government to intervene on behalf of agriculture even in peacetime and was an assurance that there would be no repetition of the events of 1921.

The government undertook to buy, at guaranteed prices, the whole domestic output of grains, potatoes, sugar-beet and fatstock; wool was later added to the list. The annual review procedure was to continue as the basis for determining prices. Farmers were urged to raise net output – i.e. gross agricultural output,

[1] Comparisons in average terms as between the agricultural sector and other sectors cannot be precise. 'Gross product' is not necessarily the same as disposable income; data on the active farm population are often defective and the comparison made here does not take into account earnings by members of farm families from non-farm employment, which may significantly redress the balance in regions where small industries, tourism, etc. are important.

Figure 11.2: Share of agriculture* in:

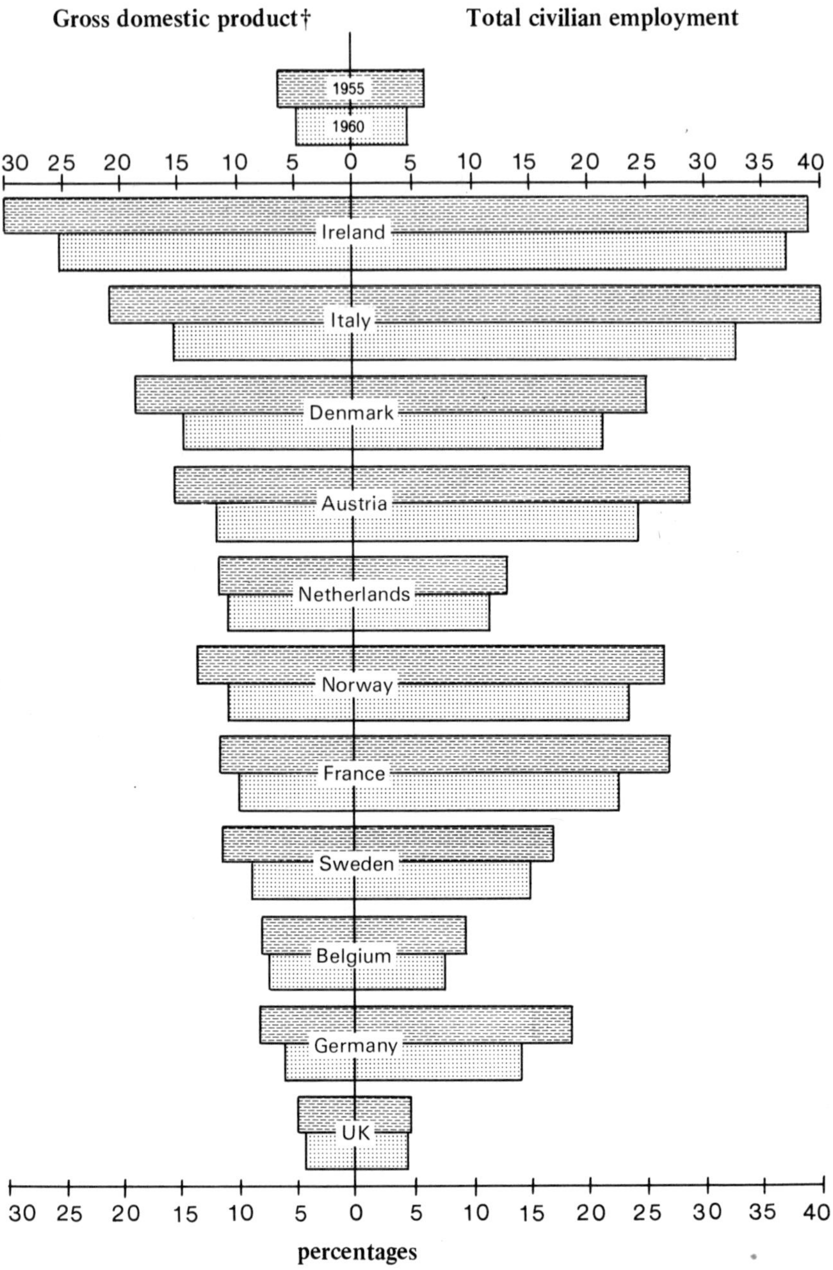

Gross domestic product† | **Total civilian employment**

1955
1960

30 25 20 15 10 5 0 5 10 15 20 25 30 35 40

Ireland
Italy
Denmark
Austria
Netherlands
Norway
France
Sweden
Belgium
Germany
UK

30 25 20 15 10 5 0 5 10 15 20 25 30 35 40

percentages

* Including forestry, hunting and fishing.
† At factor cost (at market prices for France and Germany).
Source: OECD, *Agricultural Statistics 1955–1968.*

less imports of feedingstuffs, seeds and livestock for use on farms – to 50% above the pre-war level by 1951/2 (an increase of 20% above the 1947 level).

In 1951 the Labour government was replaced by a Conservative one, which removed food controls. Imports were practically freed from quantitative restrictions; duties were generally those established in the 1930s, and the incidence of specific duties had been much reduced by inflation. However, the government maintained price guarantees to farmers, reviving for the purpose the pre-war instrument of deficiency payments which left markets free; it also restored the pre-war Marketing Boards, with the addition of some new ones. For a time, the government pursued the policy of raising food production, but around 1956 the emphasis passed to 'selective expansion'. Successive annual price reviews altered the price guarantees according to the state of the market and attempted, through price reductions and other measures, to restrain output of certain products. At the same time, the government was forced to give satisfaction to farmers in various ways, in particular by 'long-term assurances' in 1956, which limited the extent to which prices could be changed annually and seriously restricted the government's freedom of movement. As a result, it proved difficult to discourage output: in particular, production of milk came greatly to exceed the demand for liquid consumption, and the excess could only be sold at a loss for processing; egg production grew to the point where Britain became self-sufficient; output of pigmeat fluctuated but tended to rise unduly; the market for barley was over-supplied at certain periods. Market prices generally remained well below the guaranteed prices and were liable to be further depressed by increased imports, over which there was practically no control. The consequence was a heavy increase in the volume of Exchequer payments to agriculture.

In France, the first of the post-war Modernisation and Equipment Plans, for the years 1947–50, aimed to raise agricultural production in order both to meet food shortages and to relieve the balance of payments. To achieve this aim with increased labour productivity, thus releasing labour for industry, priority was given to mechanisation. In 1948, following an OEEC request to member countries to indicate how they proposed to reach balance of payments equilibrium, a revised Plan was prepared for 1948–52 which stated the aim of developing exports of cereals, meat, dairy products and sugar, and of substantially reducing imports of oils and fats. When the next Plan was being prepared in 1953, there was much discussion as to the advisability of further expanding farm production. From 1950 onwards, surpluses had made their appearance: exports of wheat and sugar had to be promoted by means of subsidies and steps similar to those adopted before the war became necessary to absorb the surplus of wine – distillation, blocking on the vineyards and uprooting of vines. The Second Plan (1954–7) nevertheless provided for a 20% increase in agricultural production, but declared that output of some products, in particular wine, alcohol and oats, should be reduced. The Third Plan (1958–61) pursued a policy of selective expansion. There was to be a further overall increase of 20% in agricultural output: in line with prospects on export markets, feed grains and livestock products were to expand most, while production of wheat and sugar-beet was to level off and that of wine and potatoes to be reduced. In fact it proved extremely difficult to restrain the production even of commodities in surplus. Not only was it politically almost impossible to reduce prices in order to discourage output, but the rapid spread of new technology throughout French agriculture enabled greatly increased productivity. In 1961 the agricultural correspondent of *Le Monde* wrote:

> Too much wheat, too much barley, maize, meat, milk, sugar, too many artichokes, apples and very soon also too many cauliflowers.... The peasants, having been encouraged by optimistic planners to raise production, are now finding that they don't know what to do with their output – this even when, in many cases, they have not even reached the planned targets. [Virieu (1961)]

From the beginning of the Second Plan, measures had been taken to give firmer price guarantees to producers. A decree of 30 September 1953 provided for the institution of a *Fonds de Soutien et de Garantie Mutuelle,* which was to support agricultural markets: it was financed in part by the State and in part by producers, as well as by levies on imports; it came into being in 1955. The markets of almost all important products were subjected to intervention in one form or another: the pre-war Wheat Board was enlarged to become the *Office National Interprofessionnel des Céréales* (ONIC), and new boards were introduced to regulate the markets of dairy products, meat and livestock, oilseeds and other products. Domestic markets were virtually insulated from outside influences by strict controls over imports, exercised for most products through import licensing, but for grains and dairy products through state-controlled import monopolies. There was also an increasing degree of assistance to exports of products in surplus through direct export subsidies and other means. In the case of wheat, a means of restraining production existed in principle in the 'quantum', under which the guaranteed price was applicable only to a standard quantity, any additional output receiving only the price realised on the export market: in fact this arrangement was not very effective in discouraging excess output. Efforts were made to deal with overproduction of wine in 1953 (blocking and distillation of surpluses and compensation for uprooting vines), and in 1959 a series of measures were introduced to stabilise the wine market, involving provision for the withdrawal of excess quantities, a buffer stock and distillation.

In spite of the increasing complexity of the price support apparatus, farm incomes remained low in relation to those of other sectors. The farm organisations became increasingly insistent in their demand for 'parity', and in 1957 the guaranteed prices for the main products were indexed on the basis of input costs, non-food retail prices and farm wages. A year later, however, de Gaulle's government abolished this system as part of its plan to reduce inflation. The farm organisations reacted with violent demonstrations and in March 1959 indexation was partially restored. The new government however sought to develop a more coherent agricultural policy with the 1960 *Loi d'Orientation.* This important act established the aim of 'parity' with non-farm incomes demanded by the farm organisations but laid emphasis on improvements in productivity and in marketing and production structures: a *Loi complémentaire* two years later gave further effect to structural policy (see chapter 14). At the same time, the various boards for supporting the markets of different commodities were merged in the *Fonds d'Orientation et de Régularisation des Marchés Agricoles* (FORMA). By now attention was increasingly turning to the prospects for French agriculture in the newly-formed European Economic Community, which appeared to offer a solution to the problems of French agriculture through increased outlets and higher prices.

In western Germany, cut off from its traditional sources of food supply in eastern Germany, a particularly serious situation was faced after the war. The worst point was reached in the winter of 1946–7, when official rations fell to around 1400 calories per head per day in the American and British zones, and to around 1300 calories in the French zone; actual supplies were often even less. Agricultural production was well below the pre-war level and capable of providing little more than 1000 calories per head. The Marshall Plan in 1948 permitted higher food imports, including feedingstuffs, making it possible to build up again the livestock sector. Still, when the Federal Republic was constituted in September 1949, shortages were still the major problem: Dr Adenauer, on taking office as Chancellor, declared that agriculture must be made more efficient and food production raised. The necessary condition for this was the assurance to producers of stable and well-balanced conditions of production and marketing, at prices which would cover the costs of well-managed average farms and at the same time permit the poorer sections of the community to meet their needs. In 1949 imports of food were still being subsidised in order to reduce prices to consumers. In 1950 and 1951, *Einfuhr- und Vorratsstellen* (Import and Storage Boards) were instituted for the main agricultural products, with the object of stabilising prices and maintaining them at levels consistent with the aims of agricultural policy: these boards controlled imports in a manner similar to the Nazi *Reichsstellen,* and could intervene on the domestic market by purchasing, selling and stockpiling. As the food situation eased and world prices declined, the Import and Storage Boards, together with other measures of import control, were used to maintain German prices at levels well above those on the world market.

Behind this protective wall, agricultural output rose, yet farmers could not keep abreast of the general increase in prosperity. Politicians, remembering the part played by agricultural distress in the rise of Hitler, were anxious to avoid any new right-wing agitation in the countryside; the *Deutscher Bauernverband* (German Farmers' Union), formed in 1948, was for its part determined to extract from the Christian Democrats, as the leading government party, an adequate price for its political support and demanded 'parity' with other sectors. At the same time, the first discussions of plans for an agricultural common market (in particular following the Messina Conference in 1955) made it clear that energetic steps would have to be taken to improve the competitiveness of German agriculture. After some controversy over the parity issue (see later section), the Agriculture Act of 8 July 1955 laid a firm statutory basis for measures to assist agriculture. Its objectives were:

(a) To ensure a reasonable standard of living for the agricultural population on well-managed farms;
(b) To raise agricultural productivity;
(c) To stabilise agricultural prices;
(d) To ensure a regular food supply at prices enabling the lower income groups to buy sufficient quantities.

The Act did not commit the government to maintaining any clearly-defined level of prices or incomes for agriculture; however, the government was required to submit an annual report to Parliament, showing in particular the development in farm incomes, and on this basis to put forward an annual plan for achieving the objectives of the Act. The annual Green Reports subsequently provided a detailed analysis of the agricultural situation, based in part on a regular study of the accounts of a sample of farms of different types and sizes. Under the Green Plans various steps were taken to support farm incomes and to improve farm structures.

Most of the other western European countries also gave income or price guarantees to their farmers in the early post-war years and later had to grapple with problems of surpluses while trying still to ensure adequate incomes to farmers. Thus Switzerland in 1947 inserted in its Constitution articles which provided a permanent legal basis for assistance to agriculture. These were implemented by the Agriculture Act of 1951, which aimed 'to maintain a large peasant population to facilitate the supplies of the country by ensuring agricultural production and encouraging agriculture, having regard to the interests of the national economy'. This Act provided the basis for subsequent legislation laying down detailed measures of support for various commodities, consisting primarily of intervention on the market and price guarantees handled by bodies responsible to the government. Imports were strictly controlled, often by the system of *prise-en-charge* or 'conditional imports', under which imports were permitted only on condition that importers purchased home-produced goods of the same kind. As a result, prices were maintained well above the world market level: although Switzerland imported more food per head than almost any other country, its agriculture remained one of the most highly-protected in the world. Special non-economic motives were largely responsible for this. Nevertheless, rising overproduction in the dairy sector necessitated measures of restraint, and it became necessary to take into account both the cost of agricultural support and the problems which would arise if Swiss agriculture were exposed to greater competition in the context of European economic integration. A report by the Federal Council in 1959, while stressing the unfavourable natural conditions under which Swiss agriculture had to operate, made it clear that the number of holdings and the size of the farm population would have to be reduced.

Norway, like Switzerland, had the problem of reconciling the social and strategic objectives of a large agricultural population with difficult natural conditions. In 1947 its general objectives for agricultural policy were laid down, including that of ensuring the farm population of a standard of living comparable with other sectors. The Agricultural Production Programme of 1955 aimed to cover requirements of livestock products from domestic production mainly based on home-produced feedingstuffs, and production of grain, fruit and vegetables was to be expanded. However, production was to be adjusted as far as possible to demand and the general aim was to raise the production of commodities then being imported rather than to produce for the export market. Agricultural markets were subjected to a far-reaching system of intervention, involving strict import controls.

Sweden also assured its farmers of a standard of living comparable with other sectors, with the special feature that food production was to cover no more than 90% of domestic requirements. The prices of the main agricultural products were maintained by a system of import monopolies and quantitative restrictions, and the difference between world prices and domestic prices was met by variable import levies. This arrangement fitted in badly with the principles applied in the rest of the economy: foreign trade played an important role and imports of manufactured goods were practically free from restrictions. An effort was therefore made to introduce a system by which changes in the world prices of agricultural products would be reflected on the domestic market, so that production would be guided along the right channels. Accordingly, in 1956 a system of 'fixed' import levies was introduced, subject to various safeguards for farm incomes. However, as these safeguards proved inadequate to maintain incomes at the desired level, the system was revised in 1959, with the result that the original aim of flexibility was to a large extent lost. At the same time, and despite a rapid movement of labour out of agriculture, difficulties were encountered in bringing production down to the desired level of 90% of domestic requirements: total production remained approximately equal to requirements, and butter and other products had to be exported, generally at a loss.

In Belgium, the price of wheat was supported while feed grains were imported at world prices for the benefit of livestock producers: the result was an increasing price differential and an expansion of the wheat area at the expense of other grains, to the point where the extra wheat production (much of which was of poor quality) could not easily be disposed of. Moreover, the relative cheapness of feed grains encouraged an expansion of the livestock sector and here too difficulties were encountered: eggs and butter in particular had to be exported with the help of subsidies. From 1957, imported feed grains were subjected to a levy in order to encourage domestic production, but livestock producers were compensated for the resulting rise in costs.

Italy, without giving formal guarantees as to price levels, sought to prevent market disturbances. For wheat in particular, the system of control instituted before the war was revised in 1948, compulsory deliveries being limited to the quantities considered sufficient to stabilise the market. Production of soft wheat began to outrun demand in the mid-1950s, necessitating subsidised exports and the price for state purchases was reduced. Rice also began to be in surplus in 1955 and 1956, and measures had to be taken to regulate the market. On the other hand, the output of products for which Italy's climate gives an advantage over most other European countries – in particular fruit and vegetables – was able to rise unchecked.

Importing countries generally were able to maintain prices and support their farmers' income by restricting imports, up to the point where their own supply began to exceed demand. Agricultural exporting countries, however, did not have this possibility and moreover were harmed by the restrictions imposed by importing countries. Denmark had been the only country which dismantled its wartime apparatus of state control and returned to private trading. Until about 1956, long-term contracts with Britain ensured outlets, at moderate prices, for specified quantities of bacon, butter and eggs. After these contracts came to an end, increasing difficulties were encountered. The British market for bacon was glutted by higher domestic production and increased exports by other countries (including Poland and Yugoslavia). Even worse difficulties were encountered for butter: by 1961 Britain was almost the only market left open and was the arena of competition between not only the traditional butter exporters but also newcomers such as France. Danish eggs also suffered badly on the British market, for the rise in Britain's own production reduced imports to negligible quantities from 1956 onwards. Denmark's trade with western Germany in live cattle and pigs expanded during the 1950s, and there was also a growth in exports of eggs which largely compensated for the decline in trade with Britain. Exports of butter to Germany, however, became unimportant.

Denmark attempted to overcome these difficulties by diversifying its exports and some increase was achieved in sales of products which formerly were of comparatively small importance: thus there were increased exports of cheese, beef and veal, and canned meat. However, the persistently low level of prices on export markets, at a time when the costs of agricultural production were rising, caused a decline in the net income of Danish agriculture in the course of the 1950s. Coming at a time when incomes in other sectors were progressing fairly rapidly, this was all the more unacceptable to Danish farmers. The result was to drive Denmark back into various forms of intervention, some of them resembling measures taken in the 1930s. An Act passed in June 1959 authorised a levy on all dairy products sold on the home market, the proceeds of which were to be distributed to the dairies in proportion to their deliveries, and ultimately to be passed on to producers; in 1961 a minimum price was set for sales of butter by dairies on the domestic market. In 1958 a Grain Marketing Act provided for guaranteed producer prices for bread grains, to be ensured primarily by a milling ratio: in 1960 the ratio was raised to 100% for both wheat and rye, thus cutting out imports. The same Act provided for minimum import prices for feed grains, to be maintained by import levies. The most drastic break with Danish traditions of non-intervention in agriculture occurred with the Agricultural Products Marketing Act of 1961. This established a 'Rationalisation Fund' for aid to farmers, introduced a subsidy on fertilisers, provided for steps to raise the home market price of pigmeat over the average export price, and authorised measures to assist the marketing of agricultural produce through the promotion of exports, research into methods for improving quality, and so on.

Danish agriculture thus could no longer be said to depend entirely on its own unaided efforts, for it now received a not insignificant degree of support in the form of direct Exchequer payments, higher prices imposed on domestic consumers for dairy products and pigmeat, and protection against imports of bread grains, feed grains and milk powder.

Other agricultural exporting countries within western Europe experienced similar difficulties. The Netherlands, after the war, continued its pre-war and wartime arrangements: the Agriculture Act of 1957 co-ordinated past legislation and introduced new provisions; agricultural markets remained subject to intervention in various forms. In 1961 Ireland took steps to organise its export marketing on a more efficient basis.

Difficulties in international trade

The growth of production, the relatively slow growth of demand and the increased degree of self-sufficiency in western Europe reduced the scope for imports of temperate foodstuffs. Figure 11.3 shows that by the early 1950s, the volume of production of foodstuffs in western Europe was well above the pre-war level, while the volume of imports had fallen. In the course of the 1950s, both production and imports rose, but imports did not make good the ground lost during and after the war. Trade between the western European countries rose faster than the net imports of western Europe from other regions: the latter, by 1960, had barely recovered to the pre-war level. This contrast is largely a consequence of the composition of trade in the two cases: meat and livestock, as well as fruit and vegetables, play an important part in intra-European trade, and demand for these commodities rose relatively fast. In the case of imports from overseas, the favourable development for these products was offset by the much less satisfactory trends for grains, sugar and dairy products.

Developments in the world prices of some major foodstuffs during the 1950s are indicated in Figure 11.4. For four of the five commodities shown, the trend was downwards. Some decline was inevitable after the high prices prevailing during the post-war shortages and the Korean War: however, a decline continued even in the second half of the decade for barley and – apart from a short-lived recovery in 1957 – for sugar. The price of butter fell sharply in 1958 and again in 1960. The price of wheat was stabilised mainly by the stocking policy of Canada and the United States. The exception to the general decline was beef: here the growth of demand enabled prices to rise steeply during the late 1950s.

Overseas exporting countries inevitably became increasingly worried at the difficulties they were experiencing. Their preoccupations were voiced particularly in the context of the General Agreement on Tariffs and Trade (GATT), the effectiveness of which, in the agricultural field, was limited by the extent to which trade was affected by measures of domestic policy as well as by direct restrictions on imports. The GATT therefore concerned itself increasingly with the effects of agricultural policies. In 1958 it published a report written by a panel of eminent economists headed by Gottfried Haberler. The foreword to

Figure 11.3: Volume of production and imports of food and feedingstuffs in western Europe: Indices, 1934–8 = 100

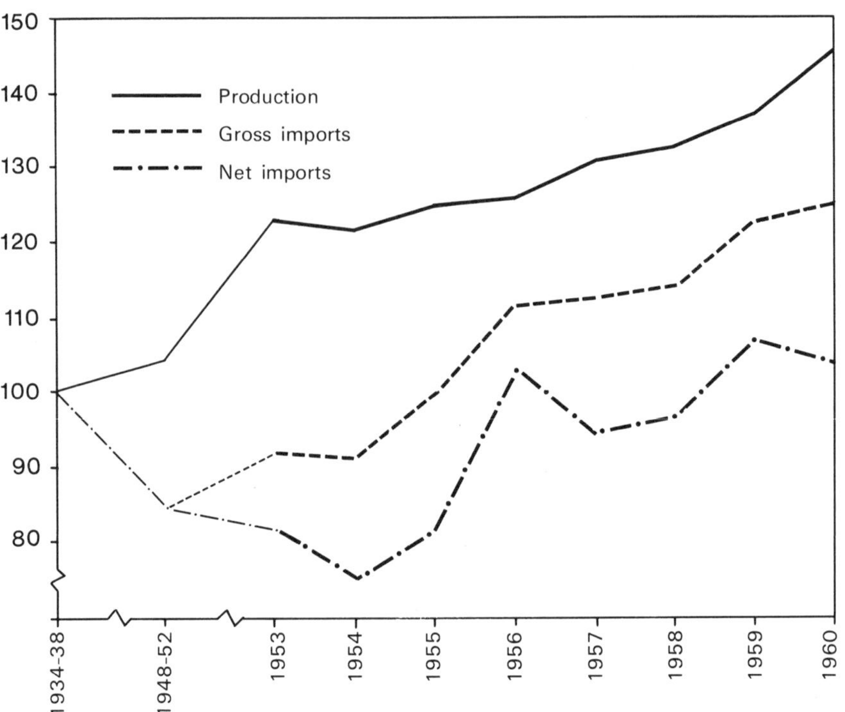

Source: FAO, *State of Food and Agriculture 1962* (and earlier issues).

this report by the Executive Secretary, Wyndham White, pointed to the disturbing elements in world trade: the prevalence of agricultural protectionism, the accumulation of surplus stocks, the sharp variations in the export earnings of primary products and the failure of the export trade of underdeveloped countries to expand at a rate commensurate with their needs. The report itself analysed trends in world trade, stressing the relatively slow growth in the volume and value of trade in agricultural products. The authors of the report considered that the main factor responsible for the unsatisfactory development of trade in temperate foodstuffs was protectionism in the industrialised countries; this protectionism appeared in the form of subsidies on domestic production and on exports, as well as of import restrictions of various kinds.

Through the operation of international agreements for a few commodities, attempts were made to restore stability to international markets. Under the

Figure 11.4: World average export unit values of major foodstuffs
Indices: 1948–52 = 100

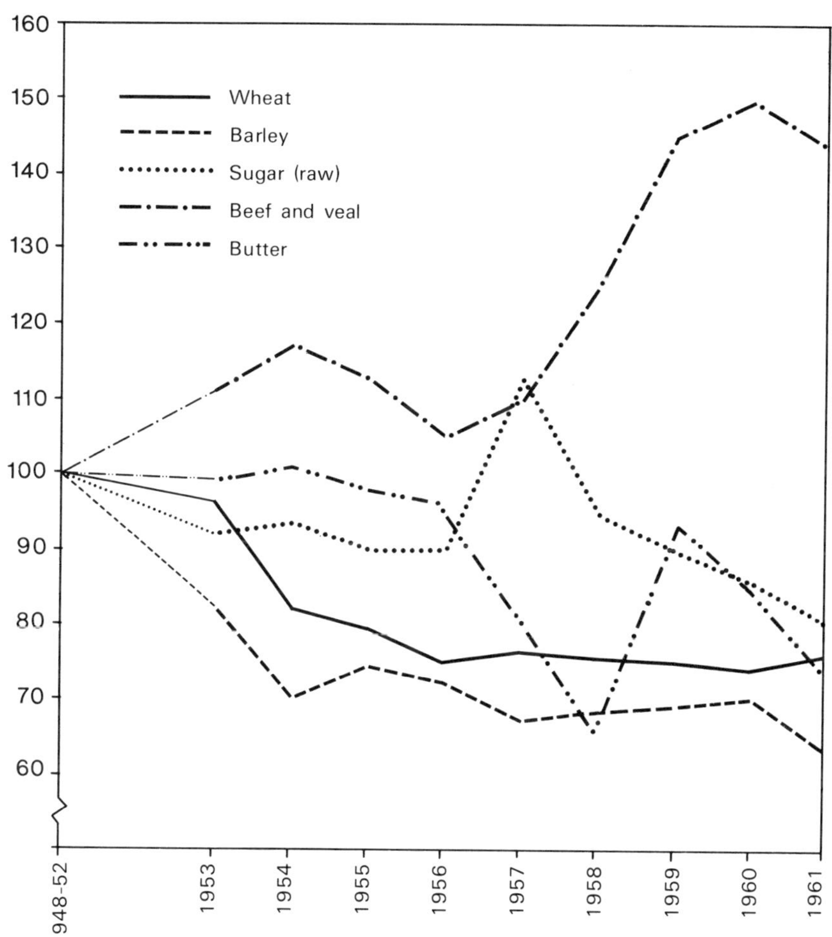

Source: FAO, *State of Food and Agriculture 1962* (and earlier issues).

International Wheat Agreements (the first post-war Agreement entered into force in 1949), each exporting country agreed to supply a stated amount at a maximum price and each importer undertook to buy a stated amount at a minimum price. These Agreements however had no power to curb the rising trend of production in exporting and importing countries. Wheat prices on the world market were kept reasonably stable during the 1950s not so much through the operation of the Agreements as by the policy of the major wheat exporters.

An International Sugar Agreement came into force in 1953. It provided for export quotas, which could be modified in order to keep prices within a given price range. As sugar production – especially in importing countries – continued to rise, there was a downward trend in market prices, together with wide price fluctuations, in face of which the export quota system proved inadequate.

Controversies

The increasing involvement of governments in agricultural support provoked much argument in most countries. In the United Kingdom in particular, there was intense debate as to the desirability of permanently maintaining, through subsidy, a large domestic agriculture. The case for doing so was based largely on the balance of payments problem; the arguments in question continued to be advanced even after Britain had recovered from the critical post-war shortage of foreign exchange. Professor E. A. G. Robinson, as an eminent economist with an influential position in Whitehall, was the most powerful exponent of this view; the basis of his argument was his pessimism as to Britain's capacity to finance increased imports of foodstuffs. Most economists however found it difficult to accept the idea that producing goods at home at a cost greater than that at which they could be purchased from abroad could be sound economics (e.g. Hallett (1959), Nash (1955), Raeburn (1958)). Probably the most thorough and well-balanced analysis of the issues was provided by McCrone (1962). He concluded that there was no good economic reason for paying costly subsidies to British agriculture either to maintain output at a high level or to encourage further expansion. However, a policy of laissez-faire aiming to allow the market mechanism to weed out the uncompetitive enterprises, would probably not be successful. Bearing in mind the experience of the late nineteenth century, he pointed out:

> The idea of forcing the inefficient producers out of production sounds plausible in theory: but in agriculture they often stay until they have ruined the other factors of production and until the job of repair and reclamation is too expensive to be worth undertaking. [McCrone (1962) page 45]

McCrone favoured measures to improve the competitiveness of the agricultural sector, including a credit policy to encourage productive investments and action to regroup uneconomic holdings.

In popular discussion in the United Kingdom, the rising cost of agricultural support gave rise to much criticism. Many large farmers became prosperous through subsidies and the charge of 'feather-bedding' was launched. Since the greater part of subsidy was paid in the form of price support, most of it went to the farmers with the biggest output who probably needed it least.

In France, there was relatively little debate, academic or public. One of the few critics was Froment (1956), who stressed the importance of reducing the costs of production and marketing and the need to move people out of agriculture. Another was Bergmann, who in 1957 analysed the principles of agricultural policy in the light of economic criteria; he criticised excessive reliance on price support, declaring that markets should be allowed to operate as freely as possible (he pointed out the advantages of a deficiency payments system along British lines), and wrote:

> The only real way of reducing the income gap is to increase the mobility of manpower.... Any measure which arbitrarily, through direct payments, raises prices and incomes would in the long run have results opposite to those which are sought for. In other words, a policy whereby prices and incomes are systematically increased would be self-defeating. [Bergmann (1957) page 20]

In general, however, agriculture in France seemed hardly to be a subject for economic analysis: its importance was generally recognised almost without question and the rest of the community seemed to accept an obligation to preserve a large and reasonably prosperous agriculture. There was widespread reluctance to admit that the agricultural population had to fall. It is true that in the still backward conditions of much of rural France, the rural exodus had adverse effects. In an important study of the agricultural income problem, Latil said that:

> In regions of general economic progress, agriculture can usually contrive to adapt to new conditions; in backward areas, either agriculture fails to progress or, if a movement of labour does take place, the land is abandoned for lack of initiative and of capital. In the first case, the emigrant from agriculture leaves behind him mechanised cultivation and modern buildings; in the second, he leaves a decaying hovel and land going to waste. [Latil (1956) page 168]

Given the great potential of French agriculture, continued production growth seemed inevitable and this could only mean increased exports. Preferential access to the markets of other European countries thus seemed increasingly to be the only way forward for French agriculture.

In Germany the farmers' demands for parity in the early 1950s had provoked substantial debate: while the Christian Democratic party favoured linking the rewards of agriculture to those of industry, at the expense of the community as a whole, most academics criticised price support policy and emphasised that the greatest need was for structural reform. The Chamber of Industry and Commerce also declared that the basic need was to make agriculture competitive, not to give it a permanently privileged position. Boerckel (1959) wrote that though the Agricultural Act of 1955 had not fully met the farmers' demand for parity and had left the income target undefined, in practice the implementation of policy, in particular through the activities of the *Einfuhr-und Vorratsstellen,* ensured price levels covering the costs of the great majority of farms, without much regard for their competitiveness. Domestic prices had thus been maintained well above import prices, so that support for agriculture was mainly at the expense of the consumer. As prices covered costs even on quite marginal holdings, there was an inducement for other holdings to expand output up to the margin. Production of rye had come to exceed demand so that the responsible Board had to stock or export the surplus; sugar production had increased; pigmeat and butter prices could be maintained only by virtually excluding imports.

After the signature of the Treaty of Rome in 1957, discussion in Germany was increasingly preoccupied with the difficult agricultural adjustments which would become necessary in the European Economic Community (e.g. Baade in 1958 – second edition 1963).

The growth of farm organisations

The post-war years saw a significant increase in the influence exerted by farmers' unions – despite the falling agricultural population. In most countries they were instrumental in obtaining income and price guarantees, and subsequently insisted on full implementation of these guarantees; in some cases they were able to secure consultative status with the government. Their relations were closest with the right-wing – Conservative or Christian Democratic – parties which were in power in several countries during the 1950s. The parliamentary membership of these parties tended to contain a disproportionately high number of farmers or others with agricultural interests, often elected with the support of the farmers' union. But agricultural policy was rarely a party issue: centre and socialist parties did not want to alienate the farm vote (even the communist parties in western Europe played down the issue of land nationalisation and based their appeal on the defence of small peasant proprietors). The Agriculture Act of 1947 in the United Kingdom, and the introduction of farm price 'indexation' in France in September 1957, were the work of socialist or socialist-led governments. The tactics of farm organisations were not restricted to the parliamentary scene: particularly in France, demonstrations sometimes leading to violence proved effective in getting government action.

In the United Kingdom, the National Farmers' Union (NFU), containing the great majority of full-time farmers, was put in a strong position by the annual price review system, at which it was recognised by the government as the official farmers' representative.[1] The government came under pressure to attune its policies to the wishes of the union. Though it insisted that the reviews were to be regarded as 'consultations' and not 'negotiations', it was often willing, during the early period of the production drive, to pay a substantial price to avoid open disagreement. Subsequently, with the attempts by the government to reduce guaranteed prices on some commodities, there was increasing difficulty in reaching an agreed settlement.

[1] The government recognised the National Farmers' Unions of England and Wales, Scotland and Ulster. The NFU for England and Wales was the biggest and played the leading role.

The strength of the NFU was reinforced by intelligent tactics. By refraining from making agricultural policy a party issue, it contrived to gain the support of both political parties. It also managed to convince politicians that the agricultural vote was an important force (it is questionable whether this was really so – see Self and Storing (1962)). In its President, Jim Turner – later Lord Netherthorpe – the NFU had a charismatic leader. It was generally well staffed: its chief economist, Asher Winegarten, was an advocate able to meet the criticisms of economists on their own grounds, making the most of the balance of payments and other arguments (cf. in particular his paper to the Agricultural Economics Society in 1960). NFU publications were well presented and well argued.

In France too the farm organisations increased greatly in strength. After the war there was an attempt by the socialists in the government to establish a single organisation in the *Confédération Générale de l'Agriculture,* but opposition by right-wing elements soon made this ineffective. The most powerful group was then the *Fédération Nationale des Syndicats des Exploitants Agricoles* (FNSEA) in the rue Scribe. This descended from the right-wing syndicalist movement that had been active in the 1930s. It was closely linked with the specialist producer associations (for wheat, sugar-beet, wine, etc.) which, as before the war, constituted effective pressure-groups. The radical tendencies of the Boulevard Saint-Germain were pursued in the *Fédération Nationale de la Mutualité, de la Coopération et du Crédit Agricole,* though this was less active in policy matters.

The *Assemblée Permanente des Présidents des Chambres d'Agriculture* (APPCA), as a public organisation, was officially neutral in political matters. In fact it also constituted a strong pressure-group, expressing views generally similar to those of the FNSEA, with which it shared the same address in the rue Scribe.

The demands of the farm organisations generally concentrated on higher prices and better markets. Mendras described their programme as follows:

> The first requirement is that family holdings must be preserved, the rural exodus restrained; here everyone is in agreement. But the measures envisaged are generally negative; positive steps on the other hand are usually expressed in vague and general terms, and they are always difficult to implement. The vital demand is of course for price support: this necessitates increased outlets, hence assistance for exports and the elimination of imports, as well as a reduction in the price of industrial products used by agriculture in order to reduce the costs of agricultural production. To allow small holdings to survive, demands are made also for additional credits to promote equipment and modernisation, and for action to disseminate the latest agricultural techniques.... Behind these demands there lies a deep concern felt by the agricultural population: the fear of overproduction and falling prices. The general agreement on this programme is in striking contrast with the rivalries between the organisations, and shows clearly that the quarrels are between different headquarters which are contending for the same body of troops. [In Association Française de Science Politique (1958) page 150]

Most of the French farm leaders were reluctant to admit that a long-term improvement in the agricultural situation could be achieved only by a thorough reform of farm structures, involving a further large reduction in the agricultural population. It was suggested by Cépède that the larger and more efficient farmers who tended to dominate the agricultural organisations did not find it in their interest to promote improvements among the more backward sectors:

> The fact is that the constant policy of this group has been to use the high costs of production which technical backwardness imposed on the greater part of French agriculture as an argument in order to obtain prices from which the most efficient farms derive a large profit. This profit being largely attributable to the difference in technical standards, the question arises whether the dominant group has an interest, in the short term at least, in reducing the difference. [Cépède (1961) pages 491–2]

The representation of agriculture in Parliament seems to have been weak in quality, if not in quantity. In a study of this question, Dogan criticised severely the standard of debate and the value of legislation on agriculture:

> An examination of the parliamentary debates leaves the impression, which is confirmed by many witnesses, that the majority of deputies from rural areas are short-sighted conservatives, incapable of seeing the problems from a national point of view ... Parliamentary initiative in agricultural matters is abundant but ineffective. [In Association Française de Science Politique (1958) pages 217–23]

The French farm organisations exerted pressure in various ways, from lobbying members of Parliament to organising demonstrations. Chombart de Lauwe (1979) has pointed out that violent and illegal demonstrations were increasingly found to be effective: the government, fearing that such unrest might lead to combined action by peasants and workers, generally made concessions.

A new tendency appeared in the late 1950s with the growing influence of the *Centre National des Jeunes Agriculteurs.* The CNJA had close links with the FNSEA but maintained an independent point of view. The leaders of the CNJA, as representatives of the young farming generation, were naturally more progressive than their elders and not so much influenced by traditional habits of thought. Their policy was determined not by a vested interest in the established order but by a concern with the future prosperity of French agriculture. Their programme thus tended to stress action that would rejuvenate French farming, doing away with the inherited pattern of uneconomic small and fragmented holdings and giving increased opportunity for young and dynamic farmers to make their way. The CNJA played a major role in the establishment of a more coherent structural policy in the early 1960s.[1]

[1] The CNJA's most prominent figure, Michel Debatisse, subsequently became Secretary-General of the FNSEA, then its President, and in 1979 was appointed Secretary of State for the Food Industries.

In Germany the *Deutscher Bauernverband* (DBV) claimed to represent 77% of all independent farmers. Most farmers voted for the Christian Democratic party (CDU): the DBV however made it clear that its support was conditional on a satisfactory agricultural policy and ensured that its views were represented within the party. In the late 1950s about one in five CDU members of parliament had farming interests and these formed a majority in the parliamentary agriculture committee. The President of the DBV, Edmund Rehwinkel, maintained direct access to Chancellor Adenauer and frequently demonstrated his ability to influence the government's agricultural policy.

In Belgium the *Boerenbond* was by far the most powerful group, closely associated with the Flemish Catholic party (CVP). In the Netherlands, the three farmers' organisations– Catholic, Protestant and non-denominational – were associated with the corresponding three farm workers' organisations in a public body known as the *Landbouwschap* (membership of which was compulsory for all engaged in agriculture even if not members of one of the six unions). The *Landbouwschap* was politically neutral, but was regarded by the government as the authoritative voice of agriculture and it developed close relations with the Ministry of Agriculture.

In Italy agricultural policy-making was mainly determined by the close links between the ruling Christian Democratic party and the *Confederazione Nazionale dei Coltivatori Diretti.* This drew its strength from the vast numbers of small peasants, especially the growers of wine, vegetables and fruit. Its leader, Paolo Bonomi, who was also influential within the *Federazione Italiana dei Consorzi Agrari* (a national federation of co-operatives with semi-official status), was a member of the central committee and the executive bureau of the Christian Democratic party, from which he could usually obtain the decisions he wanted. The other main agricultural organisation, the *Confagricoltura,* represented mainly the great landowners of the south.

Gathering clouds

This chapter has described developments in the agricultural situation and the policy responses from the Second World War until about 1960. This period saw a technological revolution in western European agriculture, permitting greatly increased labour productivity and higher output despite a massive shift of labour from farms to factories and other urban employment. After food shortages had been made up, demand for food grew less fast than its production; population growth was relatively slow and although incomes were rising quite fast, this produced a less than proportionate growth in the consumption of most foodstuffs. Inevitably food supply tended to exceed demand, and the tendency would have been to drive down prices. But to implement the guarantees given to farmers in most countries after the war, governments provided price support in a variety of ways, in most cases intervening extensively in agricultural markets (the United Kingdom's deficiency payment system gave support in a different way).

Such action, however, meant that farmers had no incentive to adjust their supply. On the contrary, they were encouraged to continue increasing their production. Governments became more and more deeply involved, the cost of support rose and the degree of protection increased.

Nevertheless the farm income situation remained unsatisfactory. Incomes and living standards on farms did increase in absolute terms. Still, with overall demand for food rising less than proportionately to total consumer income, agriculture's share in national income was falling and the transfer of manpower from agriculture to other sectors, though substantial, was in most countries barely sufficient to avoid further widening of the gap in incomes per head between agriculture and the rest of the economy. There was no significant closing of this gap (see in particular the OEEC *Fifth Report on Agricultural Policies,* 1961). As farmers' income expectations were increasingly geared to urban standards, their dissatisfaction grew. Moreover, large disparities persisted within agriculture itself, between large farms and small, and between prosperous regions and less-favoured ones. Small farms, producing relatively little for the market, gained limited benefit from price support.

The policy problem thus posed was clearly recognised at the time. Besides the writings of economists already referred to, who analysed critically the current policies of their countries, there were during this period numerous works dealing with the factors which influence the relative income of the agricultural sector and which necessitate the transfer of labour to other sectors: Clark (1957) 3rd edn.; Ojala (1952); Bellerby (1956); Latil (1956); the International Labour Office (1960).

Moreover, a series of international reports drew attention to the worsening situation. The Organisation for European Economic Co-operation (OEEC) produced five reports on agricultural policy between 1956 and 1961 which abundantly documented the growing complexity of agricultural policies and issued policy recommendations. The OEEC stressed the dangers of relying too much on price supports to implement income guarantees, and increasingly emphasised the need for action to bring about more permanent improvements in agricultural production and marketing structures, in particular the need to promote viable farms. The extent to which agricultural protectionism was restricting trade in agricultural products was pointed out by the General Agreement on Tariffs and Trade (GATT) in the 'Haberler Report' (1958). The annual reports of the Food and Agriculture Organisation (FAO), *State of Food and Agriculture,* drew attention to the growing divergence between the high-income countries of the world, characterised by over-production of food, and the less-developed countries suffering from shortages.

The Economic Commission for Europe (ECE) in Geneva produced, in 1960, the first systematic set of projections of supply and demand, relating to 1965. This report pointed out that there was still great potential for further increases in productivity, and declared:

> With the exception of one product, beef and veal ..., production is likely to increase faster than consumption ... This in turn suggests that the trend of

> agricultural prices in the near future will be downward... [Economic Commission for Europe (1960) page 96]

It predicted that the major surplus problems would arise for dairy products, wheat and sugar, and went on:

> The effects of an increasing disequilibrium in production and consumption rates clearly run counter to the general intention of governments to improve farm incomes and bring them closer to income levels in other sectors of the economy. [Ibid., page 99]

Movement out of agriculture appeared to be the only way to raise incomes per head and in the long term the only way governments could improve farm incomes was to promote this trend of falling agricultural employment.

At this time, however, farm organisations (with the notable exception of the young farmers' organisation (CNJA) in France) still regarded as anathema any suggestion that the rural exodus should be officially stimulated; they continued to concentrate their demands on price support, discouraged any suggestions that small farmers might derive little benefit from this policy, and paid little attention to proposals for structural reform. Most governments were reluctant to offend the farm organisations and their policies continued to emphasise price supports.

Against this unpromising economic background, the European Economic Community sought to initiate its common agricultural policy.

Bibliography

The bibliography for this period is extensive: selection is difficult and inevitably reflects the author's preferences. The most comprehensive sources are the five OEEC agricultural policy reports from 1956 to 1961, which include chapters on each member country; the ECE (Geneva) also carried out pertinent analysis during this period.

For the United Kingdom, the extensive debate on agricultural policy was ably resumed by McCrone (1962). In France, Froment (1950 and 1956) and Bergmann (1957) stand out among contemporary analysts; more recently, Gervais *et al.* (1976) and Chombart de Lauwe (1979) interpreted developments at length and from opposing political viewpoints. The behaviour of farm organisations in France has received particular attention: e.g. Association Française de Science Politique (1958); Barral (1968); Gervais *et al.* (1976); in English, the work by Wright (1964) remains valuable. German preoccupations before the formation of the CAP were reflected particularly by Baade (1963, 2nd edn.); Cecil (1979) has provided a useful survey in English of the post-war period in Germany.

GENERAL: MAINLY DESCRIPTIVE

Economic Commission for Europe (1954) *European Agriculture, a Statement of Problems.* Geneva.

Economic Commission for Europe (1960) *European Agriculture in 1965.* Geneva.

Economic Commission for Europe (1961) *Economic Survey of Europe in 1960* (chapter on agriculture). Geneva.

Food and Agriculture Organisation (annually) *The State of Food and Agriculture.* Rome.

General Agreement on Tariffs and Trade (1958) *Trends in International Trade* (the 'Haberler Report'). Geneva.

Organisation for European Economic Co-operation (1956–61) *Agricultural Policies in Europe and North America* (five annual reports). Paris.

Wheeler, L. A. (Nov., 1960) 'The new agricultural protectionism and its effect on trade policy'. *Journal of Farm Economics.*

Yates, P. L. (1959) *Forty Years of Foreign Trade.* London: Allen & Unwin.

Yates, P. L. (1960) *Food, Land and Manpower in Western Europe.* London: Macmillan.

THEORETICAL: ON AGRICULTURAL INCOMES AND MANPOWER

Bellerby, J. R. (1956) *Agriculture and Industry: Relative Income.* London: Macmillan.

Clark, C. (1957) *The Conditions of Economic Progress.* 3rd edn. London: Macmillan.

International Labour Office (1960) *Why Labour leaves the Land.* Geneva.

Latil, M. (1956) *L'évolution du revenu agricole.* Paris: Colin.

Martin, A. (1959) 'A comment on J. R. Bellerby's explanation of the level of income in agriculture'. *Farm Economist*, **IX** (6).

Ojala, E. M. (1952) *Agriculture and Economic Progress.* London: Oxford University Press.

Robinson, K. L. (1956) 'Political obstacles tending to retard the increased economic welfare offered by technical change in agriculture'. *Proceedings of the Ninth International Conference of Agricultural Economists.* London: Oxford University Press.

Zimmerman, C. C. (Nov., 1932) 'Ernst Engel's law of expenditure for food'. *Quarterly Journal of Economics.*

THE UNITED KINGDOM

Allen, G. (May June, 1959) 'The National Farmers' Union as a pressure-group'. *Contemporary Review.*

Blagburn, C. H. (March, 1950) 'Import-replacement by British agriculture'. *Economic Journal.*

Brembridge, P. and Briggs, E. (1955) *Agriculture and Politics – Party Records and Policies* 3rd edn. London: Conservative Central Office.

Hallett, G. (Sept., 1959) 'The economic position of British agriculture'. *Economic Journal.*

Hammond, R. J. (1954) *Food and Agriculture in Britain, 1939–45: Aspects of Wartime Control.* Stanford, California.

Imperial Chemical Industries Ltd. (1957) *Agriculture in the British Economy.* (Proceedings of a conference held in November 1956: papers by E. A. G. Robinson, E. F. Nash, J. R. Raeburn, R. C. Tress and others.)

McCrone, G. (June, 1958) 'The relevance of the theory of tariffs to agricultural protection'. *Journal of Agricultural Economics.*

McCrone, G. (1962) *The Economics of Subsidising Agriculture.* London: Allen & Unwin.

Menzies-Kitchin (1945) *The Future of British Farming.* London: Pilot Press.

Murray, K. A. H. (1955) *Agriculture.* (History of the Second World War, UK, Civil Series.) London: HMSO.

Nash, E. F. (June, 1955) 'The competitive position of British agriculture'. *Journal of Agricultural Economics.*

Nash, E. F. and Attwood, E. A. (1961) *The Agricultural Policies of Britain and Denmark: A Study in Reciprocal Trade.* London: Land Books.

Raeburn, J. R. (1958) 'Agricultural Production and Marketing'. In Burn, D. *ed. The Structure of British Industry.* Cambridge University Press.

Robinson, E. A. G. Articles in *Three Banks Review* of March 1953, March 1954, June 1958 and December 1958.

Robinson, E. A. G. and Marris, R. (March, 1950) 'The use of home resources to save imports'. *Economic Journal.*

Self, P. and Storing, H. J. (1962) *The State and the Farmer.* London: Allen & Unwin.

Williams, H. T. *ed.* (1960) *Principles for British Agricultural Policy.* London: Oxford University Press.

Winegarten, A. (Dec., 1960) 'Some reflections on the basis of international competition for the British market'. *Journal of Agricultural Economics.*

FRANCE

Association Française de Science Politique (1958) *Les Paysans et la politique dans la France contemporaine.* Paris: Colin.

Barral, P. (1968) *Les agrariens français de Méline à Pisani.* Paris: Colin.

Bergmann, D. (Oct., 1957) 'Les principes directeurs d'une politique agricole française'. *Economie rurale.*

Braibant, M. (1943) *La France, Nation agricole.* Paris.

Braibant, M. (1959) *Vocation agricole de la France.* Paris.

Cépède, M. (1961) *Agriculture et alimentation en France durant la IIe guerre mondiale.* Paris: Génin.

Chombart de Lauwe, J. (1979) *L'aventure agricole de la France de 1945 à nos jours.* Paris: Presses Universitaires Françaises.

Commissions de la Production Agricole et de l'Equipement Rural (1953) *Rapport général au Commissariat Général au Plan* (Deuxième Plan de Modernisation et d'Equipment).

Fauchon, J. (1954) *Economie de l'agriculture française.* Paris: Génin.

Froment, P. (mars, 1950) 'Le plan agricole français d'hier et d'aujourd'hui'. *Revue des sciences économiques.*

Froment, P. (7 août, 1956) 'Les problèmes actuels de l'agriculture française'. *Problèmes économiques.*

Gervais, M., Jollivet, M. et Tavernier, Y. (1976) *La fin de la France paysanne – de 1914 à nos jours.* Paris: Seuil.

Klatzmann, M. J. (oct.–déc., 1959) *Revenus agricoles et non-agricoles.* Études statistiques, INSEE.

Maspétiol, R. (déc., 1953) 'Les options de la politique agricole'. *Revue politique et parlementaire.*

Services Français d'Information (1947) *L'agriculture francaise.* Paris.

Société Française d'Economie Rurale (1959) 'L'economie agricole française, 1938–58'. *Bulletin* 39–40.

Virieu, F.-H. de (27 janv., 1961) 'Vers des surplus permanents dans l'agriculture?' *Le Monde.*

Wright, G. (1964) *Rural Revolution in France – The Peasantry in the Twentieth Century.* Stanford University Press.

GERMANY (FEDERAL REPUBLIC)

Baade, F. (1963) *Die deutsche Landwirtschaft im Gemeinsamen Markt.* 2. Auflage. Baden-Baden: Lutzeyer.

Boerckel, W. (1959) *Einfuhr- und Vorratsstellen als Mittel der Agrarpolitik.* Mainz-am-Rhein: Diemler.

Bundesministerium für Ernährung, Landwirtschaft und Forsten. *Grüner Bericht und Grüner Plan* (annual, from 1956).

Cecil, R. (1979) 'German Agriculture 1870–1970'. In *The Development of Agriculture in Germany and the UK.* Wye College, Ashford, Kent.

Decken, H.v.d. (Juli, 1954) 'Paritätssystem und Agrarstruktur'. *Wirtschaftsdienst.*

Doebel, W. (1953) 'Zwei Jahrzehnte staatlicher Agrarpreisbildung'. *Berichte über Landwirtschaft.* Heft 4.

Magura, W. (1970) 'Chronik der Agrarpolitik und Agrarwirtschaft in der Bundesrepublik Deutschland von 1945–1967'. *Berichte über Landwirtschaft* 185. Sonderheft.

Niklas, W. (1949) *Ernährungswirtschaft und Agrarpolitik.* Bonn.

AUSTRIA

Meihsl, P. (1961) 'Die Landwirtschaft im Wandel der politischen und ökonomischen Faktoren'. In Weber, W. *ed. Österreichs Wirtschaftsstruktur.* Zweiter Band. Berlin: Duncker & Humblot.

DENMARK

Nash, E. F., and Attwood, E. A. (1961) *The Agricultural Policies of Britain and Denmark: A study in Reciprocal Trade.* London: Land Books.

NETHERLANDS

Robinson, A. D. (1961) *Dutch Organised Agriculture in International Politics, 1945–1960.* The Hague: Nijhoff.

SWITZERLAND

Conseil Fédéral (29 déc., 1959) *Second rapport à l'Assemblée fédérale sur la situation de l'agriculture suisse et la politique agricole de la Confédération.* Berne.

Laur, E. (1949) *Swiss Farming.* Bern: Verbandsdruckerei.

Chapter 12

The formation of the EEC's common agricultural policy (CAP)

A Antecedents

The European Economic Community and its common agricultural policy were preceded by other ventures in integration in western Europe, more or less successful, which in various ways influenced subsequent negotiations leading to the EEC.

The Belgium-Luxembourg Economic Union (BLEU)

Already in 1922, Belgium and Luxembourg had formed an Economic Union. All trade between the two countries was freed from import duties and other restrictions, and a common tariff (corresponding generally to the previous Belgian one) was adopted for imports from third countries. The Belgian franc circulated freely in Luxembourg.

The Union was generally successful, but agriculture was its greatest problem. Farming in Luxembourg suffered from unfavourable natural conditions: the land is broken by steep hills and valleys and much of it is of poor quality; the majority of holdings were small and seriously fragmented. Moreover, since 1842 Luxembourg had been within the German tariff system and its agriculture had therefore been sheltered behind the high duties imposed during and after the Great Depression. Belgium in the 1920s had relatively little protection for its agriculture: many products, including bread grains, were free of duty and the rates of duty on other products were fairly low. It thus appeared necessary to give special assistance to agriculture in Luxembourg, and the treaty establishing the Economic Union provided that payments would be made to Luxembourg farmers, out of the customs receipts of the Union, in the amounts necessary to make up the difference between the market price of bread grains and the higher price prevailing on the protected French market (which Luxembourg farmers would have received if their country had formed a union with France instead of Belgium). Luxembourg wine was given advantages on the Belgian market, in part-compensation for the loss of free entry to the much larger German market.

During the crisis of the 1930s, Belgium – as has been seen in chapter 6 – was forced to intervene in agricultural markets in various ways. This action was apparently not sufficient for Luxembourg and further measures were introduced: a Convention of 1935 entitled Luxembourg to exercise control, by means of import licences or minimum prices, over imports of grains, meat and livestock, butter, eggs and other products. Under this arrangement, imports were authorised only in the amounts needed to complement domestic supplies.

After the Second World War, when preparations were made for the wider union of Benelux, incorporating the Netherlands, Luxembourg agriculture was still benefiting from special treatment and seemed no nearer to standing on its own feet than in 1922 when BLEU had been constituted.

Benelux

In Benelux, Belgian agriculture was faced by problems similar to those of Luxembourg agriculture in BLEU, with the difference that the Belgian market was too important to the Dutch for them to let their partners off lightly. While the Netherlands were established exporters of livestock products, fruit and vegetables, with highly organised marketing and efficiently managed holdings, Belgium was predominantly industrial and imported most agricultural products. No constructive agricultural policy had been developed, peasants had been slow to organise themselves, and many farms were too small and too fragmented to be efficient. The situation of Luxembourg agriculture in Benelux was of course even more precarious than it had been in BLEU.

The customs union between the three countries came into force on 1 January 1948: restrictions on internal trade were removed and a common external tariff was applied without any transitional period. As regards agriculture, it was agreed that each country should be free to adopt an agricultural policy which would guarantee to its producers a minimum price covering their costs of production plus a reasonable profit margin. Imports of agricultural products from other Benelux countries, as well as from outside countries, could be restricted so far as necessary to maintain these minimum prices; however, preference should be given to supplies from other members of Benelux over imports from other countries.

The Dutch soon found that these arrangements were being used to justify permanent measures of protection for Belgian agriculture. In 1950, they attempted to revise the existing provisions, but obtained little satisfaction until 1955, when it was agreed that within a year the determination of minimum prices should be subjected to independent arbitration. Further, national agricultural policies were to be 'harmonised' by 1962, so that by then free movement of agricultural produce could be permitted (still with the exception of Luxembourg). It was also agreed that an agricultural fund should be set up in Belgium and Luxembourg to assist the process of adaptation (an 'Equalisation Fund' already existed in the Netherlands).

Meanwhile, negotiations between the partners continued with the object of establishing the full Economic Union, involving the co-ordination of economic, financial and social policies and the free movement of labour and capital. Progress was held up by various difficulties, including the problem of agriculture, but in February 1958 the Treaty of Economic Union was signed. Agricultural organisations in the Netherlands, however, were still acutely dissatisfied and they were not made any happier by the fact that the Treaty was accompanied by a Transitional Convention devoted mainly to agricultural problems. This Convention authorised continued restrictions on intra-Benelux trade; further, minimum prices continued to be determined on the basis of the cost of production plus an appropriate profit margin. Experience had already shown the Dutch that this was not a satisfactory basis: it was difficult to agree on how to calculate costs of production and the arrangement tended to frustrate desirable changes involving greater specialisation and improved efficiency; moreover, the Dutch feared that this provision might form a precedent for the European Economic Community.

The Dutch agricultural organisations therefore opposed the Treaty and succeeded in having its ratification postponed. In February 1960 their government obtained some concessions from Belgium: Dutch agricultural produce would be given preference over that of other countries of the European Economic Community and the effort to harmonise agricultural policies would be intensified. With these assurances, the Dutch Second Chamber finally ratified the Treaty on 16 March 1960. But at the same time it unanimously passed a motion declaring that the system of minimum prices set out in the transitional Agreement was unsatisfactory and urging the government to ensure that the same criteria should not be adopted in the European Economic Community.

The experience of Benelux in agriculture was thus discouraging. While in other sectors a common market was achieved, trade in agriculture continued to be subject to various restrictions. The minimum price system proved much too effective in maintaining protection for Belgian agriculture; behind this barrier Belgian farmers were able to expand their production even at the expense of imports from the Netherlands. Output of butter in particular grew to such an extent that imports practically ceased, and Belgium even had net exports (heavily subsidised) at some periods; imports of fruit and vegetables from the Netherlands were frequently stopped.

The 'Green Pool' proposals

In 1950 the Consultative Assembly of the Council of Europe decided to initiate a study on how the countries of Europe could organise their agricultural markets in common. In the Special Committee set up for this purpose, the initiative was taken by France: this for the double reason that in France at that time 'European' feeling was strong – it had already found expression in the Schuman Plan for coal and steel (the 'Black Pool' – see below) – and that France had a strong interest in widening the market for her agricultural exports. A plan drawn up by the French delegate, René Charpentier, proposed the creation of a High Authority for agriculture with extensive supranational powers: it would control production, fix prices and aim at eliminating all restrictions on agricultural trade between the participating countries. The exports of these countries would be given preference over supplies from other sources, even if the latter were cheaper. 'European' prices would be determined in relation to costs of production and should be independent of supply and demand. The costs of production in different member countries would be harmonised. In a transitional period, there would be 'compensatory taxes', representing the difference between the European price and the domestic price of each member country; these would act to protect countries with high costs of production.

The supranational aspects of the Charpentier Plan gave rise to vigorous debate. A counter-proposal submitted for Britain by David Eccles ostensibly accepted the idea of an 'Authority' but stressed that it should only be intergovernmental and should be allowed merely to make recommendations to governments. Under the Eccles Plan, the 'Authority' would examine national policies for agricultural production and trade, and suggest how they could be reconciled; it would also study how participating countries could co-operate in trading with outside countries.

In spite of this opposition by Britain (supported by Denmark), the Special Committee on Agriculture adopted the Charpentier Plan, and the Consultative Assembly in December 1951 decided to go ahead with the preparation of a draft treaty along the lines indicated in the Plan.

But meanwhile the French Government was seeking more rapid action. In March 1951 it sent a memorandum to other western European countries, proposing immediate negotiations with the object of setting up a European Agricultural Community. This French proposal – known as the Pflimlin Plan after its author, then Minister of Agriculture – was based on essentially the same principles as the Charpentier Plan.

About the same time, the Dutch Government issued another set of proposals in the name of its Minister of Agriculture, Sicco Mansholt. This Mansholt Plan, while it had features in common with the French proposals, stressed the points of greatest interest to Dutch agriculture: it advocated a common market for agricultural products throughout Europe, in order to achieve the maximum efficiency of production through specialisation. An upper limit should be set to the permissible degree of protection, and this limit should be gradually reduced until a common price level was achieved. There would be a High Authority for agriculture, with supranational powers.

These various proposals led to a series of conferences on the Organisation of European Agricultural Markets, which took place in Paris between 1952 and 1954. (They were held independently of the Council of Europe, which subsequently ceased to play an effective role in European agriculture.) Fifteen western European countries were represented. Not surprisingly in view of the attitude already taken by Britain in particular, it proved impossible to reach agreement on the basis of the Pflimlin Plan. On the basic question of arrangements for trade, there was open conflict between those countries which insisted that in an organised agricultural market the member countries should give preference to each other's exports, and those which declared this inconsistent with existing trade policies. There were also marked differences of opinion about the institutional framework, in particular as to whether the proposed agricultural organisation should be attached to the Organisation for European Economic Co-operation or should be an independent and specialised body; there was growing doubt as to the desirability of the sector-by-sector approach which characterised the 'Black Pool'. There was no longer any question of a supranational body: indeed the French Government then in power had moved away from the original intentions of the Pflimlin Plan. The Dutch Government, on the other hand, was discouraged by the prospect of inadequate supranational control, which it regarded as essential if protectionist influences were to be overcome.

Seen as an attempt to set up an agricultural organisation covering all western

Europe, the Green Pool was probably doomed from the start. As a step in the gradual move towards European integration, it served a useful purpose. It represented the first serious attempt to work out the problems involved in unifying agricultural markets; the various studies that were carried out showed clearly what were the difficulties and threw light on the positions of the various countries. France's efforts received support in particular from the German Federal Republic where the desire for closer integration was shared, as well as from Belgium and other countries. Britain, on the other hand, was shown to be deeply suspicious of any arrangement tainted with supranationalism and determined to maintain its preferential arrangements with the Commonwealth. Somewhere between the two extremes was Denmark, anxious above all to preserve the outlets for its agricultural exports and thus seeing advantages both in its trade relationships with Britain and in a unified agricultural market; the Netherlands, with similar preoccupations but a firm belief in supranational principles; and Italy, who too was already allied with France and Germany in the Coal and Steel Community but who was generally willing to make compromises for the sake of a wider agreement.

These various attitudes persisted and influenced the developments which were to follow. While the limited approach favoured by the British found expression in the OEEC's work on agriculture, the attempts to create a preferential European market for agriculture were absorbed in the general movement towards closer economic integration between the six countries of the Coal and Steel Community.

The Organisation for European Economic Co-operation (OEEC)

The activity of the OEEC in agriculture before 1955 was related to its general aims of promoting post-war recovery and freeing intra-European trade. The Sixth Report of the Organisation, in March 1955, was able to record that a substantial increase had taken place in agricultural production. The trade situation was much less satisfactory. The encouragements given to member countries to remove restrictions on trade, with the help of a multilateral clearing system (the European Payments Union, operated by OEEC), had been reasonably successful in sectors other than agriculture. In agriculture, though a fair degree of liberalisation from quantitative restrictions had been attained, a large proportion of trade remained subject to state control. (Import duties were outside OEEC's competence, being subject to GATT.)

OEEC's work in agriculture took a new turn with the institution of the Ministerial Committee for Agriculture and Food in January 1955, following the breakdown of the Green Pool negotiations. As the inheritor of the moderate approach favoured by Britain, the Ministerial Committee had strictly limited functions. It had no powers to coerce governments and it was understood that its decisions, like those of other organs of the OEEC, would only be taken unanimously: any member could thus veto a proposal. The basis of its work was the principle of 'confrontation', a process of examination and questioning designed to bring into evidence those aspects of a country's policy which were likely to harm others. It was hoped that offending countries might be induced by polite persuasion to mend their ways. This was a method which could hardly be expected to resolve the serious problems already existing by 1955. The main outcome of its work was the publication between 1956 and 1961 of the five bulky reports already referred to in the previous chapter. Though it is questionable whether these had any significant effect on agricultural policies, they were the most thorough documentation so far produced on European agricultural policies and served to make clear the basic problems of western Europe agriculture at the time.

B The European Economic Community

The move to political and economic integration

It would be beyond the scope of this work to discuss in detail the political motives which led to the formation of the EEC: they were, however, vital. In the aftermath of the Second World War, *rapprochement* between France and Germany was a priority for many western European statesmen. In France, Jean Monnet, Director-General of the plan, and Robert Schuman, the Foreign Secretary, conceived a plan to make another war materially impossible and in a way which should ultimately lead to European federation: a common market in coal and steel, ensuring equal access by all participants to these strategic products, under the control of an independent High Authority. The plan eased the problem of handing the Saar back to Germany; it offered Germany international respectability once again; it was attractive to federalists everywhere who were dissatisfied with the limitations of intergovernmental co-operation in both the Council of Europe and the OEEC. Konrad Adenauer, Chancellor of the newly-constituted German Federal Republic, gave his support, and on 9 May 1950 Robert Schuman announced the plan, in which other countries were invited to join: the Benelux countries and Italy were quick to do so. The United Kingdom however refused to participate in a Community of a supranational character. The preparatory work was carried out at remarkable speed: the Treaty setting up the European Coal and Steel Community (ECSC) was signed on 18 April 1951, ratified by the six Parliaments, and came into effect the following year.

At the same time, riding on the crest of the wave, European federalists promoted the plan for a European Defence Community (EDC), a treaty for which was signed by the Six in May 1952. The Assembly of the ECSC, transformed into an *ad hoc* Assembly under the chairmanship of the Belgian Foreign Minister Paul-Henri Spaak, drafted a statute for a political authority to control the EDC; a common market was envisaged in this context. But in France, after several changes of government, the mood was no longer so favourable to European integration: in August 1954, the *Assemblée nationale* rejected the EDC scheme.

The European idea had to be relaunched. Jean Monnet, now President of the

High Authority of the ECSC, was active and the Benelux countries wanted to promote further integration. In May 1955 they presented a memorandum to the other member states of the ECSC for discussion by Foreign Ministers at a meeting which took place in Messina in June. The Foreign Ministers then adopted a Resolution (largely based on the Benelux memorandum) in which they declared their intention 'to work for the establishment of a united Europe by the development of common institutions, the progressive fusion of national economies, the creation of a common market and the progressive harmonisation of their social policies'; also to study the creation of a common organisation for developing atomic energy. An intergovernmental Committee was set up to prepare for treaties; the United Kingdom was invited to participate.

The Committee (chaired by Spaak) worked intensively from July to December 1955. British representatives were present, but declared that they were only participating as observers and withdrew when the expert work was completed. The report of the Committee, subsequently referred to as the Spaak Report, was presented in April 1956 and was approved by the Foreign Ministers meeting in Venice in May. It formed the basis of the Treaties setting up the European Economic Community and the European Atomic Energy Community which were signed by the Six in Rome on 25 March 1957 and entered into force, after ratification by the six national parliaments, on 1 January 1958.[1]

The Treaty of Rome provided in the first place for a common market, in which quantitative restrictions and customs duties on trade between member states would be removed and a common external tariff established. The Treaty also incorporated major features of economic union, including free movement of labour, services and capital, and common rules of competition (under which in particular state aids distorting competition could be banned). Member states however kept control of their economic and monetary policies.

The common market was to be progressively established over a period of twelve years, divided into three stages of four years each. Passage from the first to the second stage required a unanimous finding that the objectives laid down for the first stage had been reached: failing this, the first stage could be extended for up to two years (and implicitly, the whole process could then come to a halt if the situation was still unsatisfactory).

Reductions in customs duties between the member states and the move to the common external tariff, as well as the removal of quantitative restrictions on intra-EEC trade, were to take place in a series of steps related to the stages of the transitional period. By the end of the transitional period at the latest, the common market was to be fully established.

The Treaty set up common institutions and a decision-making procedure in which their respective roles were defined. The task of the independent Commission was to prepare proposals for policy, upon which the Council (formed of

[1] Since the merger, in 1966, of the ECSC, the EEC and Euratom, the correct title for the entity is 'The European Communities'. The plural form is however confusing. As this work is concerned with the economic aspects, reference will be made to either 'the EEC' or 'the Community'.

Ministers from the six governments and presided over by the member states in six-monthly rotation) would take decisions; implementation was then the responsibility of the Commission. The Parliamentary Assembly, composed (until direct elections in 1979) of specially designated members of national parliaments, was to be consulted by the Council on most policy decisions: until its view was obtained, the Council could not take a decision – though it was not bound to comply with the Assembly's view.[1] The Economic and Social Committee, consisting of representatives of the various professional groups (including agricultural organisations but also trade unions, industry, etc.) might also be consulted, though this consultation was generally not obligatory (the Treaty however made special provision for the ESC to be consulted on the first proposals for setting up a common agricultural policy). Finally, the Treaty set up the Court of Justice, with ultimate responsibility for interpreting Community legislation.

The Council's decisions on Commission proposals relating to the common agricultural policy were to be taken by unanimous vote during the first two stages of the transitional period. Subsequently, however, they should be taken by 'qualified majority' (the votes of the member states being weighted according roughly to their size of population). Unanimity would however still be required to *amend* a proposal by the Commission.

Agriculture in the Spaak Report and in the Treaty of Rome

The Spaak Committee had found agriculture a difficult subject – not surprisingly in view of the problems encountered in Benelux and in the abortive Green Pool negotiations – and its Report left essential questions open. All the Six, however, accepted that agriculture must be included in the common market: the Spaak Report observed that agriculture was a sector where specialisation could bring important benefits and that its inclusion was necessary to balance trade advantages between the member countries. It was clear that France, the Netherlands and Italy would not agree to open their markets to industrial goods if Germany in particular did not admit their agricultural exports. National measures of agricultural support would then have to be replaced by common regimes. The Committee, however, was unable to agree on the degree of support needed nor on the methods to be finally employed. Conflicts were already apparent on this issue between, in particular, the Netherlands, who as efficient producers wanted as free a market as possible, and the French who insisted on market organisation which would ensure adequate returns and preference over imports from third countries. The French position at this time was not very different from that of Germany: before the devaluations of the French franc in mid-1957 and late 1958 (and the revaluation of the *Deutschemark* in early 1961) French agricultural prices were nearly at the level of German ones, well above Dutch prices and still further above world market prices.

[1] 'Assembly' is the title given in the Treaty: its own preferred title, however, and that used by the Commission from the start, is 'European Parliament'. This designation is therefore used from now on.

The ambiguities of the Spaak Report were reflected in the Treaty provisions relating to agriculture. The inclusion of agriculture in the common market was confirmed: for agricultural products as for others, tariffs on trade between member states were to be removed over the transitional period of twelve to fifteen years, quotas were to be enlarged and finally disappear, and common external tariffs instituted. The development of the common market for agricultural products was to be accompanied by the establishment of a common agricultural policy; by the end of the transitional period, common organisation of markets was to be established. The Treaty however remained vague as to the nature of this organisation – it could consist of common rules of competition or co-ordination of national market organisations as well as of 'European' market organisation – and even vaguer as to the degree of support it was intended to provide.

The much-quoted aims of Article 39 were, like similar statements of objectives in national legislation, open to different interpretations. The central aim of ensuring a 'fair standard of living' for the agricultural community was carefully qualified: besides the reference to the need for 'reasonable prices' to consumers, the drafting implied that the fair standard of living was to be achieved by means of increased agricultural productivity and by increasing the *individual* earnings of persons engaged in agriculture. These provisions suggested a preference for structural measures rather than overall price support, but in subsequent practice not much attention was paid to these nuances.

As regards the transitional period, the Treaty was relatively precise: protection could be given by means of minimum import prices (a heritage from Benelux which the Netherlands had unsuccessfully opposed) and provision was made for long-term contracts for certain products; further, countervailing charges (not limited to the transitional period) could be given to offset distortions of competition. These measures, however, soon gave rise to difficulties and a different approach was finally adopted.

Towards common market regulations

It was thus for the institutions of the Community, and in the first instance the Commission, to formulate a common agricultural policy. Within the Commission, responsibility for agriculture was entrusted to the former Dutch Minister of Agriculture, Sicco Mansholt (who retained this post until 1972). The Commission undertook its work in consultation with governments and with farm organisations. In accordance with a Treaty requirement, it called together a conference of the member states to work out 'the broad lines' of the common agricultural policy, to which professional organisations grouped at EEC level were invited as observers. This Stresa Conference took place in July 1958. A highly co-operative attitude on all sides was reported: nevertheless, basic differences remained and the final resolution only outlined general principles which were carefully balanced and vague. The reports of the working parties at the Conference indicated substantial divergence, in particular between the French delegation which advocated drawing up a balance-sheet of needs and resources in order to determine what part of requirements should be satisfied by domestic production, claiming that the Treaty was based on the principle of mutual preference, and the Dutch delegation which considered that preference was a consequence and not an aim of the Treaty, while the German delegation stressed the importance of developing trade with third countries. At the request of the Italian delegation, some attempt was made to assess production trends, but only Italy itself and France actually provided forecasts. On the whole the Conference steered clear of the thorny problem of surpluses and there seemed to be some tendency to assume that within the EEC this problem would disappear.

The Commission continued its task of preparation. In September 1958 the farm organisations of the Six united in the *Comité des Organisations Professionnelles Agricoles* (COPA), with Edmund Rehwinkel, the President of the *Deutscher Bauernverband,* as its first President. In November 1959 the Commission was ready with its first proposals on which, in accordance with the Treaty, the Economic and Social Committee was consulted: in May 1960 the ESC gave an opinion in which farming interests were balanced against those of industry, trade unions and consumers. In March 1960 the European Parliament held a debate in which the principle of a common agricultural policy was reaffirmed by speakers representing all political groups and economic interests, but conflicts between farming and other interests nevertheless became apparent.

By this time parallel discussions were taking place on accelerating the dismantling of tariffs on inter-EEC trade and the introduction of the common external tariff: these led to a new conflict over agriculture, the Germans insisting that this acceleration should not apply to agriculture, while the Dutch – already dissatisfied with the lack of progress – refused to accept acceleration in other sectors unless agriculture was covered. A compromise was reached whereby the 'acceleration' would become effective on 1 January 1961 with partial application to agricultural products, and in the meantime work on the common agricultural policy should be speeded up. The Commission was invited to submit final proposals and a Special Committee for Agriculture was created to prepare for Council decisions by the end of the year.

The Commission duly submitted revised proposals on 30 June 1960, which were the object of intensive debate in all quarters in the following months, in particular because the Commission now proposed that a first adjustment of prices should be made for cereals and sugar in the 1961/2 agricultural year, involving reductions in Germany, Luxembourg and Italy, and increases in France and the Netherlands. In Germany, Rehwinkel as President of the *Bauernverband* presented a memorandum to Chancellor Adenauer in August, opposing any reduction of German farm prices and anticipating instead a general raising of prices in the future: it was only a question of time, in the *Bauernverband's* view, before French prices rose again towards the German level. The EEC must protect itself against low world prices by means of quotas and compensatory levies; there should be intervention on the domestic market at fixed minimum prices. The views of the farmers apparently prevailed on the government, despite the overall German interest in developing trade: the Minister of Agriculture, Werner Schwarz, declared that Germany could not lower its cereal and sugar prices in 1961/2 because this would mean an excessive reduction in the income of German farmers. The Belgian Government, under pressure from the *Boerenbond,* adopted a similar position. The Dutch *Landbouwschap* was more in sympathy with the Commission's aims, though it criticised some of the means proposed. In France, the FNSEA declared that the overriding goal of the CAP should be to give farmers real economic and social 'parity'; effective protection should be established against imports at artificial world prices by means of adequate import levies and also quotas in the framework of 'supply programmes' (a concept which the Commission had included in its first proposals but now rejected); effective Community preference should be ensured to bring about as high a degree of self-sufficiency as possible; outlets for farm produce should be found not only in the Community but also in underdeveloped countries. ('At a time when a world campaign against hunger is being launched, it would be inadmissible for the Common Market not to exploit rationally all its agricultural potential'.)[1] In Italy the main concern seemed to be with obtaining rapid liberalisation of intra-EEC trade for fruit and vegetables.

In October, the Parliament held a long and heated debate on the basis of a draft Resolution presented by its Agriculture Committee. Two main issues emerged. The Christian Democratic group wanted to restore the idea of controlling imports from third countries on the basis of an annual balance-sheet of supply and requirements; the President of the FNSEA, Blondelle, declared that unregulated imports from a world market dislocated by the policies of major exporters would destroy equilibrium in the Community. The socialists, and Mansholt as representative of the Commission, opposed the part of the draft Resolution which set out these ideas. The other major issue concerned the future Community price level. The Agriculture Committee's draft proposed that harmonisation should take place 'in the light of the price levels existing in the country which is the biggest consumer of agricultural goods in the Community', i.e. Germany. This was strongly supported by the Christian Democratic bloc, whose speakers claimed that high prices did not necessarily affect production and lead to surpluses; it was opposed by the socialists and by the Dutch members, as well as by Mansholt. The Resolution was nevertheless adopted (with an amendment whereby feed grain prices were excluded from the recommendation concerning prices). It was evident that the farm lobby commanded a majority in the Parliament which was opposed to the moderate approach of the Commission.

The power of decision, however, lay with the Council. In the Special Committee and in the Council itself, there was intensive discussion aiming to meet the end-year deadline. In November 1960 the Council approved certain general principles; on 20 December it adopted a more substantive Resolution declaring that a system of import levies could meet the need for a Community instrument to facilitate the transition to the common market stage. Levies on intra-EEC

[1] See *Chambres d'Agriculture, supplément au No 208,* 15 novembre 1960.

trade would be progressively reduced, subject to progress in harmonising prices and eliminating conditions which distorted competition. The relationship between levies on intra-EEC trade and those on imports from third countries should be such as to ensure 'that the member states should enjoy, on the Community market, the advantages foreseen in the Treaty' – the first explicit recognition of the principle of Community preference. Further, the agreement included a vital financial element: while proceeds from levies on intra-EEC trade would remain with the importing member state, member states would be called upon to contribute to the financing of the future common market organisation, either a progressively rising amount of the levies on imports from third countries, or direct budgetary contributions according to a 'fair distribution'.

The import levies were to take precedence over the other measures provided by the Treaty. The Netherlands in particular had become very dissatisfied with Germany's slowness in dismantling its trade restrictions and had been unable to obtain any long-term contracts, while Germany refused to remove its trade barriers so long as the Netherlands benefited from competitive advantages – including cheap feed grains.

The levy system was to apply in the first instance to cereals, sugar, pigmeat, eggs and poultry, and the Commission was asked to submit specific proposals for these products, in time for the levy system to enter into force in the 1961/2 agricultural year.

On this basis, the Council accepted that sufficient progress had been made to authorise the application to agricultural products of the 'acceleration' decision taken the previous May. On 1 January 1961 intra-EEC duties were thus reduced and the first alignment towards the common external tariff was made for industrial products.

Thus 1961 became a further year of preparation and debate. On 31 May the Commission submitted draft regulations for cereals and pigmeat, followed on 31 July by proposals for eggs, poultrymeat, fruit and vegetables, and wine. Pressures to set up common market organisations were intensifying: France, faced with increasing unrest among farmers, made it clear that it would not accept the move to the second stage of the EEC's transitional period – due on 1 January 1962 – unless sufficient progress was made on agriculture; the Netherlands took a similar position. (This conflict prevented agreement in May to implement a further 10% tariff cut that had formed part of the 'acceleration' agreement). The German Government, however, facing elections in September 1961, would make no concessions. Even after these elections, and despite several lengthy and acrimonious Council sessions between October and December, the end of the year was reached without a decision having being taken. Passage to the second stage of the Community's transitional period was threatened. The Council resorted to the device of 'stopping the clock', negotiations resumed on 4 January (still under the Presidency of the French Minister of Agriculture, Edgar Pisani), and finally, after almost continuous session, agreement was reached on 14 January; the move to the second stage of the transitional period was approved (and the relevant legislation backdated to 31 December 1961).

The principles of common market organisation

The 'package' adopted on 14 January 1962 established the method of support for the definitive common market organisation, which the Treaty had left open. In line with the Council resolution of 20 December 1960, import levies played a key role, but as one element in a more extensive system, involving much more intervention than the Commission had proposed. For cereals, target prices were to be established as a basis for determining both prices at which intervention would take place on the domestic market (by means of the various national boards now acting under Community legislation) and 'threshold prices' at the common frontier. Compliance with the threshold prices would be ensured by means of variable import levies, calculated daily by the Commission on the basis of the difference between world market prices and the threshold prices. Exports to third countries would receive export 'refunds' (or 'restitutions') corresponding to the difference between the Community market price and the world price. Import levies (taking the place of all other protective devices) would apply also to intra-EEC trade initially, but would be progressively removed as prices were harmonised: thus market unity and Community preference would be ensured. As the price paid to Germany for accepting import levies as the sole means of protection, a 'safeguard clause' was introduced under which, during the transitional period, a member state could provisionally suspend imports if its market was threatened, subject to approval by the Commission in consultation with member states in the Management Committee set up by the regulation, and with the possibility of appeal to the Council.

The cereals regulation served as the model for several other common market regimes. Those for the three 'cereal-based' products, pigmeat, poultry and eggs, adopted at the same time as the cereals regulation, provided for variable import levies (involving relatively complicated mechanisms) as the main instrument of support; for poultry and eggs no support was provided on the internal market, but for pigmeat support buying was envisaged. For all three products, export refunds were provided. When common market regimes were agreed for dairy products and rice in 1964, and for sugar in 1967 (see below), basically the same methods of price support were adopted as for cereals: target prices as the basis for threshold prices with variable import levies, intervention prices on the domestic market, and export refunds. The January 1962 package included regulations for wine and for fruit and vegetables, but these did not as yet provide for market support. When, finally, market organisation for wine was adopted in 1970, this included intervention arrangements and protection against third country imports; the fruit and vegetable regulation of 1966, however, even after reinforcement in 1968 and 1970, provided relatively flexible support.

The Community had thus embarked on a course leading it to introduce, for the major agricultural products, market intervention coupled with a protective device which effectively insulated the domestic market from the world market. The question remained as to what degree of support and protection should be aimed at. The Dutch and the French had tried to get a commitment, if not on the actual price level, at least on a timetable and on criteria for determining the level. But the Germans were adamant, and all that was accomplished in January 1962 was a commitment to freeze prices at their current level, to fix criteria for harmonisation before autumn 1962, and to begin actual harmonisation in 1963.

The other essential element of the January 1962 agreements concerned the financial provisions. The Council adopted a regulation instituting a European Agricultural Guidance and Guarantee Fund (more commonly known under its French initials, FEOGA), to finance export refunds, market intervention and structural measures. Community financing would be introduced progressively. The FEOGA contribution to the eligible expenses of market organisation would rise from one-sixth in 1962/3 to three-sixths in 1964/5. Subsequently, it would rise further until, by the end of the transitional period, all eligible expenditure would be financed by FEOGA; precise rules for this latter part of the transitional period, however, would be laid down later. Structural expenditure would 'as far as possible' constitute one-third of Community expenditure in favour of markets – an aim which proved impossible to realise.

The arrangements for financing the fund were highly controversial. The Commission had proposed that receipts from levies on imports from third countries should go to the fund. This would have put the burden on the Germans and the Dutch, who had large agricultural imports from the outside world. The conflict of interests on this point had already prevented a clear decision in the Council Resolution of December 1960; now a further compromise was adopted. Receipts from levies on third countries would go to the Community budget after the transitional period; in the transitional period, however, FEOGA would be financed *partly* by direct contributions from member states according to the Community's general budgetary 'key' (i.e. 28% each from France, Germany and Italy, 7.9% from Belgium and the Netherlands, 0.2% from Luxembourg), and *partly* by contributions based on net imports from third countries. Over the first three years, the direct contributions would fall from 100% to 80% of the total sum required, while the contributions based on imports would rise from 0% to 20%. Before the end of the third year (i.e. 1964/5), the Council would re-examine the situation and decide on arrangements up to the end of the transitional period.

Agreement was therefore not complete: difficult negotiations still lay ahead on the market regimes for the remaining products; on price harmonisation; on the financial regulations after 1964/5; and on structural policy. Nevertheless the outcome of the 'marathon' which ended on 14 January 1962 was rightly hailed as a major achievement in the development of the Community. The diverse means of support which the various countries had introduced in previous years could now be welded into a single set of instruments. Further, the principle of market unity, already implicit in the Treaty's inclusion of agriculture in the common market, was re-affirmed in the preambles to the market regulations, as was the principle of mutual preference, to be achieved through the operation of the import levies. The financial regulation, although incomplete in important respects, affirmed the principle of Community financial responsibility in the following terms (Article 2.2): 'Since at the single market stage price systems will be standardised and agricultural policy will be on a Community basis, the financial consequences thereof shall devolve upon the Community.'[1]

Thus the three principles which were constantly evoked in subsequent discussions – market unity, Community preference and financial solidarity – were embodied in these regulations, though not initially stated in these terms in any Council text. The significance which the French in particular attached to these three principles was reflected in articles written immediately after the January 1962 agreement in the *Revue du Marché Commun* (published in Paris) by the Minister of Agriculture, Edgar Pisani, by one of the chief French negotiators, Jacques Mayoux, and in a further article in November that year by an anonymous – but clearly highly-placed – author. The latter wrote:

> It is true that in the future the Common Market may no longer have such large import deficits; overall surpluses may even appear for sugar and milk. But the Community, whose agricultural policy would by then be defined by common decision, would quite naturally have to bear the consequences of that common definition; a common policy calls for common financing. [Anonymous (1962) page 426]

Consolidation of the CAP – further market organisation and price harmonisation

The first part of 1962 was a period of preparation for the implementation of the decisions reached on 14 January. Numerous Council and Commission regulations were adopted and on 30 July the common market organisations for cereals and the cereal-based products came into force. Regular decisions (in some cases daily) were now taken by the Commission on the determination of import levies, export refunds, etc. In such matters an important role was played from then on by the Management Committees consisting of representatives of member states and chaired by the Commission. The basic regulations specified the matters on which the Commission was required to consult the Management Committees; if the opinion of the Committee (expressed by qualified majority) was contrary to the Commission's intention, the latter was still entitled to take the action it proposed but was obliged to submit it to the Council, which could take a different decision (by qualified majority). As a qualified majority opposed to the Commission's intention was rarely found in the Management Committees, recourse to the Council proved rare.

This far-reaching transition worked smoothly, on the whole. In the first couple of years, the safeguard clause was invoked only on three occasions to suspend imports: once by Germany and France for certain varieties of apples, a second time by Germany for eggs, and on a third occasion by Germany for maize: in the first two cases the Commission's decisions to limit the duration of these measures or to suspend them were upheld by the Council, confirming

[1] Regulation No.25 on the financing of the common agricultural policy. This and other early regulations were republished in English in a special edition of the *Official Journal of the European Communities* in November 1972.

that recourse to the safeguard clause could be justified only in exceptional circumstances.

In the mood of optimism engendered by the progress on agriculture, the Council was able, on 15 May 1962, to take a further decision on 'acceleration', resulting in further cuts in intra-EEC tariffs on 1 July; this brought the total cut in duties on manufactures to 50% of the basic rates, while for agricultural products the total cuts amounted to 30–35%.

The cereals regulation of January 1961 had required that common prices should be fixed in time for the 1963/4 agricultural year. In November 1962 the Commission put forward criteria for price determination, which were a carefully balanced assortment of conflicting considerations and avoided any specific commitment. The stress was on the need to maintain farm incomes, but this aim was qualified: an adequate income, i.e. corresponding to that obtained in other occupations, should be sought for persons working on rationally-managed and economically viable farms. The role of prices in guiding production was emphasised, and there was discreet reference to the importance of non-price measures in supporting incomes.

But the year 1963 started in an atmosphere of crisis, provoked by General de Gaulle's veto on the UK's application to join the Community. The fact that this was a unilateral declaration, made outside the framework of the Community institutions, particularly upset the other member states and, with hindsight, can be seen as a warning of further trouble ahead. The crisis was nevertheless short-lived. In April and May, the Council was able to agree to go forward on various fronts. It would take decisions on common cereals prices by 1 July and on market organisation for the main agricultural products still outstanding by the end of the year. This progress in agriculture seemed sufficient for France to agree to a preliminary opening position for the Community in the 'Kennedy Round' of trade negotiations in GATT. The Council accepted a second 30% move towards the common external tariff on industrial products and a further 10% cut in intra-EEC duties on both industrial and agricultural products, which took place on 1 July.

Progress in the Community agricultural negotiations, however, remained slow, particularly on the issue of cereals prices. For the 1963/4 cereals-marketing year, only a small first alignment could be effected, consisting mainly of raising the lower price limits. Pressure to complete the common agricultural policy was accentuated by a warning issued in July by General de Gaulle that the Community would 'disappear' if the deadlines for settling the outstanding issues were not met. In November, Mansholt presented a new plan aiming to cut through the difficulties by harmonising cereals prices in a single step in the 1964/5 marketing year, subject to transitional compensation for producers in those member states where prices would fall (Germany, Italy and Luxembourg).

In December 1963, after another 'marathon' session, the Council reached agreement on market regulations for dairy products, beef and rice (not however on sugar). The market organisation for dairy products was, as has already been pointed out, along similar lines to that for cereals, though inevitably rather more complicated. The target price for milk would be implemented through intervention for butter and skim milk powder, and through threshold prices (with variable levies) and export refunds for the various dairy products. For rice the regime was essentially the same as for cereals. Arrangements for beef and veal were initially rather different, as this commodity was in short supply and expected to remain so: variable import levies (on top of customs duties) and export refunds were provided, and member states were authorised to intervene on their domestic markets if prices should fall; but provision for 'permanent' intervention was not introduced until 1972 in order to stimulate production, and intervention did not become significant until 1974 when, contrary to expectations, the beef market became oversupplied.

Still, however, these market regulations provided only for methods of support: the central issue of price levels remained unsolved, and the large differences in prices between the Six (see figure 12.1) made this an intractable problem. The sensitivity of the issue in Germany had been underlined when in 1962 the Commission published a study of the effects on farm incomes in Germany of a reduction in agricultural prices, carried out by a panel of eight professors (four from Germany and four from other member states). This study worked out the rate at which, to maintain the growth in farm incomes, the outflow of labour from agriculture would have to be accelerated on the assumption of reduced prices. This so-called *Professorengutachten* became a *cause célèbre*, being attacked by the *Bauernverband* as threatening the existence of 'hundreds of thousands' of German farmers and their families; in demonstrations in Göttingen, where the department of two of the professors (Hanau and Woermann) was located, the professors were personally denounced as 'grave-diggers of the peasantry' (cf. Hanau, 1971).

The German Government, moreover, was facing elections in the autumn of 1965 and was unwilling to have any cut in farm prices before that date. By May 1964 it was clear that the introduction of common prices would have to be further postponed. The most that could be decided was to fix upper and lower limits for national cereals prices in 1964/5, a decision on alignment was deferred until December 1964, and the Commission put back its target date for full harmonisation to 1966/7.

Discussions resumed in October 1964, with a new Commission paper, but Germany remained adamant. De Gaulle issued another warning to the effect that France would 'cease to participate' in the EEC if the common market for agriculture was not organised. The Community was faced with other difficult issues. The 'Kennedy Round' had been formally opened under the GATT on 4 May 1964 and the United States were pressing the Community for progress. This issue became linked with the cereal price problem, since the French were unwilling to make any moves over GATT until the Community had fixed common cereals prices. The Germans were also being urged to get the cereals prices issue out of the way in order to facilitate an agreement with the US over the proposed multilateral force (MLF), to which France was strongly opposed. Further, it was clear that France would refuse any further discussion on moves to political

Figure 12.1: Soft wheat: prices to producers, EEC support prices and world price, in units of account per 100 kg

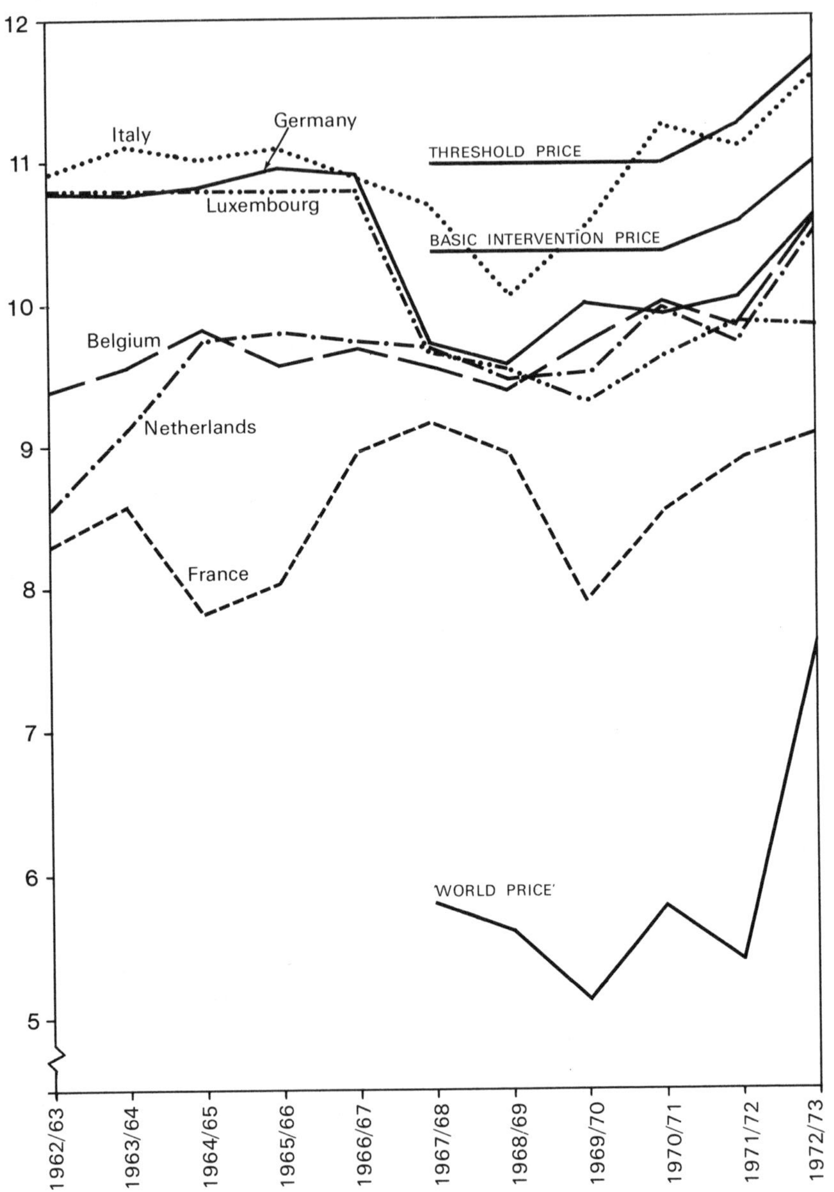

Sources: Eurostat, *Agricultural Prices* and *Yearbook of Agricultural Statistics* (various issues).

union until the main agricultural issues had been settled to French satisfaction.

At the end of November 1964 Chancellor Erhard was able to agree with the *Bauernverband* that Germany could offer to align the target price for wheat from 1967/8 onwards at a level of 440–50 DM per ton (as compared with the German target price of 475 DM), subject to financial compensation for German farmers. On this basis there was intensive debate in the Council from 10 to 15 December. Finally, the German delegation (led at the crucial point by the Minister of Economics in the temporary absence of the Minister of Agriculture, Werner Schwarz) accepted a price of only 425 DM (106.25 UA (units of account, see note on p.402)). Germany however obtained satisfaction with its request to put off the date of introduction until 1 July 1967. Further, the prices of barley, maize and rye were fixed relatively close to the wheat price – a decision which had consequences for the price structure of the cereal-based livestock products. As a special concession to Germany (and Luxembourg) where rye is grown mainly as a bread grain, the intervention price could be increased by 2.50 UA for rye for human consumption. Compensation for price reductions would be given to farmers in Germany, Italy and Luxembourg. These payments would be made on a degressive scale in 1967/8, 1968/9 and 1969/70, and totalled 280.25 million UA to Germany, 131.00 million to Italy and 2.50 million to Luxembourg. They would be paid from a special section of FEOGA and financed according to the regular Community budgetary 'key'.

In the negotiations leading to this agreement, special problems were raised by Italy, who was beginning to find that the common market was not operating to its advantage: the growth of consumer income and demand in Italy was causing increased imports of foodstuffs, while the market for Italy's main agricultural exports within the EEC had not yet come under common organisation. The Council accepted in consequence that for fruit and vegetables protection would be reinforced and that from 1 January 1966 surpluses would become eligible for Community financing; for durum wheat (only grown in the south of Italy,

Notes to figure 12.1

This diagram illustrates several of the points discussed in the chapter. From 1962/3 support prices were not allowed to move further apart: common prices entered into force on 1 July 1967, necessitating reductions in Germany, Italy and Luxembourg. Subsequently, producer prices remained fairly close together and near the 'derived' intervention prices (see below), except in Italy which was a deficit area, and in France where devaluation in 1969 reduced the price as expressed in units of account (UA).

The threshold price and the basic intervention price, both derived from the target price, were raised by monthly increments in each marketing year: the price levels shown here represent the averages over each year. The target price and the basic intervention price related to Duisburg in Germany, chosen as a major consuming area: in each producing area, 'derived' intervention prices were applied (the lowest – at Tours in France – was about 7% below the basic intervention price).

The 'world price' shown is the offer price at the common frontier, as calculated by the Commission, before payment of levy.

The unit of account was originally defined as an amount of gold equal to the value of the US dollar: its fixed relationship to the dollar ended when the latter's convertibility into gold at a fixed rate was suspended in August 1971. See also note on page 402.

Sardinia and Sicily) special arrangements were adopted, involving direct aids to make up the difference between the market price and a guaranteed minimum price; and the Council accepted that Italy's contribution to FEOGA would be limited to 18% in 1965/6 and to 22% in 1966/7, instead of being 28% as the budgetary 'key' would have indicated.

This package agreement of December 1964 once again raised hopes for the overall progress of the Community. Though the package was chiefly concerned with cereals prices, it was reasonable to hope that common prices for other farm products could also be agreed in time to enter into force on 1 July 1967 and the chances of accelerating the customs union in general were significantly improved. It was also widely believed – too optimistically as it afterwards appeared – that the institution of common farm prices would give momentum to co-ordinated action in the monetary field, as alterations in exchange rates between the member states would be inconsistent with common agricultural prices. It was also felt by the European-minded, who had already been shaken by de Gaulle's veto on UK membership and by his subsequent pronouncements, that France's stake in the well-being of the Community was now sufficient for further upsets to be avoided (events in 1965 were to show that this too was over-optimistic). In general, there appeared to be grounds for asserting that the common agricultural policy was 'the pillar of European integration'.

From the 'empty chair' to completion of the CAP

Once again, the Community entered the new year, 1965, in a mood of optimism. Discussions which had been going on for some time concerning the merger of EEC, ECSC and Euratom finally led to the signing of a Treaty to this effect on 8 April: this however was subject to ratification by the six national Parliaments. But the Community faced its most critical problem yet over the issue of financing the CAP, which led into a general conflict over the powers of the Community institutions and indeed over the very nature of the Community.

According to the financial regulation adopted in January 1962, a further decision was required by 30 June 1965. Moreover in the context of the agreements of December 1964, France had demanded that final agreement on all the remaining agricultural issues, including the financial rules, should be reached by 30 June (i.e. during the French presidency). On the financial question, the Commission put forward far-reaching new proposals, involving the creation of 'own resources' for the Community to replace financial contributions by member states. Besides agricultural import levies, customs duties should also become Community revenue once the common external tariff was introduced – in principle on 1 July 1967. The Commission pointed out that this step had been envisaged by the Treaty, and that the logic of a customs union, where duties are levied only at the external frontiers irrespective of the final destination of the goods, required that the receipts should be treated as Community revenue. At the same time, the Commission considered that the powers of the Parliament over the budget should be increased, and made proposals to this effect.

Although the French Government wanted a decision on the arrangements for agricultural finance, the concept of strengthening the powers of Community institutions was totally in contradiction with de Gaulle's policy. The French Government was particularly incensed by the fact that the President of the Commission, Walter Hallstein, disclosed the Commission's plans to the Parliament before formally submitting them to the Council. At its May session the Parliament adopted a resolution approving the Commission's proposals. The conflict came to a head in June. At a first Council session on 14–15 June, the French Foreign Minister, Couve de Murville, changed the French tactics, stating that agricultural financing could continue to be based on transitional arrangements until 1970 and that the revenue from import levies need not become Community property until then, thus deferring the question of Parliamentary control. Debate continued in a further Council session starting on 28 June until, at 2 a.m. on 1 July, Couve de Murville (still in the chair) declared that the deadline had not been met and that France would have to draw the necessary conclusions. Other delegations and the Commission felt that agreement could still have been reached. The French Government however announced that until further notice it would not be represented at meetings of the Council or its dependent bodies. (French representatives did however continue to participate in the Management Committees and some other technical groups, and the introduction of 'written procedures' enabled routine Council decisions to be taken.)

The nature of the conflict became clear with a press conference given by de Gaulle on 9 September. His real concern was about the way in which the Community institutions functioned, especially with regard to majority voting in the Council, and relations between the Council and the Commission.

To facilitate a solution, the Commission had already put forward on 22 July a memorandum in which it largely met the French objections over financing, suggesting in particular that the date for the Community to acquire its own resources could be deferred until 1970. The 'empty chair' however persisted throughout the rest of 1965, until in December France let it be known that it was willing to participate in a meeting of Foreign Ministers in Luxembourg. This 'extraordinary' session of the Council in Luxembourg took place on 17–18 January 1966 and resumed on 28–29 January, when an agreement was reached. This included a relatively mild statement on the relations between the Commission and the Council (the Commission was to establish 'appropriate contacts' with the governments of member states before adopting any particularly important proposal, and should not make proposals public before the Council was in possession of the texts). The more significant statement related to majority voting. (It will be recalled that as regards agriculture the Treaty provided for qualified majority voting from the third stage of the transitional period, i.e. from 1 January 1966.) It was now declared that where 'very important interests' were at stake, the Council would endeavour to reach solutions which could be adopted by all members of the Council. The French delegation added a declaration of its own stating that where very important interests were at stake, discussions must continue until unanimous agreement was reached; as the others did not accept this view, the Council declaration noted that there was a 'divergence of views on what should be done in the event of failure to reach complete agreement'.

This 'Luxembourg compromise' involving not a modification of the Treaty of Rome but an understanding that a vital feature of the Treaty would not be applied, meant a fundamental change in the future character of the Community. No doubt the supranational ambitions of the founders of the Community, though they had provided the necessary initial impulse, had been too ambitious: in the end, it would probably have proved impossible to impose a majority decision contrary to what a member state regarded as its vital interests. Thus de Gaulle anticipated a development which was likely to occur in any case. But now the scene was set for many further arduous Council marathons, in particular on agricultural prices, in the attempt to reach unanimous agreement. The influence of the Commission, moreover, was reduced more by the restriction on majority voting than by the other parts of the Council statement. By providing that the Council could adopt Commission proposals by qualified majority but amend them only by unanimity, the Treaty had given a pre-eminence to Commission proposals which was now significantly impaired: henceforth, the Commission representative in the Council would increasingly have to play the role of 'honest broker', being frequently compelled to amend Commission proposals in order to enable unanimity to be reached.

The agreement in Luxembourg however permitted work to resume, including that on the remaining aspects of the CAP. At successive sessions in 1966 the Council reached a number of important decisions. These provided for all customs duties on industrial goods on intra-EEC trade to be abolished by 1 July 1968, and for the common external tariff to be introduced on the same date. New market organisations were adopted for vegetable oils and oilseeds (of particular interest to Italy, and – as for durum wheat – involving direct aids); supplementary provisions strengthening the market support arrangements for fruit and vegetables were agreed (also in favour particularly of Italy); and principles for organising the sugar market were laid down. Common prices were agreed for milk, beef and veal, and sugar (entry into force in 1968), for rice and oilseeds (entry into force in 1967) and for olive oil (entry into force November 1966). Agreement was reached on the financial arrangements for the remainder of the transitional period, under which eligible expenditure for market support would be completely financed by FEOGA from 1 July 1967.

One of the most difficult commodities was sugar. The Commission had proposed free competition within the Community market, which would have promoted a shift of production from high-cost areas (notably Italy but also Germany) to lower-cost areas in northern France and Belgium. The Council however finally decided on a quota system: the full price guarantee (provided by arrangements similar to those for cereals, including a minimum price for sugarbeet) was available only for certain basic quotas (totalling 6 594 000 tons for the whole Community, somewhat greater than the current level of consumption). Quotas were fixed for each member state and allocated to sugar factories: on quantities in excess of their 'basic' quotas (or 'A-quotas') but within maximum quotas, factories had to pay a levy (this part of output being referred to as 'B-quotas'). Beyond the maxima, sugar could only be sold outside the Community, without support ('C-quotas'). This regime, including common prices, came into force on 1 July 1968 (for the previous year there had been a provisional common market organisation with different national prices): it was to apply until 1975.

Thus by 1 July 1968, when the Community achieved full customs union, the single market stage had been reached for practically all agricultural products. The main exception was wine, another commodity where the need to avoid surplus production made agreement difficult, particularly in view of conflicts of interest between the main producing countries. It was not till April 1970 that market support arrangements were finally adopted, permitting intra-Community trade to be liberalised with effect from 1 June the following year and establishing a single price system. A guide price was to be fixed each year for each type of table wine, as a basis for 'activating' prices to trigger off intervention, in the form of subsidised private storage or distillation. Imports were subjected not to levies but to customs duties and to a system of reference prices: if imports from third countries did not respect the reference prices, 'countervailing duties' could be charged. Measures were also laid down to control new plantings. A common market regulation for tobacco was also adopted in April 1970.

The Community completed its transitional period on schedule at the end of 1969, its twelfth year. A Conference of Heads of State and Government in The Hague on 1–2 December 1969 confirmed the willingness of the member states to advance towards economic and monetary union, and also to resume efforts to reach agreement on the enlargement of the Community. The Hague Conference also produced agreements of principle on the issues which had been so divisive in 1965 (Georges Pompidou had replaced de Gaulle as President of France earlier in 1969): the creation of Community 'own resources' and increases in the Parliament's budgetary powers. Later in December the Council agreed to a system giving effect to these agreements (formally adopted on 21 April 1970). From 1 January 1971, member states' contributions would be replaced by the Community's 'own resources' consisting of agricultural levies and, step by step of customs duties. To these would be added (it was hoped, from 1975 onwards) revenue corresponding to at most one percentage point of the value added tax which by then should be levied on a uniform basis of assessment throughout the Community. Member states' expenditure on agricultural market support would be directly financed by the Community budget (the FEOGA had previously acted rather as a clearing-house after expenditure had been incurred). The respective responsibilities of the Council and the Parliament in adopting the annual budget were defined. This 'financial decision' was submitted for ratification to the six national parliaments, acquiring thereby the same status as the Treaty of Rome.[1]

[1] Decision of 21 April 1970 on the replacement of financial contributions from member states by the Communities' own resources. *Official Journal of the European Communities*, No. L94, 28 April 1970.

Emerging problems

While the Community had been painfully evolving its common agricultural policy, agriculture itself had not stood still. Technological progress had been maintained, output had continued to rise in spite of the reduction of the farm labour force, and the growth rate of production had been higher than that of demand. By 1968 the Community was well above self-sufficiency in several commodities, in particular milk and milk products, sugar and wheat; apples, peaches and tomatoes were periodically oversupplied. Expenditure by FEOGA on market support was estimated to reach 2000 million UA in 1968/9. Farm structures, however, remained unsatisfactory. These factors motivated the Commission to produce in December 1968 its 'Memorandum on the Reform of Agriculture' which became generally known as the 'Mansholt Plan'.

The Mansholt Plan provoked vigorous and sometimes violent controversies: these and the content of the Plan will be further discussed in chapter 14. The debate became linked with price policy. In December 1970, the Council debated principles for the CAP in the light of the Plan. Prices for the 1971/2 marketing year could not be adopted till 25 May 1971, when the Council passed a resolution accepting certain guidelines for 'socio-structural' policy. These guidelines, aiming at the creation of farms able to provide incomes comparable with those earned in other sectors, were taken into account by the Commission in preparing its price proposals for 1972/3. The Commission considered that greater objectivity in price-fixing could thus be achieved (i.e. that smaller price increases would be required if viable farms were taken as the basis). The average price increase proposed for 1972/3 was small – 2–3%. Though the Commission submitted these proposals in June 1971, the Council did not reach a decision until 24 March 1972, shortly before the new marketing year for milk and beef. The delay was due not only to the link with the structural measures, adopted at the same time, but also to the confused monetary situation.

Already in 1969 a new threat had arisen to the barely-established common agricultural market. Throughout most of the 1960s, currency stability had lulled most of the negotiators into a false sense of security: it had been possible to fix common prices on the assumption that the relationships between the unit of account and the national currencies would remain unchanged. (As has been mentioned, it was even hoped that the existence of common farm prices would deter member states from altering their currency parities.) But in 1969, inflationary strains increased throughout the world, and in France a high rate of inflation led to a heavy deficit on current account: in August the government decided to devalue the franc by 11.11%. Strict application of the rules would have meant for France a corresponding increase in the prices of farm products covered by common market organisations: this would have run counter to the French Government's attempts to combat inflation, while from the Community's point of view it seemed undesirable to give further encouragement to farm production in France when it was already necessary to grapple with surplus production. It was agreed that most French farm prices would stay temporarily at their previous level in terms of francs, but to avoid trade distortions it was necessary to introduce compensatory amounts on trade – levied on exports and granted as subsidies on imports. In October 1969 Germany revalued the mark by 9.29%, and corresponding action had to be taken. Compensatory amounts were introduced as levies on imports (subsidies on exports to non-member countries could be granted in exceptional cases), but the single price system was to be reapplied in Germany from 1 January 1970. To compensate German farmers for the reduction in their receipts which this would involve, they could be paid subsidies (partly financed by the Community) from 1970 through 1973.

During 1971, French farm prices regained the common price level and the compensatory amounts introduced in 1969 were abolished. But new and more general problems arose. In a context of world-wide monetary instability, Germany and the Netherlands in May widened the margins of fluctuations of their currencies; after the US had suspended the convertibility of the dollar in August, the BLEU and Italy followed suit. Compensatory amounts had to be generalised. In December the Agreement reached in Washington by the 'Group of Ten' to limit fluctuations to 2.25% on either side of the official parities (the 'tunnel') restored some order to the situation. This however still meant that the spread between any two Community currencies could be as much as 4.5% at any given moment, or that over a period the relationship between them could vary by up to 9%. Consequently, in March 1972, the member states agreed to limit fluctuations of their currencies against each other to a maximum spread of 2.25% at any given moment (the 'snake in the tunnel'). For agriculture, this meant that compensatory amounts could be kept stable so long as central rates were unchanged. But the compensatory amounts introduced in 1971 remained and took on a less and less provisional aspect as time went on: only five years after a single agricultural market had been created, it was again divided into separate zones – Germany, the Benelux countries, France and Italy – between which support prices differed (when compared at market exchange rates), so that monetary compensatory amounts acting as taxes or subsidies at the frontiers between these zones (and at the common frontier) were inevitable.

The state of agriculture in the Community of Six on the eve of enlargement

The years from the Messina Conference in 1955 to the first enlargement of the Community in 1973 (the Act of Accession was signed with Denmark, Ireland and the United Kingdom on 22 January 1972) were crowded with events. Bearing in mind the diversity of national systems of agricultural support and protection outlined in earlier chapters of this book, and the extent of the adjustments required to create a single level, the introduction of common market regimes in the Community of Six was a remarkable achievement, possible only in the context of a strong political commitment. Those who participated in the early years of the Community recall the enthusiasm and sense of construction which accompanied the decisions introducing, step by step, the common agricultural policy. These decisions built up the *'acquis communautaire'* which the three new member states were required to accept, subject only to transitional derogations.

It could be asked – and was asked particularly by British critics – why the Six adopted a system which effectively insulated their domestic market from world influences and deprived consumers of access to cheaper supplies. It will however be clear from the account of earlier developments given in this book that the system introduced was fully in the tradition of support systems in the major countries of the Six. Intervention had been practised in France, for instance, since the introduction of the ONIB in 1936, while the whole system clearly owed much to the post-war German *Einfuhr- und Vorratsstellen* and in turn to the Nazi *Reichsstellen* – perhaps the ancestry can even be traced back to the Kanitz grain monopoly proposal of the 1890s. Conceivably, a system of fixed levies might have been used instead of variable ones, so that fluctuations in world prices would to some extent have been reflected on the domestic market (it has been pointed out in the previous chapter that Sweden had experimented with such a system, not very successfully), but this would probably not have met the needs of the time. There was very strong pressure for effective protection against world market prices, regarded as artificial and influenced by export subsidies given by the United States in particular. Indeed, it was fortunate for world trade that the Commission withdrew its original proposal to regulate imports quantitatively on the basis of annual balance-sheets of availabilities, although this scheme had been advocated by the majority in the European Parliament. It would have involved much more *dirigisme* and imports from third countries would have become a residual.

A system of deficiency payments, along British lines, would have been contradictory to traditions in the Six and would have stood no chance of acceptance as the basic instrument of support. (A group of economists within the Commission, led by the German commissioner Hans von der Groeben, did advocate deficiency payments, but the majority of the Commission decided to propose import levies.) As has been seen, deficiency payments (though this was not their official title) were in fact adopted for durum wheat, olive oil and oilseeds, products whose output was localised and small in relation to total requirements. But for the major products, in which the Community was nearly self-sufficient (or more than self-sufficient), the budgetary cost would have been onerous; and though economists can show that it is more equitable to finance agricultural support through progressive direct taxation than through market prices, such a system would have raised at a premature stage the controversial problem of how to finance the Community budget. Import levies, on the other hand, would provide revenue; and it was generally expected that this would more than cover expenditure.

The only country among the Six which initially had reservations about the system of market organisation was the Netherlands, which would have preferred a freer market. But the first few years had demonstrated the difficulty of getting Germany in particular to dismantle existing trade barriers: the Netherlands thus came to have a strong interest in a single instrument of protection which would replace all others.

Any system of agricultural support, however, can be more or less protective depending on the price level it provides. Because of the difficulty of reaching agreement on cereals prices, the support arrangements were in fact decided upon before the price level. Pressure from German farm interests in particular ultimately caused the price structure for wheat and feed grains, and hence for agricultural produce in general, to be fixed somewhat above the average of prices in the Six, and well above world prices. Although for the first few years common prices were kept unchanged, there was an incentive to increased production by low-cost producers throughout the Community which soon led to serious difficulties. The risks of overproduction were in fact underestimated. In 1963 (before price levels were decided) the Commission produced projections to 1970: these were in themselves quite carefully qualified – they assumed constant prices and pointed out that price increases could change the outcome – besides stressing the need to stabilise cow numbers (*Etudes* No. 10). However, these projections were announced in the Commission's *Newsletter on the Common Agricultural Policy* as heralding 'the great future awaiting agriculture in the EEC', with only discreet reference to the problems that might arise if self-sufficiency were reached.

In the first instance, however, the common agricultural market gave a decided boost to intra-EEC trade (see Table 12.1). While imports from third countries continued to rise – including those of products subject to market regulations – trade between the Six rose very much faster, particularly in the products under market regulation. Also, exports by the Six to the rest of the world – aided by export refunds in the case of regulated products – rose somewhat faster than imports from third countries.

This may be regarded as the normal consequence of a customs union. Inevitably, the introduction of common market mechanisms provoked frictions with third countries, as in the case of poultrymeat when the implementation of the

Table 12.1: Trade developments in the Community of Six

	1963	1972	1972 as % of 1963
	*million units of account**		
Intra-EEC trade			
All agricultural products	2 497	9 427	378
Products subject to market regulations	1 533	6 654	434
Imports from third countries			
All agricultural products	9 438	13 993	148
Products subject to market regulations	4 557	6 818	150
Exports to third countries			
All agricultural products	2 449	4 668	191
Products subject to market regulations	1 539	3 062	199

* In 1963, equal to the US dollar; in 1972, equal to 1.086 dollars.

Source: Commission, *La Situation de l'Agriculture dans la Communauté – Rapport 1973.*

levy system together with the growth in production provoked a 'poultry war' with the United States in 1963, in which the US threatened tariff retaliation against Community exports. In general, however, third countries including the US were able to expand their exports to the Community, even of regulated products; Community imports of animal feedingstuffs, not subject to levy, increased rapidly. The most obvious sufferer among third countries was Denmark, whose exports of livestock products to the German market stagnated while competing Dutch exports developed rapidly. Belgium too increased its exports of livestock products – particularly pigmeat and eggs – to other EEC countries.

EEC prices for agricultural products, however, remained well above world prices in practically all cases. Even allowing for the fact that the 'third-country offer prices' shown in Table 12.2 may understate world price levels and that world prices would have been higher if the EEC had imported substantially more, one could question whether the Treaty aim relating to reasonable prices to consumers was being fulfilled. But already under national policies – with the possible exception of the Netherlands – prices had been divorced from world markets. In the years after the establishment of common market regimes, food prices to consumers rose in France, the Benelux countries and Italy, but roughly in line with the overall increase in consumer prices. In the Netherlands and in France the increase was presumably due at least in part to the CAP. In Germany, food prices rose much less than other prices.

Table 12.2: Prices of major agricultural commodities in 1972/3*

	A EEC entry price†	B Third-country offer price†	C A as % of B
	*units of account*** per 100 kg*		
Soft wheat	11.7	7.7	153
Barley	10.6	7.7	137
Maize	10.3	7.2	143
Sugar	24.5	19.3	127
Beef (live weight)	76.6	68.3	112
Pigmeat	77.5	52.7	147
Eggs	65.2	41.0	159
Butter	201.1	80.8	249
Skim milk powder	67.0	46.2	145

* 1972 for pigmeat and eggs.

† The 'EEC entry price' represents the average price at which the produce was admitted after payment of levies, etc. The 'third-country offer price' is the annual average of, normally,the *lowest* offer prices referred to by the Commission in determining levies: it may not be representative of world prices.

** Equal to approximately 1.14 US dollars on average in 1972/3.

Source: Eurostat, *Yearbook of Agricultural Statistics.*

The production stimulus of the protected market, coupled with continuing technological progress in agriculture, caused substantial growth in farm output:

for the Community as a whole, output rose by about 30% between 1963 and 1973, with above-average increases in the Netherlands and Belgium, somewhat below-average increases in Germany and Italy, and virtually no increase in Luxembourg. Demand for food was slackening, and the Community's degree of self-sufficiency rose for practically all foodstuffs – beef was the main exception (Table 12.3). Wheat, sugar and dairy products had to be disposed of with the help of export refunds or domestic market schemes; fruit and vegetables were periodically withdrawn from the market. The common agricultural policy, instead of being a net source of revenue through import levies as had been hoped, became the main cause of Community expenditure: in 1972, out of total Community expenditure of 3074 million UA, the FEOGA Guarantee Section absorbed 2094 million, mainly for intervention and export refunds for cereals, dairy products and sugar and for direct aids for olive oil and oilseeds. The revenue from import levies in that year (plus proceeds from the levy on excess sugar) amounted to 800 million UA.

Table 12.3: Degree of self-sufficiency in the Community of Six

	1956–60	1971/2 –1972/3		1956–60	1971–2
	percentages			*percentages*	
Total cereals	85	98	Butter	101	116
Wheat	90	111	Cheese	100	102
Barley	84	112	Eggs	90	99
Maize	64	67	Beef and veal	92	85
Sugar	104	119	Pigmeat	100	100
Fresh vegetables	104	99	Poultrymeat	93	100
Fresh fruit (excl. citrus)	90	85			
Citrus fruit	47	46			
Wine	89	93			
Vegetable fats and oils	19	27			

Note: For milk powder, both whole and skim, big increases in self-sufficiency occurred in Germany, France and the Netherlands; as data for Italy are not available, EEC totals cannot be provided.

Source: Commission, *The Agricultural Situation in the Community.*

Despite this substantial market support, the farm income situation remained unsatisfactory. Farmers' discontent periodically erupted in demonstrations, the scene of which was now often transferred to Brussels or Luxembourg when the Council was in session. On one well-remembered occasion, protesting farmers led a cow into the Council meeting-room; in March 1971, when the Council was debating both prices and Mansholt's structural proposals, a demonstration in Brussels by some 100 000 farmers (following a COPA General Assembly) provoked violence, injuries and a death.

On average, farm incomes had risen: in real terms (i.e. deflated by the general price index) incomes per labour unit in agriculture were reckoned to have risen between 1964 and 1972 by 9% in Luxembourg, 24% in the Netherlands, 28% in

Germany, 59% in Belgium and 75% in France (data for Italy not available). But with the economic boom of the 1960s, the gap between farm and other incomes had probably increased. As observed in the previous chapter, farm income statistics are to be treated with caution, but the data in Table 12.4 are nevertheless significant. In 1970, the percentage level of farm income in relation to other incomes was calculated to range between a mere 34% in Germany and 74% in Belgium. Within each of the bigger countries, moreover, there were wide regional disparities. No doubt in many cases, particularly in Germany, these gaps were partially made up by the earnings which farm families obtained from non-farm work: part-time farming was on the increase. Nevertheless, the situation justified concern, and the central aim of the common agricultural policy hardly seemed to be fulfilled.

Table 12.4: Income disparities in 1970*

	A Agricultural sector	B Other sectors	C A as % of B
	Units of account		
Belgium	5196	7050	74
France	3479 (1986–6203)†	7849	44
Germany	2542 (1803–4904)**	7572	34
Italy	2829 (1373–4911)††	5556	51
Luxembourg	2611	7481	35
Netherlands	4522	7400	61

* The data refer to gross domestic product at factor cost per active person.
† Range: Limousin (lowest) – Région parisienne (highest).
** Range: Rheinland-Pfalz (lowest) – Schleswig-Holstein (highest).
†† Range: Molisse (lowest) – Lombardia (highest).
Source: Commission (1973) *La Situation de l'Agriculture dans la Communauté élargie* (COM(73) 1850 final).

Against this background, the tasks of the Commission and the Council became increasingly difficult. Virtually all the major policy decisions described above were taken only after protracted and bitter 'marathons' in the Council, frequently extending over several days and even nights. The original spirit of construction in the Community appeared to have been lost. The need to reach unanimity in the face of conflicting national interests tended increasingly to produce compromise solutions which satisfied no one fully; at the same time, Ministers were able to demand and obtain, as the price of their agreement, special arrangements and derogations which added to the complexity and the cost of the policy. The basic problems remained, and in the following years had to be faced by the Community of Nine.

Bibliography

A ANTECEDENTS

The work by Robinson (1961) is in fact much broader than its title suggests and is a valuable account of the role of agriculture in the early stages of European integration. The report by the Frankfurt Forschungsinstitut (1957) is also a useful and stimulating general survey.

BLEU and Benelux

Arnim, V. von (1958) *Die Agrarprobleme Belgiens unter besonderer Berücksichtigung der Probleme, die sich aus der Benelux-Union ergeben.* Gegenwartsprobleme der Agrarökonomie. Hamburg: Hoffmann & Campe.

Conix, A. (déc. 1950) 'L'agriculture belge et le protocole de Luxembourg'. *Études économiques,* Mons.

Institut National de la Statistique et des Études Économiques (1953) *Le Benelux.* Paris.

Meade, J. E. (1956) *The Belgium-Luxembourg Economic Union, 1921-1939.* Essays in International Finance', No. 25. Princeton.

Meade, J. E. (Aug., 1956) 'Benelux: the formation of the common customs'. *Economica.*

Meade, J. E. (1957) *Negotiations for Benelux, 1943-1956.* Princeton Studies in International Finance No. 6. Princeton.

Secrétariat Général de l'Union Douanière. *Benelux.* Bulletin trimestriel. (In particular Annex to No. 4 of March 1958, providing text of the Treaty of Economic Union.)

Vermeren, R. (juin, oct.-nov., 1959 et février, 1960) 'Dix ans d'application des protocoles agricoles Benelux, 1947–1957'. *Revue de l'Agriculture.* Bruxelles.

'Green Pool'

Auswärtiges Amt (Germany) (1953) Gutachten zu Fragen einer europäischen Agrargemeinschaft.

Baade, F. (1952) *Brot für ganz Europa.* Hamburg und Berlin: Parey.

Chambres d'Agriculture (15 mars, 1953) 'L'organisation européenne des marchés agricoles'. *Bulletin.* Paris.

Council of Europe. Reports of the Consultative Assembly, Second Session (August 1950), Third Session (November–December 1951) and Fourth Session (May 1952).

Council of Europe (1951) Documents of the Special Committee on Agriculture.

European Conference on the Organisation of Agricultural Markets (1952–1954) Documents.

Fédération Nationale des Syndicats d'Exploitants Agricoles (1953) *La réalisation d'une Communauté Européene de l'Agriculture.* Paris.

Forschungsinstitut der Deutschen Gesellschaft für Auswärtige Politik (1957) *Die Europäische Zusammenarbeit auf dem Gebiet der Landwirtschaft.* Frankfurt.

Fromont, P. (sept., 1953) 'Les équivoques du pool vert et les projets d'expansion

agricole européenne'. *Revue économique.*
Krumhoff, J. (1957) *Gemeinsame Wege der Europäischen Agrarwirtschaft.* Institut für Weltwirtschaft an der Universität Kiel.
Robinson, A. D. (1961) *Dutch Organised Agriculture in International Politics, 1945–1960.* The Hague: Nijhoff.
Seraphim, H.-J. *et al.* (1952?) *Probleme einer europäischen Agrarintegration.* München: Oldenbourg.

B THE COMMON AGRICULTURAL POLICY

Texts of Council regulations in the *Official Journal of the European Communities,* the Commission's monthly *Bulletin* and annual *General Report,* and other official documents, are indispensable though mostly tedious sources for study of the CAP. Contemporary comment tended to reflect national preoccupations and short-term concerns: only a few out of the many possible references have been indicated. It is symptomatic that some particularly objective and broad-based analyses of the early stages of the EEC, with considerable attention to the CAP, came from two American observers, Camps (1964 and 1966) and Lindberg (1963).

Anonymous author (nov., 1962) 'Les aspects financiers de la politique agricole commune'. *Revue du Marché Commun,* 5e année, No. 52, 422-30.
Averyt, W. F. (1977) *Agropolitics in the European Community (Interest Groups and the CAP).* New York and London: Praeger.
Camps, M. (1964) *Britain and the European Community.* Princeton University Press and London: Oxford University Press.
Camps, M. (1966) *European Unification in the Sixties – From the Veto to the Crisis.* New York and London: McGraw-Hill.
Chambres d'Agriculture, supplément au no. 208, 15 novembre 1960 (and other issues).
Comité intergouvernemental crée par la Conférence de Messine (21 avril, 1956) *Rapport des chefs de délégation aux Ministres des affaires étrangères* (The 'Spaak Report'). Brussels.
Commission of the EEC. Numerous publications and documents, including:
Bulletin (monthly).
Études – Série agriculture. N.B. No. 11 1962: *Wirkungen einer Senkung der Agrarpreise im Rahmen einer gemeinsamen Agrarpolitik der EWG auf die Einkommensverhältnisse der Landwirtschaft in der Bundesrepublik Deutschland* (the *'Professorengutachten'*). (Also in French.)
General Report (annual).
La situation de l'agriculture dans la Communauté (annual).
Newsletter on the Common Agricultural Policy (periodical).
Propositions concernant l'élaboration et la mise en oeuvre de la politique agricole commune, 30 juin 1960.
Receuil des documents de la Conférence agricole des états membres de la Communauté économique européenne à Stresa du 3 au 12 juillet 1958.

European Communities *Official Journal* (special edition in English published October 1972 covers years from 1952 onwards).

European Parliamentary Assembly. Various documents (reports by the Agriculture Committee and Resolutions).

Fennell, R. (1973) *The Common Agricultural Policy: A Synthesis of Opinion.* Wye College: Centre for European Agricultural Studies.

Hanau, A. (1971) 'Der Mechanismus der agrarpolitischen Willensbildung dargestellt am Beispiel der Getreidepreisangleichung in der EWG'. *Die Willensbildung in der Agrarpolitik.* Schriften der Gesellschaft für Wirtschafts- und Sozialwissenschaft des Landbaues, Band 8.

Houdet, R. E. (juillet-août, 1958) 'Stresa ou l'évolution de l'agriculture dans le Marché Commun'. *Revue du Marché Commun.*

Krohn, H. B. (1962) 'Agrarpolitik für Europa'. *Agrarwirtschaft.* Sonderheft 15.

Lindberg, L. N. (1963) *The Political Dynamics of European Economic Integration.* Stanford University Press and London: Oxford University Press.

Mayoux, J. (janvier, 1962) 'L'établissement de la politique agricole commune'. *Revue du Marché Commun.* 5e année, No. 43, 4-19.

OECD *Agricultural Policy Reports* (various).

Plate, R. and Woerman, E. (1962) 'Landwirtschaft im Strukturwandel der Volkswirtschaft'. *Agrarwirtschaft.* Sonderheft 14.

Political and Economic Planning. Various *Occasional Papers,* in particular:
Agricultural Policy in the European Economic Community (No. 1, Nov. 1958).
Proposals for a Common Agricultural Policy in EEC (No. 5, Feb. 1960).
Agriculture, the Commonwealth and EEC (No. 14, July 1961).

Zeller, A. et Giraudy, J.-L. (1970) *L'imbroglio agricole du Marché Commun.* Paris: Calmann-Levy.

Chapter 13

The first enlargement of the Community – Britain in Europe

On 1 January 1973 Denmark, Ireland and the United Kingdom became members of the European Community, fully accepting the Treaty of Rome and subsequent legislation (the *acquis communautaire*) subject only to transitional derogations. For the UK, this step represented the culmination of a long process during which many government leaders and a large body of public opinion had gradually come to terms with the new reality of European integration; yet this view was still contested in Parliament and in the country, as subsequent events were to show. In all stages of Britain's relations with the Community, agricultural policy played a crucial role.

Britain's special position – Commonwealth preference and deficiency payments

Political considerations were uppermost in the move to economic integration among the Six, as the previous chapter has stressed: they were also primarily responsible first for the UK standing apart from that move, yet subsequently – following a reassessment of Britain's role in the world – political motives predominated in Britain's first application to join the Community. In 1950 Winston Churchill had observed that Britain stood at the intersection of three overlapping circles – the English-speaking world, the Commonwealth and Europe. No one of these bonds could afford to be tightened to the extent that the others might be damaged. The supranational ambitions of the early moves to European unity precluded British participation. As has been seen in the previous chapter, the UK stood aside from the Coal and Steel Community and participated only marginally in the work of the Spaak Committee.

There were also strong economic objections to joining a European common market, especially one which included agriculture. These objections arose from Britain's trade links with the Commonwealth, and from the fact that Britain had adopted (in deficiency payments) a means of agricultural support quite different from that used by the Six. Until the early 1960s, the restrictions applied by Britain to imports of foodstuffs were few and generally unimportant. Apart from certain limitations on imports from the dollar area, only a few products (apples, pears and main-crop potatoes being the most important) were subject to quota. Imports from the Commonwealth were free of duty, except for revenue duties on sugar and some other items. Even in the case of foreign countries, duties were generally low; some products – wheat in particular – were admitted free of duty from all sources (Table 13.1). The main exceptions were horticultural products, subject to high specific duties at seasons of peak domestic marketing.

The policy of free or nearly-free entry for foodstuffs into the British market meant that support for British farmers had to be given for most products by deficiency payments, under which compensation was paid for any amount by which the average price obtained for all sales fell short of guaranteed prices in a given period. (All farmers received the same rate of payment, on the basis of their sales or – in the case of feed grain – their acreage: if an individual farmer could obtain a higher market price, the benefit was his.) By this method, cheap food to consumers was obtained, but with a substantial charge on the Exchequer for financing the subsidy.

The replacement of this system by one of protection at the frontier, in the context of a common market, was bound to involve increased prices to consumers. The effects on British farmers would depend primarily on differences in the levels of guaranteed prices. Effects on Britain's pattern of agricultural imports, however, would inevitably be far-reaching. As Table 13.1 shows, by far the greater part of Britain's food imports came from outside Europe, and over half from the Commonwealth; supplies from other western European countries were relatively small (bacon and butter from Denmark and the Netherlands being the major items). Participation in an agricultural common market therefore meant removing restrictions on a small part of British food imports and imposing or intensifying restrictions on a much larger part. Commonwealth countries, in the first place, would lose the preference from which they benefited on the British market. This was perhaps not so important: the degree of preference had been eroded since before the war by the rise in price levels which had reduced the incidence of specific duties. The average margin of preference on foodstuffs in 1957 was only about 6%: for the four largest suppliers enjoying

Table 13.1: UK imports of major foodstuffs in 1958, and import duties

	Sources of imports			*Tariffs*	
	Common-wealth	OEEC area	Other coun-tries	Preferen-tial	Full
		£ million		*per cent*	
Total food, beverages and tobacco	795	293	416		
of which:					
Wheat	71	13	30	–	–
Barley	21	1	6	–	10
Beef and veal*	26	1	43	–	5**
Mutton and lamb	59	1	4	–	–
Bacon	9	65	12	–	10
Butter	55	34	9	–	5**
Sugar (unrefined)	64	–	30	13†**	24**

* Fresh, frozen or chilled.
† A lower rate was granted to the colonies under certain conditions.
** Estimate based on conversion of specific duty.

Sources: Calculations by the author, based on *Annual Statement of Trade* (1958) and tariff publications.

[1] Calculations by the author (cf. *Commonwealth Preference in the United Kingdom*, PEP 1960). Average margins of preference were calculated by weighting the margin on each item (i.e. the difference between the duty on foreign supplies and that on supplies from the Commonwealth, the latter being generally nil) by the value of imports from the Commonwealth country or countries concerned. Ireland, though not a member of the Commonwealth, enjoyed Commonwealth Preference.

Commonwealth preference (New Zealand, Australia, Canada and the Irish Republic), the margin varied between 3% and 8% according to the composition of their exports.[1] But Commonwealth countries would face a barrier from which Britain's European partners would be exempt. Other overseas agricultural exporters – the United States and the Argentine in particular – would be confronted with the same barrier in place of the relatively mild restrictions to which their exports to Britain had previously been subjected.

It was therefore not surprising that Britain, in its early approaches to Europe, tried to exclude agriculture altogether or to obtain special arrangements.

The Free Trade Area negotiations

In July 1956, shortly after the Six had accepted the Spaak report as a basis upon which a treaty could be negotiated, the OEEC Council, on a British initiative, decided to study possible forms of association between the prospective common market of the Six and the other OEEC member countries, setting up a working party to examine in particular the possibility of a wider Free Trade Area encompassing the projected customs union of the Six. Such a Free Trade Area would have provided for the removal of barriers on trade between member countries – but not, in the British view, on agricultural products – while each member could have kept its own tariff on imports from third countries. In January the following year, the working party reported that such an arrangement was technically possible; in October the OEEC Council declared its determination to establish a Free Trade Area covering all OEEC countries and established a committee at ministerial level (chaired by Reginald Maudling whom the British Government had appointed as special co-ordinator on Free Trade Area questions) to prepare for a treaty. France however was without a government at this point and subsequently French objections to the scheme became more evident: both industrial and agricultural organisations pronounced against it. Differences of view appeared among the Six, the Benelux countries being the most favourable to the Free Trade Area, provided it included agriculture, while in Germany Chancellor Adenauer gave priority to political and economic integration among the Six. (On the other hand the German Minister of Economics, Ludwig Erhard, supported the Free Trade Area plan). French hostility grew after de Gaulle had been recalled to power in June 1958; the French cabinet later decided that it was not possible to create a Free Trade Area on the lines proposed by Britain.

An acrimonious period followed, particularly as the Six, having established the EEC on 1 January 1958, were due on 1 January 1959 to put into effect their first adjustments in tariffs and quotas: the other OEEC countries claimed that this would be discriminatory (and the EEC made some adjustments in consequence). The Six invited the Commission to look for solutions to the *impasse.* The Commission submitted a memorandum in February 1959 which declared that the Free Trade Area approach was impracticable. Free Trade required certain conditions, including a balance of advantages and burdens (a reference to agriculture in particular), and adequate co-ordination of economic policy; differences in external tariffs would give rise to distortions. In a second memorandum in September 1959, the Commission gave no support whatever to a wider European arrangement but stressed that solutions needed to be found in the context of the Community's policy towards the world at large. The Commission envisaged narrowing the differences between tariff and quota arrangements applicable to trade with third countries, in the context of GATT negotiations on the basis of the US 'Dillon' proposals. The Commission's approach was criticised by Germany and the Benelux countries, especially the Netherlands, for being too timid; France regarded it as excessively liberal.

The Free Trade Area idea was now dead. It received no support from the United States, to whom it would have brought trade disadvantages with no significant compensation in terms of greater political and economic stability in Europe. On the other hand the US favoured the efforts at integration by the Six and was willing to accept some economic disadvantage in consequence. When in December 1960 it was decided to transform the OEEC into the OECD (Organisation for Economic Co-operation and Development), with the US and Canada as full members, British hopes that reorganisation of the OEEC could help towards some kind of European Free Trade Area or modified customs union were disappointed.

One of the main objections to the Free Trade Area concerned the problems bound to arise from different external tariffs. Much of the discussion related to attempts to produce 'rules of origin' which would prevent goods being imported from third countries into low tariff member countries and then being shifted to higher-tariff members: inevitably, such rules would have been complicated. Agriculture was another major bone of contention. Initially, the British government wanted to exclude agriculture from the scope of the Free Trade Area, arguing that this was necessary both to preserve Commonwealth Preference and to maintain protection for British agriculture and horticulture. Other countries objected strongly, pointing out that this plan would open their markets to competition from British manufactures, while giving no advantage to their agricultural exports to Britain. In January 1958 Britain modified its position to the extent of submitting a draft Convention on agriculture. This did not require the removal of tariffs or quantitative restrictions and included only a vague reference to the possible co-ordination of national marketing organisations: its main feature was the institution of a procedure for 'confrontation' and for hearing complaints, but as decisions by the members would have been taken only by unanimous vote, this arrangement was of doubtful value.

In June 1958 the Six in their turn drew up a memorandum on agriculture. Their proposal did *not* envisage the adoption in the Free Trade Area of a common agricultural policy nor a common external tariff on agricultural products. In a first stage, certain 'very high' duties were to be lowered, with the eventual aim of abolishing barriers to agricultural trade, but there was no precise commitment. When Britain next approached the Community with a request for full membership, much less favourable conditions had to be accepted.

The European Free Trade Association (EFTA)

From the very beginning of discussions on steps to European unity, a split had been apparent between 'federalists' – essentially representatives of the Six – and 'functionalists' led by the United Kingdom. During the Maudling Committee negotiations, delegations from the UK, Denmark, Norway, Sweden, Austria and Switzerland had worked in close consultation. All these countries, for different reasons, preferred a Free Trade Area to a customs union. Following the breakdown of the FTA negotiations, these 'Other Six', joined by Portugal, embarked on a plan for a separate Free Trade Area among themselves, and the Stockholm Convention establishing the European Free Trade Association among the Seven was signed in November 1959. It provided for arrangements similar to those which Britain had originally proposed to the Six: restrictions on trade in manufactured goods between the members were to be gradually abolished (keeping in step with tariff reductions in the EEC), but no changes were required in tariffs on imports from outside EFTA. 'Rules of origin' were introduced to prevent trade diversion. There was no supranational authority.

The most difficult aspect was making the plan attractive to Denmark, as the UK and also some other members such as Switzerland were not willing to remove trade barriers on agricultural products. The Danish Government, on the other hand, had been pressed by farmers to join the EEC in order to benefit from the common market in agriculture: the German market as well as the British one was vital for Danish agricultural exports. Before agreeing to join EFTA, the Danish Government satisfied itself that Germany would not regard this as an unfriendly act and obtained an assurance (later to give trouble within the EEC) that for at least three years Denmark would receive a fair share of any enlargement of German import quotas for farm products.

Denmark finally got some advantages from joining EFTA in the form of bilateral agreements with the other members. Britain agreed to remove the duties on certain Danish exports – bacon, canned pork, canned cream and blue-veined cheese. (These products were carefully selected as being those for which there was little or no competition between Danish and Commonwealth supplies; no concession was made on butter.) The British government also gave an undertaking that it would not raise the guaranteed prices to British farmers so as to nullify the effects of the tariff reduction, and that generally it would not pursue policies likely to reduce Denmark's share in the British market. Sweden agreed to give preference to Danish supplies, and to repay to Denmark 60% of the levies imposed on Danish produce. Switzerland arranged for preferences to be given to Danish supplies of several commodities.

Britain's first bid for membership of the Community

In promoting EFTA, the British government had temporarily given up any thought of coming to terms with the Community. Soon, however, the British attitude began to shift. As the EEC progressed, EFTA seemed increasingly peripheral. The government began to re-examine its position on Europe and to seek better relations with France and Germany. The rest of the Seven were willing to have the UK undertake exploratory talks. A meeting of Commonwealth Finance Ministers in London in September 1960 stressed that Commonwealth interests should be safeguarded, but left the British government enough room to pursue its soundings. On 31 July 1961 the Prime Minister, Harold Macmillan, announced to the Commons the government's decision to open negotiations with the Community 'with a view to joining the Community if satisfactory arrangements can be made to meet the special needs of the United Kingdom, of the Commonwealth, and of the European Free Trade Association'.

Denmark and Ireland submitted to the EEC their applications for membership at the same time as Britain. The 'neutrals' – Austria, Sweden and Switzerland – had political problems, but in December 1961 they asked for negotiations with EEC on possible terms of association, followed in the course of 1962 by Spain and Portugal. In May 1962 Norway requested full membership.

Negotiations between Britain and the Six began in November 1961; by the summer of 1962 a considerable number of problems had been solved or nearly so. With few reservations, Britain accepted the general economic implications of the Treaty of Rome; the removal of internal trade barriers, the common external tariff, the free movement of labour and other aspects of economic union. Arrangements were agreed upon for imports of manufactured products from the Commonwealth, and relationships between the EEC and the less-developed Commonwealth countries were virtually settled. Arrangements for safeguarding the trade of India, Pakistan and Ceylon were decided upon. Apart from the level of the common tariff on a few raw materials, agriculture and food imports from the Commonwealth remained the only important unsolved problems.

The British government recognised from the start that its only chance of success lay in full acceptance of the agricultural provisions of the Treaty of Rome. Its policy was thus forced to come full circle, from excluding agriculture from even the relatively modest arrangements of the Free Trade Area, to endorsing a policy bound to bring about a fundamental change in the British system of agricultural support and in the treatment of food imports. (The move was perhaps facilitated by the fact that uncontrolled imports, with oversupplied world markets, were becoming an embarrassment. During 1961, in view of excess supplies of low-priced butter on the London market, the UK requested exporting countries to limit their supplies, and in the following year subjected all butter imports – including those from the Commonwealth – to quotas.) At any rate, the system of import levies which the Six had already adopted in principle, and to which they gave effect in the market regulations of January 1962, could not be called in question: Britain could only try to ensure that it would be applied in a manner such as to minimise adverse effects on British farmers and consumers as well as on Commonwealth exporters.

It was agreed without much difficulty that there should be an annual review of the agricultural situation, similar to that practised in the UK, on the basis of which the Commission would make whatever proposals might seem necessary in order to maintain a fair standard of living for farmers. There was some difficulty over the length of the transitional period for agriculture: Britain asked for twelve years, but finally accepted that for Britain, as for the Six, the transitional period would end by 1970. Britain also insisted that the system of deficiency payments should not be replaced overnight by import levies, but should be phased out through a gradual raising of market prices. The Six, having agreed among themselves to replace national measures of support by the new system without delay, wanted Britain to follow suit. When the British negotiators pointed out that this would cause a sudden rise in retail food prices in Britain, the Six suggested that consumer subsidies could be used during the transitional period – a suggestion which the British regarded as impracticable. For horticulture, Britain wanted arrangements to ease the impact of increased competition.

The issues involved in the treatment of Britain's food imports were much more significant and difficult. Early in the negotiations, the British representatives laid stress on the need for 'comparable outlets' for Commonwealth produce, by which they meant markets in the enlarged EEC comparable to those currently enjoyed in Britain. The Six were prepared to admit some degree of preference for Commonwealth supplies during the transitional period. In their view, however, this preference was to decrease over time and disappear entirely by the end of the transitional period. Commonwealth produce would then be admitted on the same terms as that from other outside suppliers. The French representatives declared that the EEC could not be expected to commit itself to maintaining specific quantities or prices after the transitional period, nor to giving more favourable treatment to the Commonwealth than to other outside suppliers.

The financial question – the use to be made of receipts from import levies – was not directly an issue, since Britain declared herself ready to accept whatever agreement the Six finally reached between themselves. However, at one stage the French asked the British to endorse a document on this question which included an interpretation upon which the Six were not yet agreed; the British negotiators refused.

Whether the outstanding issues could have been resolved is open to question. Negotiations were in fact brought to a halt by President de Gaulle, who in a press conference on 14 January 1963 emphasised the problems arising from Britain's overseas relationships: 'England is, in effect, insular, maritime, linked through its trade, markets and food supply to very diverse and often very distant countries'. Britain's entry would lead to a colossal Atlantic Community under American leadership, which would swallow up the European Community. He concluded:

> It is possible that one day England will transform herself sufficiently to participate in the European Community, without restriction or reservation and preferring it to every other connection, and in that case the Six would open the door and France would raise no objection.... It is also possible that England is not yet willing to do so, and this indeed is what seems to result from the long, much too long, conversations at Brussels.

The French representatives in Brussels accordingly refused to allow the negotiations with Britain to continue. Negotiations with Denmark and Norway also came to a halt; negotiations with Ireland had not yet begun. The other member states would have been willing to pursue the negotiations, but (under Article 237 of the Treaty) unanimity was required on this matter. The President of the Commission, Walter Hallstein, declared on 5 February 1963 that 'the chance of success was great enough to justify the continuation of the negotiations'. But several more years were to elapse before negotiations again became possible.

Britain's second bid

After de Gaulle's veto in January 1963, the British (Conservative) government made it plain that it maintained its wish to enter the Community. The Labour Opposition however criticised the government's attitude. Harold Wilson, in the Commons debate on 11 February 1963 on the failure of the negotiations, spoke scathingly of the 'new Conservative myth' that agreement had been near. In the election in October 1964, neither of the main parties had much to say about Europe: the Conservative manifesto recognised that there was no question of further negotiations in the immediate future, while the Labour party declared that 'the first responsibility of a British Government is still to the Commonwealth'. The election, fought mainly on domestic issues, was won by Labour and Harold Wilson became Prime Minister.

British interest in the Community revived from 1965 onwards, for a number of reasons. EFTA was well on the way to achieving full Free Trade among the members, but still appeared lacking in permanence; Denmark in particular remained keen for a settlement with the Community. Within the UK, the current of opinion in favour of joining the Community which had developed during the first negotiations was still flowing strongly: economic justifications were now stronger, as the prospects for expanding trade with the Commonwealth appeared less favourable than those for trade with the EEC if access could be obtained. Under the Conservative government, Britain had moved further away from the traditional policy of free imports of foodstuffs. Besides the butter import quotas introduced in 1962, agreements on grain imports had been concluded in April 1964 with the four major suppliers – the United States, Canada, Argentina and Australia – involving a system of (relatively low) minimum import prices; under a Bacon Market Understanding, Britain was trying to regulate supplies to the British market in consultation with exporting countries; a quota system for meat had been sought, though only an advisory group had been set up with the exporting countries concerned. Finally, in the political field, it had become evident that Britain had counted too much on a 'special relationship' with the United States.

Such considerations finally led to the decision by the Labour government, announced to the Commons by Harold Wilson on 2 May 1967, to renew the bid for membership. President de Gaulle's immediate response was discouraging, but Wilson said that Britain would not take 'no' for an answer. However, at a press conference on 27 November 1967, de Gaulle declared that what was needed was 'a radical transformation of Great Britain, to enable her to join the Continentals'. Economic considerations played a greater part in his reasoning than in January 1963: Britain's chronic balance of payments deficit and the international role of sterling were major obstacles in his view. As in 1963, the rest of the Six wanted to pursue negotiations with Britain, but were unable to act in the face of the French refusal. The whole issue had to be postponed for another two years.

The accession of Britain, Ireland and Denmark

After Georges Pompidou had been elected President of France in June 1969, the French position on Britain's accession was somewhat modified. French policy on the Community was now expressed in the triple aim 'completion, reinforcement and enlargement'. Adoption of all the measures necessary to permit passage on 1 January 1970 to the common market's final stage and extension of Community activities would create conditions in which France could envisage the enlargement of the Community. The Hague 'Summit' of 1-2 December 1969 made it possible to move forward on all fronts.

The Six subsequently agreed among themselves on two basic principles for the negotiations on enlargement: the candidate countries must accept the Treaties and all subsequent Community legislation, and any problems of adjustment must be solved by transitional arrangements, not by changing existing rules. When negotiations opened on 30 June 1970 (just after elections in the UK had returned the Conservatives to power), the four candidate countries accepted these principles. The British delegation confirmed that it accepted the principles of the common agricultural policy, but said that there were problems of adjustment which did not affect the principles.

Agreement was reached on all issues after nineteen months of negotiations, and the relevant acts were signed on 22 January 1972, providing for the entry into the Community on 1 January 1973 of the UK, Ireland, Denmark and Norway. National parliaments ratified these acts during 1972, except in the case of Norway: the Norwegian public rejected the terms of entry in a referendum. In the Irish referendum 83.1% of the votes were cast in favour, in the Danish referendum 63.5%. No referendum was held in the UK: the results of the Commons debates are referred to below.

Under the Act of Accession, the general rule was that customs duties between the new member states and the Community would be scaled down to nil by five annual steps of 20% each (the last on 1 July 1977); the new members' tariffs on imports from third countries were to be adjusted to the common tariff over the same period. For agricultural and horticultural products, there were differences in the rate of tariff alignment to which significance was attached during the negotiations but which were in fact minor. The main difference was that for horticultural products the rate of alignment was delayed by a year. The arrangements for adjusting price levels were more significant: for all products subject to price-fixing under the CAP, the new member states would adjust their prices to the common level in six annual steps. By 1 January 1978 at the latest the common prices were to be applied. In the meantime, remaining price differences were to be offset by compensatory amounts ('accession compensatory amounts', not to be confused with 'monetary compensatory amounts'). From 1 February 1973, the CAP market regimes would be applicable in the new member states, and all measures incompatible with these regimes (such as import quotas) were to be abolished.

In the negotiations with the UK, issues relating to domestic food and agriculture proved relatively easy to solve – much more so than in Britain's first bid. The British Conservative party had become disenchanted with deficiency payments because of their high budgetary cost and their unpredictability: indeed the Conservative government, soon after it had been elected in 1970, began replacing deficiency payments with 'minimum import price' schemes implemented through import levies, with the aim of shifting over completely in a few years, independently of EEC entry. As EEC guaranteed prices were on average higher than British prices, there remained the problem of the impact on consumer prices: the British White Paper of 1970 estimated that retail food prices might rise by 18-26%, causing a 4-5% increase in the over-all cost of living. The UK therefore requested a tolerable period of adjustment, but found the transitional arrangements just described to be acceptable. (The 1971 White Paper reckoned that food prices would rise 2.5% per annum and the cost of living 0.5% per annum in consequence over the six-year transitional period.) Among British farmers, opinion had swung in favour of entry as they realised that, with the main exception of horticultural products, they were likely to get higher prices under the CAP, particularly for cereals. (For horticultural products, where the tariff was the main instrument of support, a slower rate of tariff adjustment was provided for.) The NFU was particularly attached to the annual price review in view of the role which it could play in this procedure: the UK was able to accept that existing Community procedures would be adequate, after obtaining confirmation that these procedures would include 'appropriate contacts' with farm organisations. There was some difficulty over the support arrangements for pigmeat and eggs, where the British government felt that its support arrangements ensured better market stability, but the British delegation finally stated that Community arrangements might be adequate while the Community declared that it shared the British concern as to the need for stability. The status of the Milk Marketing Boards and their methods of pricing were left unclear (this matter was settled only in 1978). Hill farming grants were an important aspect of British policy: the Six acknowledged this problem and in 1975 the enlarged Community adopted a programme of aid for less-favoured areas which took the place of the British national scheme.

Fisheries were a more difficult issue, but one which is outside the scope of the present work. The Community's fisheries regime, involving in particular free access for vessels of member states to each others' fishing grounds, were a major reason for rejection of the terms of accession by the Norwegian public, and proved to be a serious source of discord in Britain's relations with the Community after the decision, at The Hague in October 1976, to extend national fishery zones to 200 miles.

The issue of Britain's relations with the Commonwealth, which had seemed so intractable in 1961–3, now broke down into a limited number of relatively technical problems. The UK was bound to accept that in general it could no longer grant Commonwealth Preference, and indeed that *Community* preference would have to be substituted: the relatively slow growth of Commonwealth trade and the erosion of reciprocal Commonwealth preferences facilitated this reversal in the British position. Still, the problems which remained were difficult enough. Britain obtained recognition from the Six of the particular difficulties which New Zealand would face in view of its dependence on the UK market. Lamb was not yet a serious problem, since the Community had not adopted a common market organisation, though the UK had to accept the common tariff of 20%. But New Zealand exports of dairy products would be hit by the import levy regime. The Community agreed to guarantee a market for certain quantities of New Zealand butter and cheese in the UK market: the quantities would be reduced to reach in 1977 80% of the initial level for butter and 20% for cheese. Thereafter there would be no concession on cheese; for butter, the Council would have to decide (unanimously) on 'appropriate measures to ensure the maintenance after 31 December 1977 of exceptional arrangements'. The problem of New Zealand butter became another recurrent issue in Britain's relations with the Community. Following the Labour government's 'renegotiation' in 1974–5, discussed below, the concession was extended to the end of 1980.

Commonwealth sugar exports were a particularly difficult item. Under the Commonwealth Sugar Agreement (CSA), Britain had undertaken to import specified quantities of cane sugar at a guaranteed price: this arrangement was vital not only to the developing Commonwealth countries concerned – Mauritius, Fiji and certain Caribbean islands – but also to the British sugar refining industry. France, with its own strong interest in producing and exporting sugar, was unwilling to give Britain any firm guarantee for the future. The issue became linked to the general question of relations with associated developing countries, currently defined by the Community of Six under the Yaoundé Convention. This was due to expire in 1975, and in 1974 Britain could opt out of the CSA. It was agreed that Commonwealth countries in Africa, the Caribbean, the Pacific and the Indian Ocean could enter into a new association with the enlarged Community (this took shape subsequently in the Lomé Convention) and that the question of their sugar exports would be dealt with in this context. This in itself was insufficient for the Commonwealth sugar producers and the British refiners: the British delegation consequently demanded 'bankable assurances' that the developing Commonwealth countries would still be able to sell their crops in the UK from 1975 onwards. The formula finally proposed by the Six gave no quantitative assurance, but declared that the Community *'aura à coeur'* the interests of the developing Commonwealth sugar exporters (the English version read: 'It is the firm purpose of the Community to safeguard the interests ...'). Despite doubts by the interest groups as to the validity of this assurance, it was accepted by the British government. Subsequently, under the Lomé Convention of 1975 (negotiated at a time of shortages and high prices on the world sugar market), the Community undertook, for an indefinite period, to purchase at guaranteed prices – to be negotiated annually 'within the price range obtaining in the Community' – specific quantities of cane sugar from the ACP countries: these quantities totalled 1.4 million tons in terms of raw sugar, which was as much as the ACP countries had asked for, and the amount actually taken up totalled approximately 1.3 million tons. Australia (which had been the largest sugar producer in the CSA) got no concession, so that its former place on the UK market was available to be filled by EEC domestic supplies.

A particularly significant and complex issue was that of Britain's future contributions to the Community budget – a general question which had important agricultural implications. The British authorities estimated during the negotiations that on the basis of the Community's 1970 financial regulation, the UK would end up paying 31% of the budget (because of the relatively high incidence in the UK of agricultural import levies and customs duties on imports from third countries), and receiving back only 6% (largely because the UK would get little benefit from the CAP). This problem, much discussed and analysed in British circles, was a major criticism of the Community and the CAP: it gave weight to the popular feeling that in joining the Community the UK would be subsidising 'inefficient French farmers'. A settlement was reached, after arduous negotiations, on the basis of a complex formula under which Britain's contributions were limited up to 1979; no formal safeguard was given beyond 1979, though the British White Paper of 1971 on the outcome of the negotiations quoted a Community declaration that if 'unacceptable situations' were to arise, 'the very survival of the Community would demand that the institutions find equitable solutions'. This temporary settlement did not solve the issue, which remained in subsequent years a major British complaint, being taken up again in the 'renegotiation' by the Labour government and yet again in 1979 by the Conservative government under Margaret Thatcher.

The negotiations with Ireland and Denmark led to no special difficulties over agriculture. Ireland had every interest in following the UK into the Community and gaining access to the Community market for its agricultural exports: its dependence on the UK market was excessive and diversification was desirable. Denmark too could only gain in agriculture; in fact Denmark did not need a transitional period but it accepted this in view of the UK's difficulties. Norway, having very high agricultural prices, would have had problems – a protocol was adopted recognising that special arrangements might be needed even after the transitional period – but as has been seen above, the Norwegian public's verdict was in any case negative.

As regards the other EFTA countries, all asked for discussions with the Community which led to a series of Agreements, to enter into force on 1 January 1973, at the same time as the accession of the three new member states (Norway obtained a similar Agreement slightly later). These Agreements all provided for free trade in industrial products, according to the same schedule as under the Act of Accession, and for rules of origin to avoid deflections of trade: the extensive Free Trade Area unsuccessfully discussed in 1956–8 thus came into being, but in the meantime the Community had become firmly established. As regards agriculture, the Agreements in general simply affirmed the principle that the contracting parties were prepared, while respecting their agricultural policies, to foster the harmonious development of trade in agricultural products, and set up a procedure for examining any problems that might arise.

Controversies in Britain: 'Renegotiation'

Britain's second application to join the Community had been made by a Labour government under Harold Wilson in 1967 and carried to a successful conclusion by a Conservative government under Edward Heath. In the British election of June 1970, little was said about the EEC: preparations for negotiations were already well advanced under the Labour government and neither side at that point wanted to rock the boat. The newly-elected Conservative government took over negotiating briefs prepared for Labour Ministers, apparently with little change.

But joining the EEC was still unpopular in the country at large. A Gallup poll in April 1970 showed 59% opposed to entry and only 19% in favour. After the June election, the trade unions began to take a much clearer anti-EEC stand: a resolution hostile to entry, put forward by the powerful Transport and General Workers' Union at the party conference in September, was narrowly defeated. The TGWU later demanded a nation-wide referendum on the issue. Harold Wilson's European policy came under attack from other Labour leaders: John Silkin in January 1971 put down an anti-EEC motion in the Commons; Peter Shore denied that there had ever been any commitment in the Labour cabinet beyond negotiations to find out the terms of entry; Michael Foot, Barbara Castle and Douglas Jay were other prominent anti-Marketeers; in May 1971 James Callaghan openly challenged Wilson on the issue in a bid for the party leadership. By the summer of 1971 the party had swung against entry; at the party conference that October, Callaghan gave a warning that a future Labour government would re-open negotiations, particularly over the CAP and the unanimity rule. The Conference returned a huge majority against entry 'on Tory terms'.

In the Commons debate in October 1971, the Labour Party officially opposed entry. However, 89 Labour MPs defied the party 'whip', 69 voting in favour and 20 abstaining; on the Conservative side (where a free vote was allowed), 39 voted against entry and 2 abstained. The outcome was 356 votes for, 244 against. In subsequent votes on the implementing legislation, the government obtained much narrower majorities.

The public remained unenthusiastic. An Opinion Research Centre poll in December 1971 indicated that among Conservative voters 57% were in favour of entry and 29% against, among Labour voters only 24% were in favour and 64% against. Concluding an analysis of these political developments, Kitzinger observed:

> The government succeeded in their primary aim. They obtained parliamentary approval of the principle of entry by a large majority, and they secured – by a hair's breadth – the passage of the implementing legislation. But they failed to obtain the 'full-hearted consent' of the British people before entry. [Kitzinger (1973) page 397]

Just over a year after Britain's entry, Labour was again in power, as the result of an unexpected election in February 1974 provoked by domestic troubles. The party was now committed – in particular by its election manifesto – to renegotiating the terms of membership and to calling a nationwide referendum on the results. On 1 April the Foreign Secretary, James Callaghan, made a statement to the Council of the Community declaring that the Labour government opposed membership on the terms previously negotiated and quoting passages of the manifesto. The objectives of renegotiation included 'major changes in the Common Agricultural Policy, so that it ceases to be a threat to world trade in food products, and so that low-cost producers outside Europe can continue to have access to the British food market'. Referring to the trade problems of Commonwealth and developing countries, Callaghan specifically mentioned sugar and New Zealand butter. Another major aim was to achieve 'new and fairer methods of financing the Community Budget', for otherwise, after the transitional period, Britain would be paying considerably more than its fair share of Community GNP, and receiving a relatively low share back.

The demand for renegotiation was received with dismay by the other member states – France in particular refused as a matter of principle to consider that there could be such a thing as renegotiation, and the term itself was avoided in Community documents. However, progress in a number of specific areas, combined with external factors, helped to facilitate a settlement.

Labour's first Minister of Agriculture, Fred Peart, obtained several concessions in Agricultural Council sessions during 1974. The British government's concern to keep down food prices to UK consumers, at a time of rapid general inflation, was partly met at a Council session in March that year, when in the context of the price decisions for the coming year, special arrangements (intended to be temporary) were made whereby the UK was exempted from 'intervention buying' for beef and authorised to pay consumer subsidies for butter. In July the Council agreed that the UK, instead of intervention on the beef market, could apply a system of slaughter premiums (subsequently referred to as 'the Peart premiums'). In October, when the world sugar price had risen well above the Community price, the Council agreed to meet any shortfall in Britain's sugar requirements over the coming year with subsidised imports from the world market (the current world price was over £400 a ton, Australia had sought a long-term contract at £190, and under the scheme Britain was guaranteed supplies at £156).

The world wheat price too was above the Community target price following the massive Russian grain purchases in the second half of 1972. Until early 1975, when the world price fell again, the common market mechanisms went temporarily into reverse as export *levies* were applied, so that market prices in the

Community were kept down and supplies to consumers assured. Further, as sterling had floated downwards, the market exchange rate had diverged from the 'green rate' for converting Community prices (in units of account) into sterling, so that Britain was receiving from the Community budget substantial import subsidies in the form of monetary compensatory amounts, also to the benefit of consumers.

When the newly-instituted 'European Council', consisting of Heads of State or Government, met in Dublin in March 1975 with Britain's 'renegotiation' high on their agenda, there remained only two substantial issues: arrangements for New Zealand dairy produce after 1977, when the transitional derogations of the Act of Accession would expire, and Britain's budgetary contribution. On New Zealand, the European Council, in a carefully-worded statement, accepted that New Zealand butter exports would continue after 1977 in quantities which, 'depending upon future market developments', could remain close to those of 1974 and those envisaged for 1975 (it was not until June 1976 that these quantities were actually fixed), and at prices which would be subject to periodic review; these assurances were valid up to 1980. For cheese, it was noted that the exceptional arrangements would expire at the end of 1977 and no further guarantee was given, beyond a general expression of willingness to co-operate with New Zealand in promoting 'an orderly operation' of world markets for dairy products.

The budget issue was the most fundamental. The basis of a solution was provided by the Heads of Government when they met in Paris on 11–12 December 1974, in a declaration recalling the statement made during the accession negotiations by the Community to the effect that 'unacceptable situations' would require 'equitable solutions', and asking the Council and the Commission to set up a 'correcting mechanism' to prevent the development of situations 'unacceptable for a member state and incompatible with the smooth working of the Community'. The Commission subsequently suggested that such an 'unacceptable situation' implied a combination of an adverse economic situation (relatively low GNP per capita, relatively low GNP growth and balance of payments deficit) and a disproportionate contribution to Community financing. On this basis the Heads of Government, at the Dublin summit, agreed upon a detailed 'financial mechanism' which was subsequently adopted by the Council. This solution was hailed by the British government as a 'major improvement' in the budgetary terms of British membership (though the criteria were such that the UK did not in fact benefit until the conditions were changed in 1980).

Harold Wilson and his government were then able to recommend to Parliament and the nation that Britain should stay in the Community (cf. White Paper of 1975). Several Labour Ministers however dissented: Tony Wedgewood Benn, Barbara Castle, Michael Foot, Judith Hart, Peter Shore and John Silkin issued a statement saying that the CAP would still 'ensure dear food for our people when it is cheaper elsewhere', that the budget system remained 'intrinsically unfair', that Britain had an 'appalling trade deficit' with the EEC, and above all that the Community represented a threat to democratic self-government in Britain.

The outcome of the referendum, held on 5 June 1975, was a surprise even to the pro-Marketeers: on a turn-out of 65% of the electorate, 67.2% voted in favour of staying in the Community, 32.8% voted against. Commentators however doubted that the result really meant a strong commitment to Europe: reluctance to envisage another change, it was suggested, may have been the main motive (cf. Butler and Kitzinger, 1976). A referendum held *before* Britain joined the Community might well have given a very different result. It was generally felt, however, that the referendum had settled the issue once and for all, both in the country at large and within the Labour Party. Labour members now took their seats in the European Parliament, and British trade unions were represented on the Economic and Social Committee. However, some of the major issues which had divided Britain and the rest of the Community remained alive.

Continuing issues

The priority given by the Labour government to keeping down farm prices in the interests of consumers – particularly after John Silkin had become Minister of Agriculture in 1976 – was a major element in the annual review of common agricultural prices. The British attitude to common prices was even more restrictive than the Commission's 'prudent price policy' from 1977 onwards (see next chapter). Nevertheless, successive devaluations of the 'green pound' enabled British farmers to obtain price increases above the Community average. This was a period of considerable friction between Britain and her partners, intensified by conflict over fisheries policy and other matters. Britain's refusal to participate in the European Monetary System, which entered into force in March 1979, disappointed the other member states.

The Conservative government, returned to power in May 1979, demonstrated its willingness to improve relations with Europe. Under the new Minister of Agriculture, Peter Walker, the British attitude to farm prices appeared to be more flexible, and the attitude to the CAP generally more positive, although the problems arising from surpluses and the resulting costs continued to be stressed.

Of the various Commonwealth issues which had played so big a role in earlier negotiations between Britain and the Community, there remained above all the problem of New Zealand's farm exports. Britain's ties with New Zealand remained strong; agricultural exports to the UK – of butter and lamb in particular – were still the mainstay of the New Zealand economy, despite attempts at diversification.

The Dublin concession on New Zealand butter expired at the end of 1980. In the difficult CAP context at that time – which is discussed in the following chapter – and especially in view of the oversupplied state of the Community's own market for dairy products, it proved impossible to reach agreement on any further long-term commitment: only a temporary extension was accepted and the issue was left in suspense.

Britain's traditional imports of mutton and lamb from New Zealand, Australia and some other countries complicated the establishment of a common market regime for 'sheepmeat'. The price on the open British market was well below that on the protected French market (returns to British farmers being supported by deficiency payments). Under a ruling of the European Court of Justice (the 'Charmasson' case of 10 December 1974), derogations from Treaty rules arising from national market organisations were no longer admissible after the end of the transitional period, but France defied a Court injunction of September 1979 to remove the barriers excluding British lamb from the French market (while imports from Ireland were admitted) – the first serious challenge to the Court's authority in the history of the Community. The French insisted that a common market organisation giving adequate guarantees to their sheep farmers must first be set up. This was achieved only in October 1980, after difficult negotiations to reconcile the interests of all concerned (the outcome was described as 'a patchwork of national demands'). The content of this agreement is described in the following chapter. Under it the UK gained a higher price guarantee to its sheep farmers than under the domestic scheme, and transfer of the cost of support to the Community budget.

The central issue in Britain's relations with the Community was still that of the budgetary contribution, and this was brought into the foreground by the Prime Minister, Margaret Thatcher. The European Council at the end of April 1980 failed to reach agreement: a critical situation then arose since the UK was blocking a farm price increase until it obtained a satisfactory deal on the budget, while France threatened to introduce national farm support measures on 1 June if common price increases were not agreed by then. Crisis was averted at the last minute (on 30 May) when the Foreign Affairs Council reached agreement on a budgetary arrangement whereby the UK's net contribution would be limited, enabling the Agriculture Ministers in parallel session to adopt the 1980/81 farm prices.

This budgetary settlement modified the previous criteria of the 'financial mechanism' in such a way as to permit the UK to benefit from repayments in 1980 and 1981, and provided for Community finance for up to 70% of the cost of special investment programmes in regions of the UK, also in 1980 and 1981. For 1982, 'the Community was pledged to resolve the problem by means of structural changes', with the aim of preventing the recurrence of 'unacceptable situations' for any member state. The Commission was to make proposals in this respect by end June 1981. Since the imbalance between Britain's budgetary payments and receipts was primarily due to the large share in the budget of the CAP, from which the UK derived relatively little benefit, the UK was bound to keep up its pressure for action to contain the cost of the CAP.

A new situation arose in the course of 1980, as the spectacular recovery of sterling reversed the relationship of the market exchange rate and the 'green rate': 'positive' monetary compensatory amounts were now applied – i.e. acting as *levies* on imports into the UK – and reached 12.1% by the end of the year. The government was under pressure to *revalue* the green rate, so as to reduce these m.c.a.s and bring down consumer prices: in these circumstances, price increases to British farmers depended on increases in common prices. This however conflicted with the general British aim of keeping down the cost of the CAP.

Ireland and Denmark

Most of this chapter has been about Britain's adjustment to Europe. A brief comment may be made on the experience of the two other countries which joined the Community together with the UK.

For Ireland, where the agricultural sector accounts for nearly a fifth of national income and which exports the greater part of its produce, the advantages of joining the EEC were clear. Irish agriculture benefited markedly from increased outlets and higher prices: indeed the downwards float of sterling – to which the Irish punt was tied until 1978 – provided additional scope for increasing Irish farm prices through adjustments in the 'green rate' (see Sheehy, 1980). The Irish consumer interest, inevitably, was not served by this development and the fight against inflation was made more difficult; but in the Irish case agricultural prosperity made an important contribution to the economy as a whole. By 1980, however, Irish agriculture was being hit by rising costs, while there was no further scope for raising prices through the 'green rate'.

Denmark's agricultural interests too benefited from the wider Community market and Community preference. However, though agriculture and agricultural exports continued to make an important contribution to the economy, the agricultural share in the gross domestic product was down to about 7% by the early 1970s. Further, though Danish agricultural exports within the Community expanded rapidly after 1973, Denmark was not able to catch up with the very substantial advantage which the Netherlands, its main competitor, had gained in the Community of Six, while the value of French agricultural exports had overtaken those of Denmark (see figure 13.1). The 1970s were a difficult period for Danish agriculture, hit by rising production costs, high land prices and especially by high interest rates, while in the EEC generally agricultural markets were increasingly oversupplied. The common agricultural market remained vital to Danish agriculture, but Danish agricultural circles became increasingly preoccupied with the disruption of the common market and of common conditions of competition through monetary compensatory amounts (especially the 'positive' German m.c.a.s) and through national aids (see Kofoed, 1979).

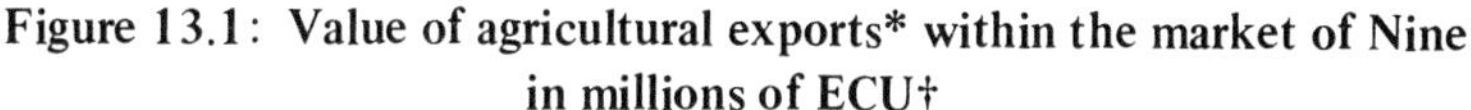

Figure 13.1: Value of agricultural exports* within the market of Nine in millions of ECU†

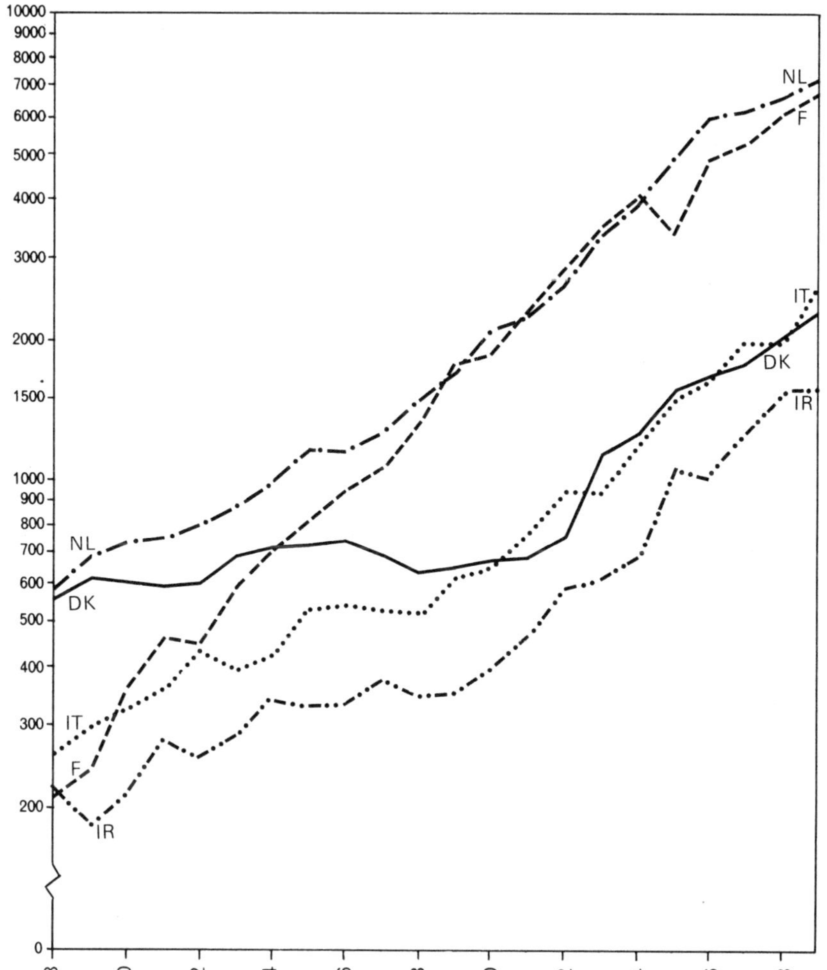

* Food, beverages and tobacco.

† An explanation of the various units of account used in the Community is given in a note at the end of this book. This particular source, however, gives a confusing definition of the ECU, and uses conversion factors different from those in other publications. Here the ECU was calculated as equal to the US dollar up to 1970, subsequently as equal to the 'basket' EUA; in 1979 it was worth $1.37.

Source: Eurostat, *Monthly external trade bulletin*, special number, 1958–79 (Table 7b).

Bibliography

The issue of Britain's relations with Europe has produced a vast flood of literature which continues up to the time of writing; only a small selection is mentioned here. On the early stages, the PEP series is useful, together with Camps (1964 and 1966) and Lindberg (1963) – see bibliography to previous chapter. Marsh and Ritson (1971) provided a convenient survey of the issues at the time of the final negotiations. The terms of access were analysed by Young (1973) and Puissochet (1974); see also the Commission's *General Report* for 1971 and of course the Act of Accession itself (including its Protocols and the attached Declarations). Kitzinger (1973) and Butler and Kitzinger (1976) provided useful insight into British political attitudes on Europe.

British Government White Papers, the Commission's *Bulletin* and *General Reports,* the press and the specialised press agencies – European Report, Europe and Agra-Europe, are important sources for this whole period and especially for recent developments.

BRITAIN AND EUROPE

British Government White Papers:

(1970) *Britain and the European Communities: An Economic Assessment.* Cmnd. 4289: HMSO.

(1971) *The United Kingdom and the European Communities.* Cmnd. 4715: HMSO.

(1974) *Renegotiation of the Terms of Entry into the European Economic Community.* Cmnd. 5593: HMSO.

(1975) *Membership of the European Community: Report on Renegotiation.* Cmnd. 6003: HMSO.

(1975) *Food from our Own Resources.* Cmnd. 6020: HMSO.

(1979) *Farming and the Nation.* Cmnd. 7458: HMSO.

Butler, D. and Kitzinger, U. (1976) *The 1975 Referendum.* London: Macmillan.

Commission of the European Communities:

Bulletin (monthly).

General Reports (annual).

(February 1959) *First Memorandum to the Council of Ministers concerning the establishment of a European Economic Association.* COM(59)18 rev. 2.

Second Memorandum. In *Bulletin* 4–59.

Denton, G. *et al.* (1975) *The Economics of Renegotiation.* Federal Trust.

Economist Intelligence Unit (1960) *The Commonwealth and Europe.* London.

Kitzinger, U. (1973) *Diplomacy and Persuasion – How Britain joined the Common Market.* London: Thames & Hudson.

Lardinois, P. J. (December, 1961) 'The United Kingdom and the European Economic Community (a Continental view-point).' *Journal of Agricultural Economics.*

Marsh, J. (1979) 'U.K. Attitudes to the CAP'. In Tracy and Hodac *eds. Prospects for Agriculture in the European Economic Community.* College of Europe, Bruges.

Marsh, J. and Ritson, C. (1971) *Agricultural Policy and the Common Market.* Chatham House/PEP European Series no. 16.
National Farmers' Union (1961) *Agriculture in the Community.*
National Farmers' Union (1966) *British Agriculture and the Common Market.*
National Farmers' Union (1973) *A Review of the Common Agricultural Policy.*
Political and Economic Planning *Occasional Papers.* Especially:
no. 2 by Camps, M. (1959) *The Free Trade Area Negotiations.*
no. 3 (1959) *Agricultural Policies in Western Europe.*
no. 13 (1961) *Food Prices and the Common Market.*
no. 14 (1961) *Agriculture, the Commonwealth and EEC.*
Puissochet, J.-P. (1974) *L'élargissement des Communautés Européennes.* Paris: Editions techniques et économiques.
Warley, T. K. (1967) *Agriculture: The Cost of Joining the Common Market.* Chatham House/PEP European Series no. 3.
Young, S. Z. (1973) *Terms of Entry – Britain's Negotiations with the European Community, 1970–1972.* London: Heinemann.

DENMARK

Agricultural Council of Denmark (1977) *Agriculture in Denmark.* Copenhagen.
Kofoed, N. A. (1979) 'Prospects for EEC Agriculture'. In Tracy and Hodac *eds. Prospects for Agriculture in the European Economic Community.* College of Europe, Bruges.
OECD (1974) *Agricultural Policy in Denmark.* Paris.

IRELAND

Baillie, I. F. and Sheehy, S. J. *eds.* (1971) *Irish Agriculture in a Changing World.* An Agricultural Adjustment Unit Symposium. Edinburgh: Oliver & Boyd.
OECD (1974) *Agricultural Policy in Ireland.* Paris.
Sheehy, S. J. (1980) 'The impact of EEC membership on Irish agriculture'. *Journal of Agricultural Economics* **XXXI** (3), 297–310.

Chapter 14

CAP adjustment in the context of 'stagflation'

The first enlargement of the European Community in 1973 coincided with the end of the post-war economic boom. Until then, the problems of agriculture had been discussed in terms of adjustment to economic growth: how could farm incomes keep pace with the rise in living standards in the urban economy? This preoccupation was underlined by an OECD report in 1965 entitled *Agriculture and Economic Growth,* which pointed out that rapid and sustained growth in gross domestic product had been attained in most OECD countries since 1950, and stressed that adaptation of agriculture to current and future economic conditions implied further movement of people out of agriculture, encouragement to appropriate use of technology and capital within the sector, improvements in farm structures, and adjustment of output to economic requirements.

Such adjustments were difficult enough. The problems existing around 1960 and described in chapter 11 grew no simpler during the following decade: farm incomes still lagged behind other incomes, disparities within the farm sector remained substantial, markets were even more oversupplied. Still, in a context of economic growth, rational solutions could be advocated: an accelerated shift of manpower from agriculture to other activities could be expected to benefit both the agricultural sector and the economy as a whole; farm people leaving rural areas could hope to find other jobs; expansion of industry into rural areas helped to provide employment and extra earnings for farm families; growth in consumer incomes stimulated demand for high-value foodstuffs such as meat.

The economic climate from 1973 to 1980 was a very different one. No doubt the boom was bound to slow down in any case, as underlying economic and social tensions made themselves increasingly felt. But the major shock to the Western economy was the energy crisis, starting in 1973 when the Arab oil-producing countries, in the wake of the Arab-Israeli War, first restricted supplies and then sharply raised prices: the average c.i.f. cost of oil to importing countries rose from about $1.70 per barrel in 1970 to around $4.15 in 1973, to about $10.75 in 1974, and by 1980 was over $30 per barrel. Inflation in western Europe accelerated into double figures, while GDP growth in real terms slackened and became negative in 1975: the new phenomenon of 'stagflation' was born and proved very difficult to shake off. Unemployment rose from 2.5% of the labour force in the Community in 1973 to over 6% in 1980; balance of payments deficits, due largely to the burden of high oil prices, added to the difficulties. Monetary instability was inevitable: while the dollar and the pound sterling fluctuated, only the Swiss franc and the Deutschemark – benefiting from relatively low rates of inflation – remained strong, and frequent parity changes occurred among the EEC currencies.

In such circumstances, agricultural adjustment in general was bound to become much more difficult, and the organisation of common agricultural markets in the Community was subjected to special problems. The farm sector, like other branches of the economy, suffered cost increases, which squeezed farmers' incomes causing them to demand price increases which could not easily be met in view both of the need to contain inflation and of the state of the markets; demand for some foodstuffs slackened (especially for beef), creating new market problems; employment opportunities outside agriculture were severely restricted and structural change in agriculture was hampered. In the Community, the assumption of fixed exchange rates, upon which the common market organisation had been based, was undermined, necessitating compensatory measures at the frontiers contrary to the objective of free movement of goods within the Community. The cost of market support became a major burden, pushing the total Community budget up to near its statutory limit. The 'reform' of the common agricultural policy came to be urgently advocated from many quarters.

These developments will be looked at more closely below.

Trends in the economic state of agriculture after 1973

During the 1970s, many of the earlier trends continued. Gross agricultural product continued to rise in terms of current value, though the share of agriculture in national income continued to fall. By the end of the decade, agriculture in the Community accounted for only about 4% of gross domestic product (barely above 2% in the United Kingdom). Agriculture's share of total employment also continued to fall, to just under 8% by the end of the decade in the Community. The difference between these two proportions gives an indication of the average *per capita* income gap between agriculture and other sectors, although, as previously observed, such comparison of averages can be misleading and also takes no account of earnings by farm families from non-farm sources (see below).

The technological revolution in farming continued, causing higher productivity in terms of both land and labour, and increased output. It would be beyond the scope of this work to refer to all the technical changes which caused this development (see, however, Thiede (1979), who analysed the process in terms of mechanical, biological and organisational advancement and observed that the potential for yield increase had constantly been underestimated). The wheat yield per hectare in the Community increased at an average rate of about 1.4% per annum between 1973 and the end of the decade; the milk yield per cow by about 2.3% per annum. Wide variations in yields between regions suggested that there was still considerable scope for further improvement through wider application of existing technology, quite apart from the continued introduction of new technology. Total agricultural output in the Community rose by some 1.3% per annum, while agricultural employment fell by about 3.0%; labour productivity had thus risen at a rate of some 4.4% per annum.

The growth of food consumption was limited (Table 14.1). Population growth in the Community slowed down from a rate of 0.8% per annum in 1958–72 to barely 0.2% in the latter half of the 1970s. Income elasticities of demand for most foodstuffs were low or even negative.

Higher self-sufficiency was the inevitable outcome (Table 14.2). The effects on trade with third countries will be discussed in Chapter 15: it is sufficient here to note that for most temperate foodstuffs the Community reduced its volume of imports during the 1970s, and in several cases increased its exports – the

Table 14.1: Trends in total human consumption of major foodstuffs in the Community of Nine

	'1973/4' – '1977/8'		'1974–8'
Common wheat	+ 0.6	Fresh milk products (excl. cream)	+ 0.5
Durum wheat	– 2.2	Cheese	+ 3.1
Potatoes	– 2.3	Butter	– 1.2
Sugar	– 1.0	Eggs	+ 0.8
Wine	– 0.8	Total meat	+ 1.0
Vegetables	+ 0.7		
Fruit (excl. citrus)	– 0.7		

Source: Commission *The Agricultural Situation in the Community, 1980 Report.*

main exception to this trend being the growth in imports of animal feedingstuffs (especially soya and cassava). World prices were, at most periods and for most products, well below Community levels: output in excess of domestic requirements could therefore be exported only with the help of costly export 'refunds'; other schemes to relieve the market were also expensive.

Agriculture, like other sectors, was hit by the rise in costs arising directly and indirectly from the energy crisis. The cost of fuel for agricultural machinery and for heating buildings was the most directly affected from 1974 onwards (figure 14.1), but other farm inputs also rose in price. Overall, the cost of agricultural requisites increased about two and a half times during the decade of the 1970s. Farmers suffered a cost-price squeeze especially during the initial impact of the crisis in 1974. In terms of the Community average, the prices received by farmers appeared to catch up with input costs by 1976, but the average is misleading: input prices rose faster than product prices in all member states of the Community except Italy and possibly Ireland, where 'green rate' changes (see later

Table 14.2: Degree of self-sufficiency in the Community

	'1973/4' %	'1977/8' %		'1974' %	'1978' %
Total cereals	92	91	Dairy products:		
Wheat (total)	104	105	fat content	. .	108
Barley	106	106	protein content	. .	108
Maize	56	52	Butter	97	111
Sugar	90	117	Cheese	103	103
Fresh vegetables	94	93	Whole milk powder	216	310†
Fresh fruit (excl. citrus)	78	76	Skim milk powder	147	110
Citrus fruit	40	42	Eggs	100	100
Wine	100	99	Beef and veal	95	96
Vegetable fats and oils	. .	25	Pigmeat	100	100
Oilseeds	. .	27*	Poultrymeat	103	105
Oilcakes	. .	6*	Sheepmeat and goatmeat	63	65

* '1976/7'.
† '1977'.

Sources: Commission *The Agricultural Situation in the Community.* Eurostat *Yearbook of Agricultural Statistics.*

Figure 14.1: Producer prices and input prices in the Community: Indices, 1970 = 100

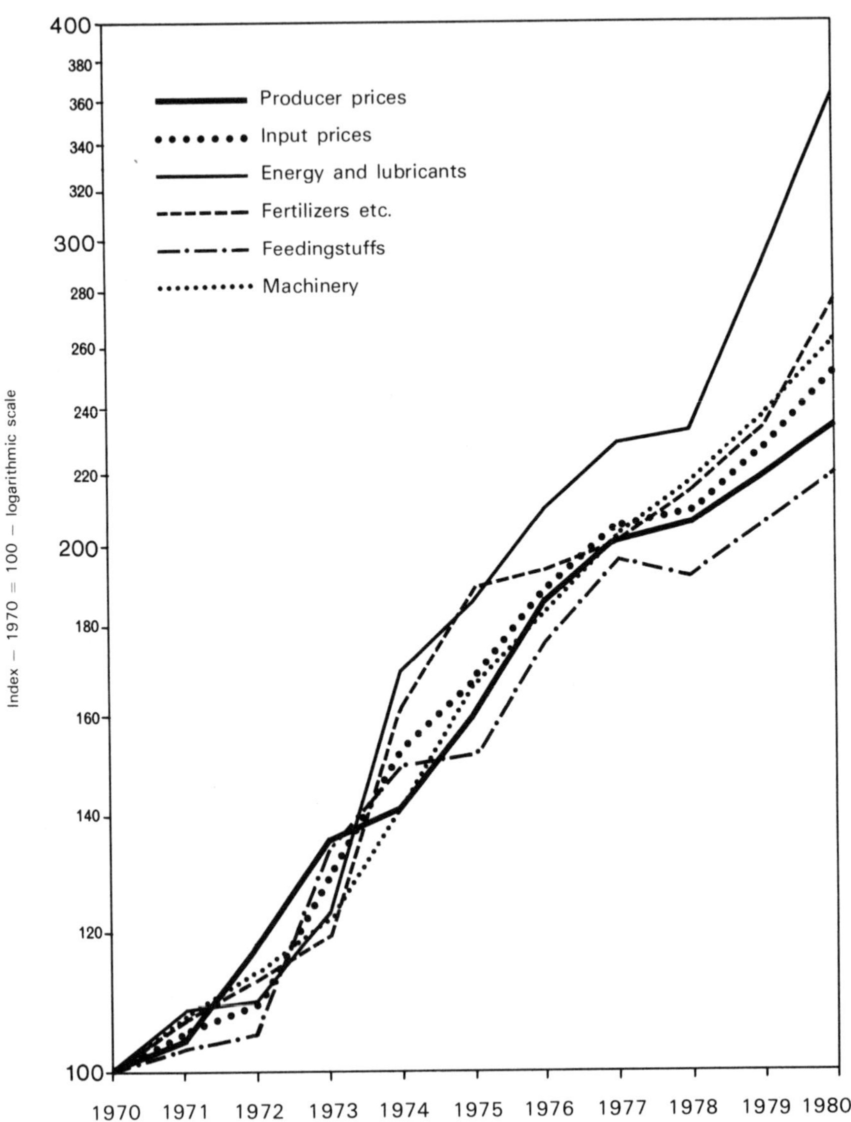

Source: Commission *The Agricultural Situation in the Community.*

section) pushed up product prices. High interest rates – well into double figures except in Germany and Benelux – particularly hit farmers who had invested heavily in farm improvements or land purchase. Preliminary data for 1980 indicated a further worsening of agriculture's 'terms of trade', with a renewed surge of energy prices.

In most countries, 'value added' per person occupied in agriculture in 'real' (constant price) terms dipped in 1974, and by 1979 had barely regained its previous level (major exceptions were again Ireland and Italy) (Table 14.3). A deterioration probably occurred in all countries in 1980.

Table 14.3: Indices of gross value added in agriculture at factor cost *per capita*, in real terms*: 1967–9 = 100

	1967–9	1973	1974	1979
Ireland	100	152	140	185
UK	100	139	131	121
Denmark	100	124	122	126
Netherlands	100	122	104	113
Belgium	100	154	127	134
Luxembourg	100	129	113	111
Germany	100	127	112	120
France	100	167	159	165
Italy	100	136	134	159
Community	100	143	133	144

* Deflated by the implicit price index of gross domestic product. See also notes to Tables 14.4 and 14.7 as regards the employment data used to obtain these *per capita* results.

Source: Eurostat *Sectoral income index 1979*. 'Rapid information note' of 18 February 1980

The rate of outflow of manpower from agriculture slowed down after 1973 (Table 14.4). In most countries the reduction in farm numbers also slowed down: in the nine Community member states, the total number of farms was falling by 4.0% per annum at the end of the 1960s and by only 2.3% per annum in 1970–75; a further slackening probably occurred in the latter half of the decade.

In spite of substantial changes in farm structures, with substantial reductions in the number of farms in the smaller size groups, inadequacies in farm size structures remained serious in many regions. Figure 14.2 indicates the contrasts between the Community member states in 1975. The influence of the historical developments described in this book remained evident, with a markedly better farm structure in the United Kingdom than in other countries. The proportion of small farms was still extremely high in Italy. Fragmentation too remained serious and consolidation work made slow progress (cf. section on structural measures below).

An extensive survey of farm structures in the Community in 1975 provided much useful information. Table 14.5 shows that in several countries – especially Italy, but also Belgium, Germany, France and even Denmark – a substantial

Table 14.4: Annual rates of reduction of agricultural employment*

	'1968–73'	'1973–8'
	%	%
Ireland	– 3.4	– 2.6
UK	– 3.5	– 1.9
Denmark	– 4.8	– 1.2
Netherlands	– 2.6	– 1.7
Belgium	– 6.4	– 3.6
Luxembourg	– 5.3	– 4.5
Germany	– 4.9	– 3.9
France	– 5.3	– 3.6
Italy	– 4.7	– 2.4
Community	– 4.8	– 3.0

* Including forestry and fisheries. The series from which these data are derived (cf. Eurostat *Social Statistics*) relate to the numbers of persons recorded as having their main occupation in agriculture, etc. (with the exception of the Netherlands where the data are expressed in 'man-years'). Consequently, they do not take account of changes in the amount of time worked, which may result in particular from changes in the incidence of part-time farming. (Contrast Table 14.7.)

Source: Commission *The Agricultural Situation in the Community.*

Table 14.5: Distribution of farms by number of labour units in 1975*

	less than 1	1–2	2 and over	Total
	%	%	%	%
Ireland	26	49	25	100
UK	11	45	44	100
Denmark	29	51	20	100
Netherlands	18	58	24	100
Belgium	39	49	12	100
Luxembourg	15	29	56	100
Germany	34	44	22	100
France	29	47	24	100
Italy	59	27	14	100
Community	43	38	19	100

* One labour unit equals 2200 hours of work per year for the farmer or his wife (or statutory working time for farm labourers).

Source: Eurostat *Farm Structures Survey 1975.*

number of farms did not provide enough work for one man. From Table 14.6 it appears that the proportion of farmers engaged full-time on their farms ranged from over 70% in the Netherlands and the United Kingdom down to 16% in Italy. Many farmers – though by no means all those with inadequate farms – had other employment. The extent of income obtained by farmers and members of their families from work off the farm was not generally known, but was certainly significant in areas near towns and also in rural areas to which small-scale industry had penetrated, as in much of south Germany. Many agricultural holdings in such areas had in fact become little more than weekend pursuits for city-

Figure 14.2: Distribution of farms by size group in 1975

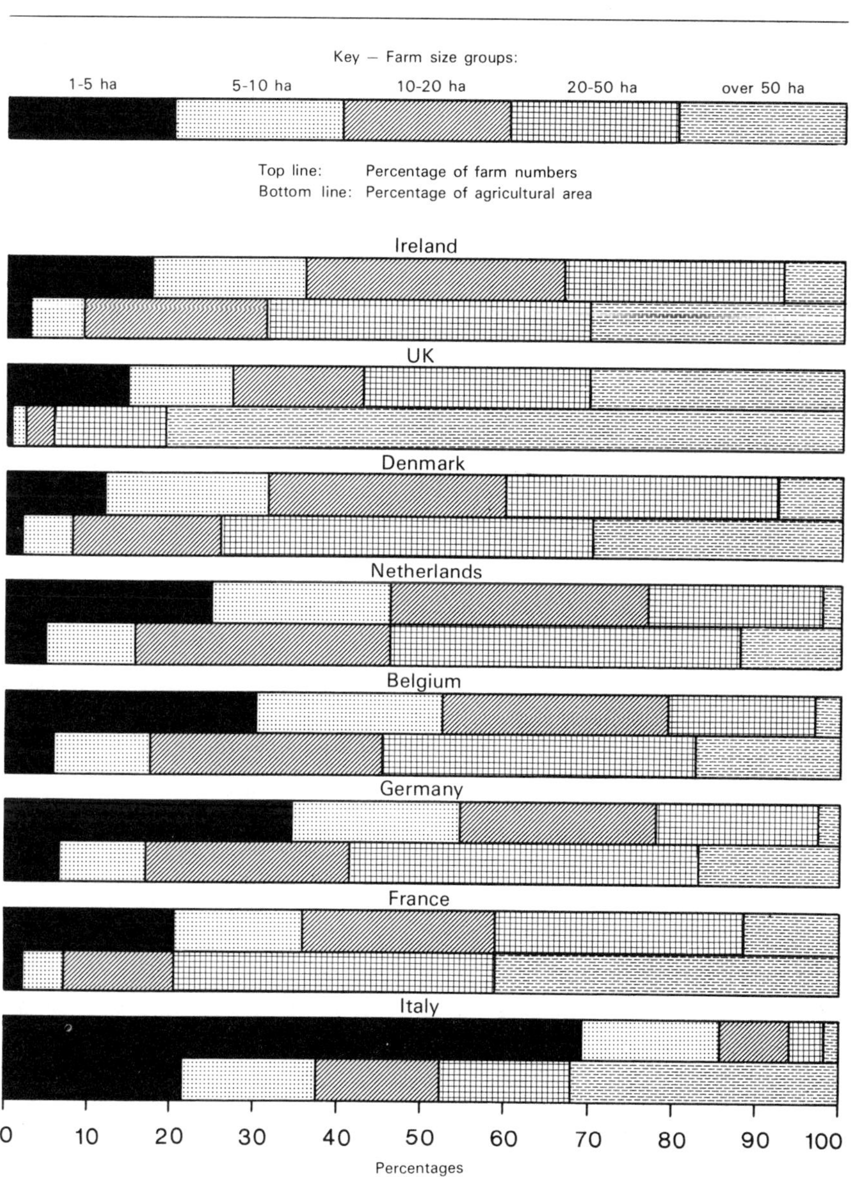

Source: Commission *The Agricultural Situation in the Community, 1979 Report.*

dwellers.[1] Still, in remote rural areas such as the south of Italy, the centre of France, the western fringes of Ireland or Scotland, natural handicaps, poor farm structures and the lack of alternative employment combined to create difficult problems. The proportion of elderly farmers – which was high on average in each Community country, as Table 14.6 shows – tended to be particularly high in such regions.

Throughout the Community, the proportion of elderly people working in agriculture was much higher than in other occupations. About one farm in five was run by a farmer over 65; these tended to be the smaller farms, but still accounted for about 13% of the agricultural area. The retirement of elderly farmers could thus release a substantial amount of land.

Table 14.6: Full-time and part-time farmers, and elderly farmers, in 1975 as percentages of total number of farms

	Farmers occupied full-time %	Farmers with other employment %	Farmers over 65 %
Ireland	56	..	25
UK	72	23	18
Denmark	60	21	18
Netherlands	71	19	10
Belgium	57	24	12
Luxembourg	70	23	24
Germany	47	43	9
France	52	20	18
Italy	16	30	29
Community	37	21*	21

Note: The above categories are not mutually exclusive – farmers can be defined as full-time yet have another occupation, and in either of these groups they can be over 65. Also, the group of farmers with other employment does not necessarily coincide with the category of farms in the previous table providing less than full-time employment.

* Excluding Ireland.

Source: Commission *The Agricultural Situation in the Community, 1979 Report.*

An analysis of trends in productivity in the agricultural sector, based on total output and estimates of the total input of factors of production, was made for each member state of the Community by Behrens and de Haen (publication due 1981). This indicated (see Table 13.7) that in the Community as a whole, between 1970 and 1976, the reduction in labour was offset by increases in other inputs (feedingstuffs, machinery, fertilisers, etc.) and by productivity gains, which together permitted output growth of nearly 2% per annum. The main increases in output were in the Benelux countries and – after its accession to the EEC – in Ireland. These countries both increased their inputs and had above-average increases in productivity. Denmark had hardly any output growth, but a big outflow of labour and high gains in productivity. In Germany output increased, with a substantial fall in inputs (particularly labour) and a big increase

[1] See also OECD's survey (1978) of part-time farming in several countries.

in productivity. France had moderate growth in output and productivity, with a small net increase in inputs. Italy had the lowest growth in productivity, associated with a low rate of decline in labour input together with limited use of capital and other inputs. The authors observed that national shares of agricultural production in the EEC had shifted in favour of the Benelux countries, and of Ireland after her entry into the EEC, and that – if Ireland was excepted – the productivity gap among the member states appeared to be increasing. In their view, this underlined the difficulties of a 'harmonised income-orientated price policy', since different national rates of price increase would be required to achieve a uniform rate of growth of farm incomes. Finally, they pointed out that there was no indication that productivity growth was slowing down and an extrapolation of the trends indicated a continuing rate of growth of production of about 2% per annum.

Table 14.7: Average annual rates of change of agricultural production, factor input and total factor productivity, 1970–76

	Production %	Input %	Productivity %
Ireland	3.6	1.0	2.6
UK	1.1	– 0.2	1.3
Denmark	0.2	– 1.9	2.1
Netherlands	4.2	1.8	2.4
Belgium-Luxembourg	3.3	0.7	2.6
Germany	1.6	– 0.9	2.6
France	1.7	0.2	1.5
Italy	1.2	0.3	0.9
Community	1.9	0.1	1.8

Note: The calculation of total factor input raises statistical and methodological problems. For labour input, which is a major element, the authors used data from the *Farm Structures Surveys* which are expressed in 'labour units' (see Table 14.5). This method shows in some cases a different rate of change from national employment data (as in Table 14.4). In particular, it indicates a much lower rate of decline in the labour input in Italy, a somewhat lower rate in France, and a higher rate in Germany. Probably the data used here reflect reality more accurately.

Source: Behrens and de Haen (1981).

Agricultural policy predicaments

In the circumstances just described, the task of agricultural policy became even more difficult: how to pursue the traditional aim of ensuring a fair income to farmers, while curbing surplus production and avoiding excessive cost, avoiding also additional inflationary pressures upon the economy – this at a time when further structural adjustment, still badly needed, was impeded by the general state of the economy. Within the Community, different rates of inflation and monetary instability compounded the difficulties; yet regional imbalances required redress.

(A) PRICE AND MARKET POLICY

(i) Common prices and national prices

The scope for increases in support prices was inevitably limited. In the Community, the Commission, in making its annual proposals, was bound to take increasingly into account market situations and the general state of the economy. The 'objective method', intended to indicate the price increases needed to ensure that 'modern' farms (i.e. with incomes comparable to other sectors) should continue to earn comparable incomes, allowing for cost increases on the one hand and estimated productivity growth on the other, would generally have led to much bigger increases than could be contemplated on other grounds. The Council's annual negotiations on the Commission's proposals for prices and related measures became increasingly difficult and protracted. In the end the Council of Agricultural Ministers, always under strong pressure from national farm organisations and from COPA, tended to decide upon somewhat higher price increases than the Commission had proposed. In 1974 the steep rise in farm costs caused the farm organisations to demand an additional price review, causing a record overall increase in common prices of about 14% in the 1974/5 agricultural year; in the following two years the average increases were about 10% and 8% respectively. At this stage the Commissioner responsible for agriculture was Pierre Lardinois – like Mansholt, a former Dutch Minister of Agriculture. The new Commission which took office in January 1977, with Finn Olav Gundelach as the Commissioner for agriculture, emphasised the need for a 'prudent' price policy. For 1977/8 it proposed only a 3% increase: the Council decided on changes amounting to 3.9% in that year, to 2.1% in 1978/9, to 1.2% in 1979/80 (the Commission had proposed a 'freeze') and – after particularly arduous negotiations lasting until May 1980 – to 4.8% for 1980/81.

While political constraints prevented any overall reduction in nominal support prices in order to discourage excess supply, there was in principle some scope for adjustments in price *relationships*. In practice, this possibility too proved limited (see Table 14.8). The main surplus problem was milk, and the milk price was the most sensitive of all, being regarded as virtually equivalent to a minimum wage for small farmers. When prices were fixed for 1979/80, no increase was made in the milk price, while other prices were increased by 1.5%. In 1980/81, however, a further increase had to be granted, in conjunction with an increase in the 'co-responsibility levy' (see below), and on the whole, over the period from 1972/3 to 1980/81, the target price for milk was raised only slightly less than the average for all products. The intervention price for butter was however increased substantially less than that for skimmed milk powder. Above-average increases were given for beef until 1974/5, but subsequently the difficulties on this market, too, limited the extent to which beef production could be encouraged as an alternative to milk. Among the cereals, durum wheat received the most encouragement up to 1976.

An evaluation of the Community's price policy during this period, however, needs to take account of two factors which worked in opposite directions. On the one hand, the evolution of common prices in 'units of account' was misleading: the unit of account, originally defined in terms of gold but subsequently calculated by reference to the stronger currencies within the 'snake' (see Chapter 12), became increasingly overvalued in terms of the average of currencies of member states.[1] This meant that for most member states, the increase in support prices in national currencies, converted at the 'green rates' (see next section), was greater than the increase in common prices in units of account. The extent of this difference is indicated by the two lines (a) and (b) of Table 14.8.

Table 14.8: Changes in target prices etc. in units of account* in nominal terms

	1972/3 – 1976/7 *indices*	1976/7 – 1980/81 *indices*
Soft wheat	134	116
Barley	132	117
Durum wheat	165	111
Sugar-beet (minimum price)	139	111
Olive oil	148	111
Table wine (average of guide prices)	137	113
Milk		
target price	142	110
intervention price for:		
butter	124	108
skim milk powder	169	110
Beef (guide price)	158	112
Pigmeat (basic price)	139	115
All products under common market regulations†		
(a) Expressed in units of account	143	113
(b) Expressed in national currencies**	162	140

* In the price series upon which these indices are based, prices up to 1978/9, originally expressed in UA, were converted into ECU by the factor 1.208953. This maintains the same relationship over the period as if prices had been expressed throughout either in UA or in ECU. See explanations of units of account at the end of the book.

† Including minor products not listed.

** Average weighted by shares in total value of output in 1974-6. The faster increase reflects the progressive overvaluation of the UA.

Source: Commission, various reports.

On the other hand, price increases in nominal terms were offset by inflation to varying degrees between member states. The real price increases allowing for these two factors – together with the effects of adjustment to the common prices as regards the three new member states – are shown in Table 14.9. In the

[1] The extent of this overvaluation was shown in 1979. When it was decided henceforth to express common farm prices in 'ECU' without any change in national prices or m.c.a.s, it was necessary to adjust nominal common prices before the 1979/80 price-fixing by a factor of 1.208953. If common prices had simply been relabelled in ECU, either all national prices would have been reduced, or all green rates would have had to be realigned, causing a substantial increase in the incidence of positive m.c.a.s and decrease in the incidence of negative m.c.a.s. (See explanation of units of account at the end of the book.)

first four years of the period, farmers in the United Kingdom, Ireland and Italy were able to keep well ahead of inflation which, however, caught up with them in the next four years. In Germany the relatively low rate of inflation enabled farm prices to rise in real terms in the early period, but not subsequently. All other member states experienced reductions in real prices, particularly from 1976/7 onwards.

Table 14.9: Changes in target prices, etc. in national currencies, in real terms*

	1972/3 – 1976/7 *indices*	1976/7 – 1980/81 *indices*
Ireland	A 107	91
	B 120	97
United Kingdom	A 97	82
	B 133	91
Denmark	A 95	97
	B 98	97
Netherlands	95	91
Belgium	98	92
Luxembourg	100	90
Germany	110	91
France	97	95
Italy	116	89
Average†	A 104	91
	B 108	92

A Theoretical changes in the absence of 'accession' adjustments.
B Real changes allowing for 'accession' adjustments.
* Target prices converted through 'green rates' and adjusted by the implicit GDP price index.
† Weighted by shares in total value of output in 1974-6.
Source: Commission and Eurostat data.

Price policy, taking account of green rate changes and varying rates of inflation, thus affected farmers in the various countries of the Community in different ways; the impact was not clearcut. On the whole, price policy did not appear to have restrained production growth significantly. Even under the 'prudent' price policy from 1976/7, when farm incomes were undoubtedly squeezed, production continued to rise. Adjustments made to commodity programmes, outlined in section (iii) below, also did not succeed in keeping down expenditure on support. By the end of 1980, the Commission was looking for a new approach, described in the final section of this chapter.

(ii) The 'agri-monetary' complication

Chapter 12 has already described how currency changes among the Six, starting in 1969, necessitated the introduction of monetary compensatory amounts on agricultural trade. With increased monetary instability after 1971, the agri-monetary problem became a major issue within the common agricultural policy.

In the case of the strong-currency member states – the Benelux countries and especially Germany – successive revaluations would in strict orthodoxy have called for corresponding changes in the 'green rates' at which common farm prices were converted into national currencies, the consequence of which would have been to reduce national farm prices proportionately (the unit of account being worth less in national currency). To avoid this, these member states accepted only partial or delayed adjustments in their green rates, and the difference between the green rates and the market rates had to be offset at their frontiers by 'positive' monetary compensatory amounts: between 1973 and 1980, the German m.c.a. ranged from 12% to 7.5%, the Benelux m.c.a. from 3.3% to 1.4%.

The opposite situation prevailed in the United Kingdom and Ireland (sharing a common currency until 1979), in France, and in Italy. In these countries, currency depreciation led to overvalued green rates and 'negative' m.c.a.s. 'Phasing-out' proved more acceptable in this situation, where it implied *increases* in national farm prices. France eliminated its m.c.a. in 1975 and again – after further falls in the franc which led in early 1978 to an m.c.a. of over 20% – in 1980. Italy behaved similarly. Ireland also sought to eliminate its m.c.a. by green rate devaluations as quickly as possible. British policy under the Labour government up to May 1979 was (as has been observed in the previous chapter) more consumer-oriented and green rate devaluations were delayed. With downward fluctuations in the pound, the British m.c.a. ranged between about 40% and 25% from late 1976 to early 1979. In 1980, as a result of several green rate devaluations followed by the remarkable recovery of sterling, the United Kingdom found itself with an *undervalued* green rate requiring a *positive* m.c.a. (which rose to 12.1% at the end of the year).

The aim of eliminating border charges within the common market was thus frustrated (only Denmark systematically eliminated m.c.a.s by immediate green rate adjustments). The introduction in March 1979 of the European Monetary System (EMS) offered the hope of greater stability among the currencies of the participating countries, without however excluding the further parity changes which different rates of inflation seemed to make inevitable; but the United Kingdom refused to participate in the EMS rules limiting currency fluctuations and providing for intervention on exchange markets.

For the common agricultural policy the issues were serious, and the Commission repeatedly urged the member states to phase out their m.c.a.s as quickly as possible. The persistence of green rates which did not reflect economic reality appeared increasingly as an element of distortion: Germany in particular came under growing pressure to remove the advantage its farmers obtained from unduly high national prices. Moreover, decisions on green rates began to have almost as much significance as the annual determination of common prices, and seriously complicated the latter. Insistence by France in particular on a commitment to phase out m.c.a.s caused delay in the introduction of the whole EMS in early 1979. Eight member states – the United Kingdom did not subscribe – agreed that any new m.c.a.s to appear after the entry into force of the EMS should be dismantled within two years, but such reductions must not result in reductions of prices expressed in national currency. One consequence was additional pressure for common price increases in order to give leeway for green rate revaluations. This underlined the assumption which had previously been implicit in the price-fixing process: that nominal support prices could not be reduced.

Finally, another problem arising from the agri-monetary situation was that of cost. M.c.a.s, if negative, acted as a subsidy on imports or a tax on exports; if positive, as a tax on imports or a subsidy on exports. In the situation prevailing between 1973 and 1979, the cost to FEOGA of subsidising food imports into the United Kingdom and Italy, or exports by Germany, substantially exceeded the revenue gained from taxing exports by France and Italy or imports by Germany. On the other hand, the increasingly large *positive* British m.c.a. in the latter part of 1980 was a source of revenue for the FEOGA.

(iii) Adjustments in commodity programmes

In view of the constraints on price policy, the Community had to rely largely on adjustments to its commodity market regimes in order to bring markets into better balance. At the same time, support was reinforced for a number of products, mainly as a result of pressure by Italy and France for better balance between 'northern' and 'southern' produce and for counterparts to trade concessions made to other Mediterranean countries. Concern over the increased competition that would result from entry into the Community by Greece, Portugal and, above all, Spain also played an increasing role. Moreover, the need at each annual review of prices and related measures to produce a 'package' acceptable to all Ministers, in view of the practice of unanimity following the 'Luxembourg compromise' of 1966, meant that frequently additional measures or derogations were granted at the request of particular delegations.

All these changes and innovations made the common market regimes increasingly complex and only a brief review will be given here (for more details see works referred to in the Bibliography). Table 14.10, showing FEOGA expenditure on the various support schemes and also the share of each commodity in both total expenditure and total output, helps to give a general perspective. The developments of the Community's market regimes also need to be seen against the background of world market developments, which are discussed in the next chapter.

CEREALS

The basic elements of the system have been outlined in chapter 12. The main adjustments were the abolition at the end of 1976/7 of different regional intervention prices, and from the beginning of that year the progressive introduction of a single intervention price for all cereals including feed wheat. The cost of the regime had been increased largely by the Community's growing surplus of wheat qualifying for intervention as milling wheat, while large quantities of feed grain were being imported. The change was intended to encourage greater use of low-quality wheat as livestock feed and to discourage the overproduction of milling wheat. The threshold price for milling wheat remained substantially higher than

Table 14.10: FEOGA expenditure on market support, etc., by commodity, and shares of the various commodities in total agricultural output

	FEOGA expenditure			Distribution of total agricultural output
	1974	1979		1979
	*million EUA**		%	%
Cereals	383	1 564	16.1	11.6
Export refunds	67	1 185		
Intervention and aids	316	379		
Rice (mainly refunds)	1	43	0.4	0.3
Sugar	106	940	9.7	2.7
Export refunds	8	685		
Intervention etc.	98	255		
Wine (mainly distillation)	41	62	0.6	6.4†
Fruit and vegetables	58	443	4.6	10.9**
Export refunds	15	35		
Intervention – fresh produce	43	123		
Processing aids	–	285		
Olive oil (aids and intervention)	109	404	4.1	0.8
Oilseeds (aids)	11	202	2.1	0.4
Tobacco (mainly premiums)	166	225	2.3	0.4
Dairy products	1 258	4 527††	46.6	19.5
Export refunds	362	2 088		
Intervention and aids	896	2 439		
Skim	499	1 671		
Butter	259	645		
Beef and veal	322	748	7.7	15.8
Export refunds	54	270		
Intervention and premiums	268	478		
Pigmeat (refunds and intervention)	70	105	1.1	12.1
Poultrymeat and eggs (refunds)	18	80	0.8	7.4
Other products	51***	372***	3.8***	11.7
All agricultural products	2 594	9 715	100.0	100.0
'Accession' compensatory amounts	346	1		
Monetary compensatory amounts	154	708		
Total FEOGA Guarantee Section†††	3 094	10 434		

* The 'basket' European unit of account (EUA) replaced the original UA in the Community budget in 1978: data for 1974 have been converted into EUA at the rate 1 UA = 1.04714 EUA.

† Of which 3.7% quality wines without common prices.

** Of which 6.3% without common prices.

†† Allowing for receipts of 94 million EUA from the co-responsibility levy.

*** Includes export refunds on processed foodstuffs.

††† Excluding fishery products (1 million EUA in 1974 and 17 million EUA in 1979).

Sources: Commission, *Financial Reports on the EAGGF* and *The Agricultural Situation in the Community, 1980 Report.*

that for other grains. (The new system was given the confusing name of the 'silo' or 'cathedral' on account of a diagrammatic representation of the various price levels.) A higher intervention price was still provided for wheat recognised under special tests (which gave rise to some difficulty) to be of bread-making quality: the original intention was that full intervention buying should not be provided, but this was subsequently modified. The intervention price for rye was, by 1980/81, still above that of feed grains (although there were large intervention stocks) and a premium was paid for rye of bread-making quality – concessions mainly to Germany where rye production was important. Rye was to become subject to the single intervention price for feed grains only in 1982/3.

Growing self-sufficiency in cereals meant that the cost of the regime depended largely on the world market price, which determined the size of the export refund that had to be given. (On the other hand, low world prices also meant high import levies: these however were treated as receipts for the general budget, while export refunds were a cost for FEOGA).

A derogation from the normal system of import levies continued to be made on behalf of Italy, a lower rate of levy being charged on seaborne imports of feed grains on which Italy had become highly dependent for the development of its livestock sector. This derogation was justified on grounds of higher handling charges at Italian ports.

Refunds continued to be granted to Community starch producers to compensate them for the higher cost of cereals on the Community market; the Council did not accept a Commission proposal to abolish this aid.

Durum wheat for pasta, in which the Community was not self-sufficient, was treated practically as a separate product, with higher prices and with direct aids to producers in Italy – the latter being a substantial cost item, but regarded by Italy as one of the elements which helped to redress the balance of advantages as between 'northern' and 'southern' produce.

SUGAR AND ISOGLUCOSE

Sugar remained the only common market regime with a quota system. As has been observed in chapter 12, at the beginning the system was primarily intended by the member states to prevent the rapid shifts in the location of production which free competition would have induced; efficient producers, however, could profitably make full use of the B-quota, on which a levy was fixed annually subject to a maximum. When the regime was reviewed in 1974, at a time of shortages and high prices on the world market (see chapter 15), the quotas were raised and the maximum levy was reduced from 40% to 30%. This permitted growth in output and in net exports, requiring increasingly expensive export refunds as world prices fell again. In 1979 the sugar regime cost over twice as much as the receipts from the levy on producers (not accounted for in FEOGA, being treated as revenue for the general Community budget). A record harvest in 1979 led to the production of 12.3 million tons of sugar, 1.4 million tons of which was produced outside the 'maximum' (A plus B) quota.

In 1980 world prices again rose above the Community level and export *levies* were applied. By this time the 1974 regime had already been temporarily extended and discussions on a new regime had started. The Community's undertaking to import 1.3 million tons of sugar from the ACP countries annually was a factor stressed by the sugar interests, who felt that the cost of re-exporting this amount should be considered as overseas aid rather than as FEOGA expenditure. The Commission's proposals for the new regime aimed to ensure that while the cost of disposal of Community sugar corresponding to preferential imports remained a Community responsibility, the cost of other exports or disposal should be borne by the producers themselves. Some reduction in the maximum quotas was proposed, with A-quotas remaining unchanged but a reduction in the B-quotas of those member states who had not fully used these quotas: the overall outcome would be 9.1 million tons of A-quotas, 2.1 million of B-quotas (instead of 2.5 million previously), giving a maximum quota of 11.2 million. As domestic consumption amounted to about 9.5 million tons, and allowing for the import of 1.3 million tons of raw ACP sugar (equal to about 1.1 million tons of refined sugar), this could lead to exports amounting to some 2.8 million tons. The Commission also proposed to introduce a levy of 2.5% on A-quota sugar, thus spreading 'producer co-responsibility' more widely. Discussion in the Council on these proposals was continuing at the end of 1980.

Sugar production costs in Italy were relatively high and production was tending to decline. To offset this, Italy was regularly authorised to grant additional aid for domestically-produced sugar, up to a certain maximum tonnage. Similar authorisation was given to France for the benefit of its overseas *départements* (DOM).

The development of isoglucose, made mainly from maize starch, as a competitor for sugar in the food industry, led to the introduction in 1977 of a quota and levy system. After the first scheme, which applied a levy to all output, had been held discriminatory by the European Court following appeal by the manufacturers, a system parallel to that for beet sugar was substituted.

WINE

The wine regime was one of the most complicated under the CAP, though not particularly expensive in relation to its share in total agricultural output. FEOGA expenditure was mainly required for aid for private storage and for the various distillation schemes. Production of 'table wines' (as opposed to quality wines) was however growing faster than consumption, and in some years the market imbalance was serious, causing agitation especially among the wine-growers of southern France. Moreover, a conflict of aims arose in the context of the Community's 'Mediterranean policy' in view of the interest of the Maghreb countries in wine exports. Market support through distillation schemes was reinforced on several occasions.

To remedy the problem of basic disequilibrium, the Commission presented an 'Action Programme', the main features of which were adopted by the Council in December 1979. Besides additional measures of market management (including provision for possible prohibition of marketing below a minimum price – a provision demanded by France as a safeguard against competition from cheaper Italian wines), this programme provided for a series of measures to control wine-growing potential. A basic element was the classification of land according to its natural suitability for wine production. The existing prohibition of new planting of table wine was extended up to 1985; premiums were made available for temporary (eight years) or permanent cessation of wine-producing, with additional compensation to producers who then gave up all agricultural work; and aid was provided for restructuring vineyards to improve quality.

FRUIT AND VEGETABLES

The market regime for fruit and vegetables was intended not to support prices permanently as a high level but – in combination with a relatively liberal import policy – to provide safeguards against periodic slumps in the Community market. The main instrument to this effect was provision for withdrawal of produce from the market by producer organisations, with compensation from FEOGA subject to certain conditions. Imports were subject to the common tariff and – for a number of products – to a system of 'reference prices': in effect minimum import prices, imports being subject to 'countervailing charges' if they failed to respect these reference prices. Under 'safeguard clauses', imports could be temporarily banned. The Community's trade and association agreements with Mediterranean countries required tariff concessions to these countries in the fruit and vegetable sector. To meet concerns by Italy and France, additional support measures were introduced from 1975 onwards, including reinforcement of the protection given by the reference price system, 'penetration premiums' to assist marketing of Italian citrus fruit, import certificates for 'sensitive' processed products, a minimum import price for tomato concentrate, and aids to processors of tomato products, canned peaches and prunes, conditional on minimum prices being paid to producers. These processing aids became a substantial extra cost item for FEOGA.

OLIVE OIL AND OILSEEDS

Another important item for Italy, which was expensive for FEOGA in relation to its importance in total Community agricultural output, was olive oil: here too a direct aid to producers was the main element in the regime. The market price for olive oil – though protected by threshold and intervention prices – had to be set low enough to enable olive oil to compete with substitute oils (vegetable oils and oilseeds being imported into the Community at low or zero rates of duty). The direct aid provided extra revenue for producers. It was fixed in advance for each marketing year and was intended to represent the difference between an estimated 'representative market price' and a higher 'production target price'. Under a new system which came into effect in 1979, an additional 'consumption aid' was introduced, payable (to bottlers) at variable rates during the marketing year to take account of changes in the price of competing oils.

The olive oil regime – especially the production aid – was one from which Greece expected to gain particular benefit on entry to the Community; it was also one likely to become exceedingly expensive on Spanish entry.

Direct aids to producers were also paid for rapeseed (or colza) and sunflower seed. Unlike the olive oil aid, these were variable depending on differences between target and market prices. Expenditure on these aids increased substantially. A subsidy to encourage the development of soyabean production was also introduced in 1974 after a temporary US embargo on exports.

TOBACCO

This was another item which was minor in terms of overall Community agricultural output – being produced mainly in Italy and France, with smaller amounts in Germany and Belgium – yet which accounted for a disproportionate share of FEOGA expenditure. Expenditure was mainly on premiums to buyers to compensate for the high price level of Community produce relative to the world market. Through relatively small increases in prices and aids, expenditure was kept roughly constant over the period, falling in relative terms.

BEEF AND VEAL

Until 1973, beef was regarded as unlikely to require support. A system of 'permanent' intervention was adopted in 1972 with the aim of *encouraging* production. A drastic slump in the market in 1974, resulting both from an increase in the number of cattle coming on the market and from a fall in demand due to the economic recession, caused the Community to suspend import licences for most categories of cattle, calves, beef and veal. When this system was abolished in 1977 and import licences were again issued freely, the variable import levy system was modified in such a way as to reinforce protection when the Community market price was low: the result was that from 1977/8 import levies were at relatively high levels (amounting in some cases to almost 100% of the purchase price on world markets), and imports were practically limited to schemes under which concessionary levies or duties applied.

Even with this drastic limitation of imports, pressures on the domestic market required substantial intervention buying, on the basis of the provisions adopted in 1972. Besides intervention, various types of premium were authorised: a slaughter premium (in effect a deficiency payment, for the benefit of the United Kingdom), a calving premium (to encourage production in Italy, still a deficit region) and a 'suckler cow' premium to encourage producers to rear calves rather than sell milk.

Continued growth of production in excess of demand caused the Community to become a net exporter of beef in 1980, with costly export refunds, while intervention stocks reached a high level.

'SHEEPMEAT'

The background to the establishment of this regime has been described in the previous chapter. Unlike all other market organisations, it included no protective measure other than the external tariff (this being 'bound' under GATT), and the duty was reduced from 20% to 10% as the counterpart for voluntary export restraint agreements by New Zealand, Australia and other suppliers – conceded with difficulty by Australia in particular, who feared that provision for export refunds in the new regime could lead to subsidised Community exports to third countries. Premiums (a form of deficiency payment) would be paid to offset the difference between the Community market price and the 'reference price' in each producing region. The reference prices were initially highest in Italy and France, lowest in the UK, but were to be aligned over four years. On French insistence, provision was made for intervention: member states could choose between premiums and intervention.

As this regime was only introduced towards the end of 1980, no expenditure was incurred by FEOGA for 'sheepmeat' up to this date. Subsequently, the premium arrangements and intervention would inevitably become an additional cost item: budgetary allowance was made for 264 million EUA in 1981.

PIGMEAT, POULTRY, EGGS

Among the livestock products, pigmeat, poultry and eggs were minor cost items, limited FEOGA expenditure being required for export refunds and – in the case of pigmeat – for occasional intervention.

DAIRY PRODUCTS

By far the biggest problem of the CAP was the dairy sector. FEOGA expenditure in support of the dairy market rose steadily during the 1970s, amounting to nearly half of total expenditure on market support – although milk production accounted for under 20% of total agricultural output (see Table 14.10). The problem was a particularly sensitive and controversial one, because of the large number of producers involved, many of them small farmers with little alternative to milk production. Output rose steadily, well in excess of total consumption of dairy produce within the Community. Profitable export possibilities were limited, except for cheeses and other specialities. Food aid provided a limited outlet (see next chapter). Commercial exports of skim milk powder required large export refunds in view of a depressed world market; virtually the only outlet for butter was the USSR, and there were strong political objections (particularly from the UK) to selling butter to Russians below Community prices.

Various schemes were implemented for subsidised disposal on Community markets. Several butter subsidy schemes were applied, to the particular benefit of the UK in view of the previous low level of prices in that country and the need to avoid the fall in consumption (with a shift to margarine) that would have resulted from a steep price increase to Community levels; a 'social butter' scheme provided cheap supplies to people in receipt of social assistance; special 'Christmas butter' sales were carried out; butter from intervention stocks was also made available at reduced prices to the armed forces, to non-profit-making organisations, and to the food processing industry.

As farmers, instead of separating milk on the farm, feeding the skim milk to livestock and sending only the cream to the dairies, increasingly delivered whole milk for separation in the dairies, the latter found themselves handling increasing quantities of skim milk. A scheme was introduced in 1976 requiring skim milk powder to be purchased from intervention as a condition for import of vegetable protein products for use in manufacturing compound animal feeds and for obtaining certain aids: this was ruled discriminatory and excessive by the European Court. From 1977 onwards subsidies were granted for the use of skim milk powder in pig, poultry and calf feeds.

The various disposal schemes helped to reduce the size of intervention stocks of butter and skim milk powder, but at considerable cost.

Farm organisations, stressing that the imbalance on the Community dairy market was largely due to the free import of fats and oils (which permitted cheap manufacture of margarine) and of substitute protein feeds such as soya, demanded higher duties or taxes on these competing products. These demands – which would have required difficult exercises in 'deconsolidation' of duties bound under GATT – were strongly opposed by manufacturing interests.

The Commission regularly stressed the need to bring milk production into line with demand, which could only be done by basic structural changes. Already in 1969 schemes had been applied to encourage the slaughter of cows on small farms and non-marketing of milk (to give preference to beef) on larger farms; a further 'beef conversion' scheme operated in 1974. In 1976 the Commission proposed an 'Action Programme' involving various measures. A new scheme of premiums for non-marketing of milk and for conversion to beef was introduced the following year. By the end of 1980 the scheme had probably taken nearly 1½ million cows out of producing milk for dairies (about 6% of the total number of cows in the Community), though it was difficult to say how far the changes would have occurred in any case, and also how far reductions in milk cow numbers on some farms were offset by increases in others: the total number of cows fell slightly, but this reduction was more than offset by the continuing rise in yields.

Given the political impossibility of reducing the milk price, the main thrust of action from 1977 onwards was the 'co-responsibility levy' imposed on milk deliveries to dairies. Whether this, in its original form, significantly discouraged output is questionable; it did however have a financial effect in that it was set against FEOGA expenditure in the dairy sector, being used (in consultation with producers) to finance a school milk programme and other schemes to expand outlets.[1] The rate of levy was modest: 1.5% in 1977/8, 0.5% in 1978/9 and in 1979/80 (it will be recalled that there was no increase in the target milk price in the latter year) and 2% in 1980/81. In the 1978 and 1979 financial years the levy brought in respectively 156 million and 94 million EUA (as compared with FEOGA expenditure on the dairy sector of 4 527 million EUA in 1979). For 1980/81, the Commission proposed a much higher 'superlevy' on increments in deliveries to dairies: this was not accepted by the Council, though it agreed that a supplementary levy would be applied in 1980/81 if milk deliveries in 1980 grew by more than 1.5% over 1979. The prospect of a high rate of levy provoked requests for exemptions on grounds which varied according to the national interest of each member state; already farmers in 'less-favoured areas' were permitted to pay less than the full 2% levy.

The Community was by no means alone in facing major problems in the dairy

[1] The co-responsibility levy was treated as 'negative expenditure' under FEOGA. This was in contrast to the sugar levy, which was a receipt for the general Community budget, and a derogation to the general budgetary principle that revenue should not be automatically allocated for a specific purpose. The practice came under criticism from the Court of Auditors (1980).

sector: the problems were inherent in western European farm structure and in demand trends. Austria in 1980 introduced a system under which the State assumed full financial responsibility for milk deliveries only up to 115% of domestic use, the excess being marketed at the expense of producers. Individual producers who went beyond their own target volumes, fixed at 93% of their deliveries in 1976–8, received a much lower price for the excess (AS 1.67 per kg as compared with AS 3.66 for deliveries within the target): the scheme was thus akin to a quota system. Switzerland introduced production quotas on milk in 1977 and revised the system in 1979: under the later arrangements, if a producer's milk deliveries exceeded his individual quota – based on deliveries in 1975/6 and subject to a maximum of 8500 kg per hectare for farms up to fifteen hectares – he had to pay a levy of SF 0.40 per kg on the excess, as compared with the basic price of SF 0.79. Milk producers in mountain zones were subject not to this quota arrangement but to a system of maximum deliveries per hectare (fixed at levels well above those normally attained).

(B) STRUCTURAL POLICY

Previous chapters in this book have emphasised the relative neglect of measures to deal with the basic causes of maladjustment in the farm sector. This theme was pursued in a series of OECD reports: in particular, the study *Low Incomes in Agriculture* (1964) pointed out that low incomes were associated with the small size of business on many farms, resulting from the insufficient mobility of farmers out of agriculture in past periods. Consequently there was an excessive labour supply in relation to the land resources available. The report declared:

> Policies so far pursued in most OECD countries have not been based on a sufficiently close analysis of the farm income problem. Emphasis has been laid on general measures of assistance, mainly through price support. These measures, while alleviating the farm income situation in the short run, are not of a nature to bring about the necessary long-term adaptation. Such an adaptation would be assisted by more selective action aimed at low-income farms. [OECD (1964) page 11]

(i) Early initiatives

The consolidation of fragmented holdings was the earliest type of measure – it has been seen that in Germany efforts in this direction were begun even in the early nineteenth century. But inheritance practices in many regions continually aggravated the problem, and fragmentation remained serious in several countries of continental western Europe.

Consolidation was a major preoccupation of the first Minister of Agriculture of the German Federal Republic, Wilhelm Niklas. In 1952, writing to the President of the Republic in support of a proposed Federal law to promote consolidation, he declared that in parts of south and south-west Germany (especially where in the past Franks and Swabians had settled), inheritance customs had led to such a degree of fragmentation that mechanisation was impossible. He described the situation in a Bavarian village where, in an area of about 1 000 hectares, there were no less than 35 000 plots. 'One can no longer speak of fields but of elongated handkerchiefs, where in ploughing, even if the peasant walks in the middle, his cow has two feet on the neighbour's land'. He said that earlier plans for consolidation over 70–80 years must be accelerated and carried out in 10–20 years.

However, consolidation schemes – usually accompanied by new roads, drainage systems, etc. and sometimes by removing farmsteads from congested villages to the consolidated farmland – still proved costly and extremely slow to implement. The German Ministry of Agriculture estimated in 1972 that about 40% of the agricultural area still had to be consolidated (some of which had previously been consolidated but had become split up again through divided inheritance): this amounted to nearly ten million hectares. The rate of implementation of consolidation schemes during the 1970s was about 200 000 hectares per annum. In France, by the late 1970s, nearly a third of the agricultural area still needed consolidation, or about ten million hectares: at the current rate of progress, this would take another 25–30 years. The ecological effects of consolidation schemes, where they involved removal of hedges which had acted as wind-breaks and as shelters for wild-life, had to be taken increasingly into account.

Consolidation schemes, moreover, did not necessarily create viable units if the total area of the farms concerned remained small.

The possibilities for more far-reaching action were already shown by some countries during the 1950s: Sweden and the Netherlands were the first in the field. In Sweden, County Agricultural Boards had extensive powers to promote restructuring, through intervention on the land market, provision of state-guaranteed loans, advisory work, etc. The Netherlands tended to concentrate efforts on designated areas, combining land consolidation, improved water management, reallocation of farmland, compensation to farmers who gave up their farms, and retraining facilities.

In France a new departure in favour of structural reform took place on the basis of the *Loi d'Orientation* of 1960 and especially the *Loi complémentaire* of 1962. The latter resulted from the combined efforts of a young and dynamic Minister of Agriculture, Edgar Pisani, and of the pressure exerted by the young farmers' union, the CNJA (*Centre National des Jeunes Agriculteurs*), then led by Michel Debatisse (see chapter 11). Under this legislation, public bodies, the SAFER (*Sociétés d'Aménagement Foncier et d'Etablissement Rural*) were established to buy and resell farmland with the aim of improving structures, and special retirement aids were granted to elderly farmers.

(ii) The development of the common structural policy

Within the Community, structural policy took longer to organise than market policy, and remained a much less significant element in terms of FEOGA expenditure. The provision of the basic financial regulation of 1962 under which structural measures should receive up to a third of FEOGA expenditure on market support was never fulfilled. However, Community funds for 'individual projects' were made available from 1964 onwards. By the time this scheme (Regulation 17/64) was terminated in 1979, it had provided a total of two billion units of account for over seven thousand projects, involving a total investment of nearly ten billion UA. The projects – usually submitted not by individual farmers but by co-operatives, etc. – concerned improvement of production structures (land improvement, irrigation, drainage, etc.) and of marketing structures (dairies, slaughter-houses, packing and processing of fruit and vegetables, wine-making etc.).

The 'Mansholt Plan' presented by the Commission in December 1968 (in the form of a memorandum) broke new ground. Structural policy was seen as a necessary adjunct to the common price policy. The problems of oversupply and the need for basic adjustments were recognised. The long-term programme 'Agriculture 1980' set out a new approach to price policy, under which prices were again to play their true economic role of guiding production. Structural policy would aim to create 'modern production units' through selective investment aids: such farms should reach 80–120 hectares of cropland, or 40–60 cows, with similarly high targets in other lines of production. About five million people should be helped to leave agriculture during the 1970s by early retirement schemes or by retraining. The land thus released would be made available for amalgamating holdings or for afforestation. Since the creation of the 'modern production units' would accelerate production growth, the agricultural area in the Community should be reduced by at least 5 million hectares in the period 1970–80 (it amounted to about 70 million hectares in 1968). The Community dairy herd should be reduced by about 3 million cows by 1976, by slaughter and by encouraging a shift to beef production.

The Mansholt Plan provoked violent opposition in farming circles throughout the Community. Attention was concentrated on the proposals for reducing labour and land. Mansholt was attacked in public debate on numerous occasions. In France in particular, bitter reactions were provoked by the Mansholt Plan, as the debate took place against a background of concern about the prospects for the farm community, faced by acute problems of transition from a peasant-type, small-scale farming pattern, depending almost exclusively on family labour and making little use of purchased inputs, to larger-scale, mechanised and capital-intensive units. A pessimistic analysis of prospects for French agriculture (the 'Vedel Report') was made at this time: it expected structural reform to be insufficient and pressure on farm prices to continue.

In April 1970 the Commission presented a set of specific proposals on the basis of its earlier memorandum. However, as has already been seen in chapter 11, the debate over the structural proposals became intertwined with the Council's negotiations over prices, and it was not until May 1971, after particularly violent farmers' demonstrations, that the Council agreed on a 'package' including 'guidelines' for socio-structural policy; in the following year three schemes were adopted on this basis. This outcome was both less ambitious and less selective than the Mansholt Plan, and the link stressed by Mansholt between structural adjustment and action to reduce surpluses was much weakened: in particular, the Commission's proposal for aids to encourage shifts of land from farming to forestry, recreation, etc. was omitted. However, provision was made for 'common measures' (the term laid down in the financial Regulation No. 729/70) for investment aids for farm modernisation (Directive 72/159); for payments to outgoers in the form of annuities (or lump sums) to elderly farmers, or premiums (not eligible for Community finance) to younger ones (Directive 72/160); and for promotion of socio-economic guidance and training (Directive 72/161).[1]

The investment aids under Directive 72/159 – in the form of interest rate subsidies or capital grants – were made available for 'farms suitable for development', able to reach an income comparable with other occupations for 'one or two' labour units (the Commission had proposed a minimum of two labour units). Selectivity in the provision of credit was aimed at, in that farms not meeting these criteria could in principle get credit from national funds only on less favourable terms, though an important exception was made which was intended to be temporary but which by 1980 was still being applied in some member states. National aid could also supplement Community aid for farms with development plans, mainly for land improvement and buildings.

Subsequently, other common 'socio-structural' programmes were adopted. Directive 75/268 provided income aids or 'compensatory allowances' – based usually on the number of livestock units per farm – to farmers in mountain, hill and other less-favoured areas. This gave effect to the commitment obtained by the United Kingdom during its entry negotiations, but was of benefit to other member states too: the definition of less-favoured areas covered not only the hill areas of the UK but also large parts of Ireland, Italy and France, as well as some areas in Germany and the Benelux countries. The terms of Directive 72/159 were made more favourable for farmers in these regions.

Regulation 355/77, to promote improvements in processing and marketing, was designed as a 'common measure' to replace the marketing aids previously given as 'individual projects' under Regulation 17/64. As experience with the previous scheme showed that applications would be numerous, a financial ceiling of 80 million UA per annum was imposed.

A significant new development in structural policy was the introduction from 1978 onwards of a series of measures for designated regions. An element of regional differentiation existed already in Directive 72/160 in that the rate of FEOGA contributions to the compensatory income aids was set at 65% for Ireland and parts of Italy, instead of the general rate of 25%. Similarly Directive 75/268 provided initially 35% of expenditure for Ireland and Italy instead of 25%. The important 'Mediterranean package' introduced from 1978 onwards – which included measures in favour of Ireland and Northern Ireland – was intended to support specific types of improvement corresponding to the needs

[1] The legal form of the 'Directive', as opposed to a 'Regulation', left scope for flexibility in implementation under national legislation. On the other hand, this proved a hindrance and cause of delay in Italy, where legislation was required in each region. In official numbering, the year of enactment is given first for Directives, second for Regulations.

of the regions concerned. Measures adopted by the end of 1980 – with rates of FEOGA contribution up to 50% – included irrigation in the Mezzogiorno and in Corsica; the development of an agricultural advisory service in Italy; reafforestation and development of rural infrastructures in Italy and the south of France; support for producer groups in Italy, parts of France and Belgium (for certain products); drainage and farm improvement in the west of Ireland; a cross-border drainage scheme in the north of Ireland; and promotion of sheep-farming in Greenland. To promote marketing and processing in the Mezzogiorno and the south of France, funds amounting to 40 million UA per annum were added to the 80 million per annum already provided under Regulation 355/77. The maximum rate of aid under Directive 75/268 to farmers in less-favoured areas was raised, and the rate of FEOGA contribution increased to 50% in Ireland and Italy. Further proposals still before the Council at the end of 1980 included livestock development in Italy; farm improvement in Northern Ireland; 'integrated' rural development projects in the Western Isles of Scotland, in the French *département* of the Lozère and in the Belgian province of Luxembourg. Proposals for the French overseas departments and finally for certain regions in Germany had been added to the original package.

In presenting these regionalised schemes, the Commission was motivated both by the general aim of shifting the balance of the CAP more in favour of the Mediterranean regions and by the realisation that the existing structural policy was making little impact on the less-favoured areas. Apart from the fact that Italy was very slow to apply the socio-structural directives, the selective criteria for farm modernisation under Directive 72/159 meant that in areas where structures were particularly bad, few farms could qualify for assistance. The Commission therefore submitted proposals – also still on the Council table at the end of 1980 – with the aim of making the farm modernisation scheme under Directive 72/159 more flexible, and of increasing the impact of Directive 72/160 for the cessation of farming, in particular by increasing the rate of annuity to outgoing farmers.

(iii) Results of the common structural policy

The impact of the 1972 Directives was mixed: on the whole, it was limited by comparison with the adjustment aims of the Mansholt Plan and even by comparison with the more modest ambitions of the 1972 package itself. It initially prove difficult to spend even the relatively small sums provided for the FEOGA Guidance Section under Regulation 729/70, which had fixed a 'ceiling' of 285 million UA per annum, raised to 325 million on the enlargement of the Community in 1973. However, the number of farm development plans approved under Directive 72/159 rose steadily, reaching by the end of 1978[1] a total of 107 000; among farms over ten hectares in the Community, about one in twenty

[1] Data on implementation of the structural Directives (see annual Commission reports) become available only with some delay: by January 1981, the latest data described the situation at the end of 1978. The implementation reports need to be studied in conjunction with the financial data in the annual FEOGA reports, which also provide slightly more up-to-date figures (as in Table 14.11).

had by then presented a plan. The impact varied widely as between the member states: in terms of the number of beneficiaries and in relation to the agricultural area, the most extensive use of the scheme was made in the Netherlands, Belgium and Denmark, while use appeared moderate in Ireland and Germany, and low in the UK. France got off to a late start; Italy even later and then only in a few regions. In terms of the amount of farm investment under the scheme, the picture was rather different, with a comparatively high impact in the United Kingdom, where the 'development farms' were large and invested heavily; in the Netherlands too the volume of investment was relatively high; in Belgium and Ireland, on the other hand, both the average size of farms in the scheme and their investments were small. (The figures in Table 14.11 for Directive 72/159 are not entirely comparable between countries, because the UK and Ireland gave aid as outright capital grants rather than as interest rate subsidy payable over several years.) In contradiction to the original aims of the scheme, most of the development plans did not involve an increase in area but only intensification of output. About half the plans aimed at increased numbers of cattle.

Table 14.11: Expenditure on common socio-structural measures*

	1978	1979	IR	UK	DK	NL	B	Lux	D	F	IT
	million EUA		*Distribution of expenditure, 1974–9 percentages*								
Farm modernisation (Dir. 72/159)	44.7	85.7	8	29	9	13	1	–	35	4	1
Cessation of activities (Dir. 72/160)	0.3	0.7	8	4	–	1	2	–	82	3	–
Socio-economic guidance and vocational training (Dir. 72/161)	3.3	2.2	4	3	1	0.1	4	–	19	62	7
Less-favoured areas (Dir. 75/268)	58.4	93.2	15	38	–	–	3	0.8	15	26	2
Marketing and processing (Reg. 355/77)†	102.9	133.6	6	8	4	5	3	0.2	14	18	41

* Data relate to period of 'eligible' expenditure by member states (FEOGA reimbursement is made later).

† Including, in 1979, additional funds earmarked for the Mezzogiorno, Languedoc-Roussillon, etc. 'Commitment' appropriations: payments in these years were much less.

Note: This table does not include all FEOGA Guidance Section expenditure. In particular:
(a) Payments continued to be made for 'individual projects' under Reg. 17/64.
(b) Various structural schemes were related to product sectors, the most important being premiums for the non-marketing of milk and conversion to beef.
(c) The 'Mediterranean package' would make a budgetary impact in later years.

Source: Commission, *Financial Reports on the EAGGF, Guidance Section.*

It had been intended that encouragement to outgoers under Directive 72/160 should support farm modernisation under Directive 72/159 by providing extra land for enlargement. This aim was largely frustrated, except to some extent in Germany, because of the limited success of Directive 72/160. Annual applications under the scheme reached a peak of 15 700 in 1975 and then fell off. By the end of 1978 there were about 46 000 beneficiaries altogether, almost all in France and Germany. Some 650 000 hectares had been freed under the scheme and had been used to enlarge nearly 100 000 other farms – but usually the land was split up, and of the other farms only 15% were carrying out a development plan. In France especially, very few of the claims for outgoers' payments met the 'structural' conditions necessary for Community financing, partly because of France's late start with Directive 72/159 (hence France's low share in terms of FEOGA expenditure and Germany's relatively high share). The outgoers were almost entirely elderly farmers taking the annuity, very few younger men being interested in the premium offered by the scheme (no doubt because of the poor job prospects outside agriculture, together with the inadequacy of the payment). Denmark opted out of Directive 72/160 entirely; Italy failed to apply it in spite of the large potential benefit. FEOGA expenditure was negligible.

Under Directive 72/161, a few hundred 'socio-economic counsellors' had been trained, mostly in Germany and the Netherlands, and by the end of 1978 about 110 000 farm people had followed agricultural training courses, mostly in France. In terms of promoting movement out of agriculture, the impact of this scheme too was small.

The structural schemes which consistently attracted most applications were the old Regulation 17/64 for 'individual projects' of improvement of production (mainly land improvement) and of marketing structures, and Regulation 355/77 which in part replaced the former. In both cases, applications for aid exceeded the sums available, even in the case of Italy (though in the latter case not all the projects were in fact carried through). These schemes contributed to improved efficiency, but not to structural adjustment in the sense envisaged by the Mansholt Plan.

Directive 75/268 provided useful social benefits in less-favoured areas. By the end of 1978, some 390 000 farmers were receiving the compensatory allowance, most of them in Ireland, France and Germany – though the highest share of FEOGA expenditure went to the United Kingdom, where the highest rate per livestock unit was paid and the average size of cattle or sheep herd was relatively large. Italy began to apply the Directive only in 1978, though here again the potential benefit was considerable.

The key to structural adjustment was land mobility, and this problem was not solved. A scheme to encourage long-term leasing had been part of the original Mansholt Plan, but was withdrawn after encountering stiff opposition: land tenure remained a matter for national legislation. In several countries farmers were suspicious of tenancy – although most farm enlargements took place through renting additional land. In general farmers wanted to own their farms, but with rising land prices and high interest rates, this was increasingly difficult.

The alternative path to improving farm income – intensification of output on existing area – was hardly consistent with the general policy aim of adjusting supply to demand. Moreover, as has been seen, about half the development plans under Directive 72/159 related to cattle, and it appeared that most of these aimed at milk production. Directive 72/159 included a special premium to encourage beef cattle or sheep rearing: only about one in ten of the farmers pursuing development plans took this option, most of these being in the United Kingdom and a few in France. Elsewhere, the amount of land available even to 'development farms' hardly permitted the extensive use of land involved in rearing livestock for meat, while milk production offered a higher return per hectare.

The relationship between structural and market policy became a major and difficult issue. Aid for buying cows under Directive 72/159 was suspended in 1977. In the context of the 1980 price package, the Council agreed in principle to restrict investment aids for dairy production, and to modify previous restrictions concerning pig production: different views on the interpretation of these agreements, however, prevented them being enacted by the end of 1980. The concern with effects on supply also affected the formulation of some of the regional measures already referred to, since they too, while pursuing the legitimate aim of redressing imbalances within the Community, would inevitably add to output.

The cost of the Community's structural policy was minor by comparison with market support, but became an issue in view of the extra expenditure foreseen for the 'Mediterranean package' and the other regional measures likely also to be incurred if the 1972 Directives were made more flexible and rates of aid generally increased. The basic rate of FEOGA contributions had been fixed at 25% of expenditure in the 1972 Directives, subject to higher rates in less-favoured regions; higher rates of contribution were however provided under the regional measures. The Guidance Section ceiling of 325 million UA per annum was raised in 1979, in view of the 'Mediterranean package' and other measures, to 3600 million EUA for the five years 1980–84 (a total sum still substantially less than current annual expenditure on market support for dairy products alone); as it was provided that this sum should not prevent the adoption of additional measures, it was questionable whether in fact it constituted a 'ceiling'. By the end of 1980, estimates of the cost of implementing all the measures already adopted and those still before the Council (including measures in fisheries) indicated that the amount of 3600 million UA would be fully used up in 1980–84.

(iv) Environmental considerations

The climate of opinion within which structural policy had to be developed was, by 1980, significantly different from that of the 1960s. The assumption of continuing economic growth had been rudely shaken; with growing urban unemployment, the security of farm work seemed some compensation for lower earnings. Growing preoccupation with the environment also made its impact: while living standards in the countryside steadily improved – except perhaps in the least-favoured regions – urban congestion and pollution grew worse. It was doubtful whether further large transfers of population from country to town would produce added welfare. Policy aims in terms of farm structure itself also came into question, with increased attention to agriculture's contribution to pollution: intensive livestock units were treated with growing suspicion and subjected in several countries to environmental legislation. Sweden – which as has been seen was a forerunner in structural reform – decided under new guidelines in 1977 to keep the arable area at its current level (accepting that this would mean some surplus production) and, while continuing to promote adequate-sized farms, aimed to discourage further enlargement of already well-established larger units. Sicco Mansholt himself, in his retirement, was converted to the 'small is beautiful' view and to 'closed-cycle', organic farming on ecological principles (Mansholt, 1979).

The full implications of such trends in thought remained to be worked out. The basic problem of excess agricultural resources and their consequent relatively low remuneration remained largely unsolved. The farming population in general still had income expectations geared to the rest of the economy, and could hardly be expected to accept relatively low standards of living for the sake of ecological principles.

The CAP, the Community budget and the European Parliament

As has been seen in chapter 12, the Council adopted on 21 April 1970 a vital Decision – subsequently ratified by national parliaments – creating 'own resources' for the Community budget. These resources were to consist, in the first place, of receipts from the import levies on agricultural products, the sugar levy and receipts from customs duties; in the second place, of a share of the value added tax (VAT) levied in member states, to the extent necessary to balance the budget but subject to a maximum of 1% of the VAT base. (This VAT contribution became operative only under the 1979 budget, after a uniform basis of assessment had been introduced – in the meantime, member states made financial contributions based on their gross domestic product.)

The Community's budgetary procedures, as defined (following amendments) by Article 203 of the Treaty of Rome, were complex. Following presentation of a 'preliminary draft budget' by the Commission, and establishment of a 'draft budget' by the Council, powers of decision were divided between the Council and the Parliament. In the case of 'compulsory' expenditure (i.e. held to result 'necessarily' from the Treaty or derived legislation – a definition which included FEOGA Guarantee expenditure), a proposal by the Parliament to *reallocate* funds stood unless the Council took action against it, while a Parliament request to *raise* expenditure required positive approval by the Council; in the case of non-compulsory expenditure (including notably the Regional and Social Funds and overseas aid), Parliament could propose increases – subject to maxima laid down each year by reference to developments in Community GNP, national budgets and the cost of living – which went through unless *modified* by the Council. In the budgetary procedure the Council acted by qualified majority. It was for the President of the Parliament to declare the budget finally adopted.

In June 1979 the European Parliament – previously formed of members delegated by national parliaments – was directly elected. The new Parliament was anxious to use to the full its powers in the budgetary field; moreover, strong elements were critical of the level of agricultural expenditure. A conflict with the Council arose over the 1980 draft budget, as the Parliament on the one hand demanded increased spending on the Regional and Social Funds and other items of 'non-compulsory' expenditure, and on the other sought to gain control of spending under the FEOGA Guarantee Section by the device of transferring funds to the reserve. These so-called 'Dankert modifications' (after the rapporteur of the Parliament's Budget Committee, Piet Dankert), were rejected by the Budget Council, which however – followed by the Agricultural Council – expressed 'sympathy and understanding' for the Parliament's move over agricultural spending and declared that the Council shared the Parliament's preoccupation over the need to deal with the financial consequences of persistent agricultural surpluses. The significance of the issue was underlined by the need for a supplementary budget for 1979 to cover increased FEOGA expenditure. In December, the Parliament accepted the supplementary budget after receiving explanations from the Commissioner, Finn Gundelach, on dairy export policy in particular, but rejected the 1980 budget by a large majority. Under Treaty rules, the Community's expenditure, from 1 January onwards, was limited to one one-twelfth of the 1979 budget in each month.

In March 1980, however, the Parliament was unable to give an opinion on the level of farm prices for 1980/81, beyond stating that the average increase of 2.4% proposed by the Commission was inadequate. The Parliament's Agriculture Committee supported the increase of 7.9% demanded by COPA; the Liberal and Christian Democrat groups suggested a compromise of 5%, but no majority could be found for any figure. The Parliament was also unable to give a view on the proposed milk 'superlevy'. A basic conflict over the CAP was apparent between political groups but also between nationalities, with resistance to changes in the CAP principally among French Gaullists, the Irish members, German Christian Democrats and Liberals, and some British Conservatives (including Sir Henry Plumb, former President of the National Farmers' Union, who was chairman of the Parliament's Agriculture Committee).

A revised 1980 budget was not adopted until July, with increases going some way to meet the Parliament's demands over 'non-compulsory' expenditure, but also with increases in appropriations for FEOGA to take account of the 4.8% price increase and other decisions which the Agricultural Council had taken in the meantime.

The draft 1981 budget provoked a further dispute between Parliament and Council, as the Parliament again sought to increase expenditure on the Regional and Social Funds, energy development and overseas aid. The procedures it used were regarded as unconstitutional by France, supported by some other member states. Madame Simone Veil, President of the Parliament, nevertheless declared the 1981 budget adopted.

These tensions brought into the foreground the problems arising from the growth in agricultural expenditure. The cost of export refunds on grain, sugar, dairy products, beef and other items had risen from a low point in 1974, when world market prices for grain and sugar were high, to nearly 5 billion EUA in 1979; as has been seen in Table 14.10, for dairy products alone, export refunds plus intervention amounted to about 4.5 billion EUA. In the 1981 budget, total expenditure by the FEOGA Guarantee Section accounted for 12.9 billion EUA out of a total of 19.3 billion. This high share of agriculture in the total partly reflected the comparative lack of development of other Community policies. But the growth in agricultural spending was bringing the total budget perilously close to the ceiling imposed by the 1% limit on the VAT contribution. The 1981

Table 14.12: The Community budget

	1980	1981
	million EUA/ECU	
Appropriations*		
Agricultural policy	11 803	13 338
Guarantee Section	11 486	12 870
Guidance Section	317	468
Fisheries policy	64	48
Regional Fund	403	619
Social Fund	701	620
Food aid	396	369
Other development aid	246	239
Repayments to UK	848	1 388
Other expenditure	1 721	2 707
Total appropriations	16 182	19 328
Revenue†		
Agricultural import levies	1 520	1 902
Sugar and isoglucose levies	504	571
Customs duties	6 000	6 274
Other	902	330
Total above revenue	8 926	9 077
Revenue required from VAT	7 256	10 251
Total revenue	16 182	19 328
Yield from max. 1% VAT†	9 910	11 510
as % of revenue required	0.73%	0.89%

* Appropriations for payments as finally adopted by the European Parliament at the end of 1980.

† Preliminary estimates.

Source: 1981 Budget of the European Communities (*Official Journal, no. L378 of 31 Dec. 1980*).

budget as amended by the Parliament called on VAT contributions at the rate of 0.89%, and the Commission estimated that funds were bound to run out in the early 1980s. Under the 1970 Financial Decision, however, the 1% ceiling could only be raised by a procedure involving a Commission proposal and Council approval, and ratification by all national parliaments.

The budgetary problem was further complicated by the United Kingdom's demands for fairer arrangements: as has been pointed out in the previous chapter, the agreement of 30 May 1980 on the UK's budgetary requests included the commitment to resolve the long-term problem of budgetary imbalance by means of 'structural changes', and the Commission was requested to make proposals to this effect by June 1981. Keeping down the cost of the CAP was crucial to any solution of these problems.

'Reform of the CAP'

(A) ECONOMIC ANALYSES

Many studies of the CAP were made during the 1970s. Economists tended to emphasise that the root of the problem was the dual role expected of price policy: being used to support farm incomes, prices could not satisfactorily guide production in a situation of oversupply. Some reports (e.g. House of Lords, 1980) simply advocated a progressive reduction of producer prices in real terms. A report by the expert committee to the German Economics Ministry (*Deutscher Bundestag,* 1980) advocated placing an upper limit on expenditure, progressively abolishing intervention on domestic markets and export subsidies, and relying for support simply on protection at the common frontier. Many economists, recognising the need to soften the impact of price reductions on farm incomes, advocated direct income payments: this approach was advocated by panels of economists in 1970 (Atlantic Institute) and in 1973 (Wageningen Memorandum), and developed in more detail by, for example, Van Riemsdijk (1973), Bergmann (1975), Koester and Tangermann (1977). Some economists began to explore the scope for national policies to operate within the CAP (Britton, 1976) and even to suggest more precisely that the Community member states should regain freedom to fix national prices, with compensatory mechanisms at the frontier (Marsh, 1976/7); a joint paper by two British and two German economists (Heidhues *et al.*, 1978), while stressing the need for common prices, came to the conclusion that the system of green rates provided a useful element of flexibility and should be regarded as a long-term policy instrument. Memoranda attached to the above-mentioned House of Lords report, by Marsh (1980) and by Tangermann (1980), made suggestions to the effect that common prices should be lowered to levels more in line with international prices, while governments should have scope to decide what level of direct income payments would be appropriate for farmers in their country. While the studies so far referred to were mostly the work of German, British and Dutch economists, an assessment of different approaches, also tending to advocate a combination of lower prices with direct income compensation, was made in 1980 by two French economists, Mahé and Roudet.

On the whole, reduced support with direct income aids (not to be confused with deficiency payments or direct product aids) was clearly the preferred solution of agricultural economists. The writings referred to, however, did not always take sufficiently into account both the political and the budgetary and legal constraints in the Community context. Direct income aids already existed in the Community (and other countries) for farmers in less-favoured regions: however, using them more generally, as at least a partial substitute for price support, was a different matter. Apart from the fact that generalised use of direct payments was opposed by farm organisations, who preferred farmers to get their income through the market, it was generally admitted that they would, at least initially, raise budgetary costs. If financed by the Community, this would tend to aggravate political problems already arising from transfers between member states through the Community budget (Germany and the UK being likely to be the main net contributors). Moreover, as the total Community budget approached the 1% VAT ceiling, little scope was left for such initiatives. The administrative problem of handling direct payments to millions of small farmers throughout the Community tended to be under-rated. If financed nationally, and particularly if scope were admitted for different national rates of aid or levels of price, distortions of the conditions of competition would arise which would almost certainly be in contradiction with Treaty rules (Article 92). Further, national financing would be no solution for member states (Ireland and Italy) with relatively weak economies but large and structurally-deficient sectors. A more fundamental difficulty was how to ensure that direct payments would in practice *substitute* for price support and would not simply be added. In fact, as has been seen early in this chapter, real farm prices underwent a substantial decline after 1976 in all the Community member states (a point which was disregarded by some of those who advocated price reductions). In a highly inflationary economy, the establishment of a clear criterion for price policy, even in the absence of political considerations, was not an easy one. Further analysis of these various problems appeared necessary.

Other approaches received on the whole less favour among economists. Quota systems were generally regarded as likely to lead to excessive rigidity in farm structures. Switzerland, as has been seen, introduced individual farm quotas for milk: the difficulties associated with this were underlined by Rieder (1980), who pointed out in particular that as over-capacity would remain, so would the need to maintain the quotas. However, a system of degressive prices by *tranches* of production on each farm, similar to the French *quantum* for wheat before the CAP, was advocated by Mahé and Roudet (1980) and in a report by some French socialists in the European Parliament which was drafted by the former French Minister of Agriculture, Edgar Pisani (see Agra-Europe, 1980). Such a system would reduce the benefits of price support for the larger farms. As has been seen, the only quota system so far applied in the Community was that for sugar, where national interest conflicts tended to push up the total quota. With quotas, as with direct income aids, solutions which would have been difficult even at the national level were subject to additional complications in a Community of Nine. As with direct payments, further analysis seemed desirable, for instance into the extent to which the limitations of quotas might be mitigated by making them negotiable between producers (a solution advocated already in 1973 by OECD, taking into account experience with milk in Canada, tobacco and cotton in the United States).

Structural policy received on the whole less attention. Probably most economists were aware of its limitations as a means of bringing about basic adjustments in a period of stagnant economies. The growth of part-time farming (see OECD, 1978) was a development which provoked some thought, particularly in Germany. Already in 1972, Priebe pointed out the advantages of a 'multi-functional agrarian structure with mixed farms, varying sizes of farms and the combination of several occupations and incomes'. This could lead to less intensive production and ease the surplus problem.

(B) THE POLITICAL LEVEL

At the political level too, demands for reform of the CAP were increasingly widespread and insistent by the end of 1980, primarily in view of the budget situation. In the United Kingdom the CAP had always been criticised by practically all quarters except perhaps the NFU. Now it was increasingly under attack in Germany too, especially in the Socialist Party and the trade union federation (*Deutsche Gewerkschaftsbund*); as has been seen, proposals for reform were put to the Economics Ministry by its expert consultative committee. Much then depended on the attitude of the Minister of Agriculture, Josef Ertl. France's traditional interest in the CAP remained, but here too there was increased recognition of the need for adjustments, as in a report by the *Commissariat du Plan* (1980); even among farmers' representatives, there appeared to be acceptance of the principle of some degree of participation by producers in the costs of market support, though stress was laid on the need to reinforce Community preference especially by controlling imports of cereal substitutes and of vegetable fats and oils (see Agra-Europe (1980) *La PAC: Qui veut Quoi?*). The Community's Economic and Social Committee was preparing a report on the CAP with numerous contributions from Committee members.

Discussion of CAP reform at the Community level was not new. Indeed the Commission had examined possibilities for improvements in a series of reports: *Improvement of the Common Agricultural Policy* in 1973; *Stocktaking of the Common Agricultural Policy* in 1975 (an exercise demanded by the German government following the exceptional mid-season price increase granted to farmers in September 1974); 'Action Programmes' for milk in 1976 and for wine in 1978; and finally *Reflections on the Common Agricultural Policy* in December 1980. The cost-saving adjustments to the market regimes referred to earlier in this chapter mostly originated with one or other of these reports: in particular, the move to a single intervention price for cereals, the adjustments in the price ratio between butter and skim milk powder, the introduction of a milk co-responsibility levy, the measures to adjust wine production to market trends. In addition, as has been seen, the Commission from 1976 onward stressed the need for a 'prudent' price policy.

The Commission's *Reflections* of December 1980 (see *Green Europe*, 1980, no. 13) were presented as an element in the general study of 'structural changes' in the Community required of the Commission by the end of June 1981. The Commission stressed that the re-examination should not call into question common financial responsibility for common policies, nor the three basic principles upon which the CAP had been developed – which it described as 'freedom of trade and Community preference, the creation of market organisations based on common prices, and the sharing of the cost of this common policy'. The Commission considered that for political, financial and administrative reasons, it was not possible to envisage a radically different model for the CAP than the support of market prices, though it did not rule out direct income support for problems of a special regional nature or concerning particular commodities.

But solutions had to be found for the difficulties encountered by the CAP.

In particular, under the existing market regimes, the rise in expenditure on milk, beef and processed fruit and vegetables could no longer be kept under control, and production of cereals, sugar and wine was tending to rise faster than consumption. Price support, moreover, had tended to aggravate disparities within agriculture, benefiting mainly the larger farms and the more prosperous regions.

The Commission rejected 'two-tier' financing whereby Community responsibility would be limited and extra expenditure borne by member states, possibly even with variations in the degree of support between member states. This, in the Commission's view, would create distortions, make impossible a sound common policy and do nothing to remedy production imbalances.

The solution advocated was based on the introduction of 'a new basic principle: co-responsibility or producer participation'. As has been seen in a previous section of this chapter, a 'co-responsibility levy' had already been introduced for milk; but a significant extension of this principle was now envisaged. Any production above a certain volume – to be fixed taking into account the Community's domestic consumption and its external trade – should be charged fully or partially to the producers. This new principle should be introduced as a permanent feature and should be a general rule. The Commission had not, by the end of 1980, made specific proposals as to how this principle should be applied for each product – that was left to the new Commission due to take office after 1 January 1981 in the context of the price proposals for 1981/2 – but it recalled that the Council had already accepted that there would be a supplementary levy on milk in 1981/2 if deliveries to dairies in 1980 were more than 2.5% higher than in 1979 (which proved to be the case). For sugar, co-responsibility had always existed in the form of a levy. For cereals, either a levy could be imposed or prices could be reduced – the Commission preferred the latter, as livestock producers would benefit and Community prices could be brought closer to world prices. For products such as processed fruit and vegetables, co-responsibility could take the form of a ceiling on the quantities eligible for aid; this could also be the solution for olive oil. For beef, intervention mechanisms could be made more flexible. For tobacco, the quantities eligible for premiums could be limited. For other products, such as wine, production restraints could be regarded as a form of producer participation.

A counterpart to these restraints on Community producers would be a check to the excessive rise in imports of certain livestock feedingstuffs (especially cassava and soya), while the growth of Community agricultural exports should be promoted, in particular by means of long-term contracts. 'If the Community is to remain open to the rest of the world, there must be a balance. If it is to import agricultural produce it must also have the means to conduct an export policy.'

In structural policy, action on behalf of less-favoured regions should be pursued and the Council was urged to act on the proposals to adapt the general socio-structural directives.

*

*   *

At the turn of the year from 1980 to 1981, it was beginning to appear that the budgetary problem would be less acute in 1981 than had been thought. Stocks of dairy produce had been reduced by the various disposal schemes; high positive British m.c.a.s were providing an unexpected financial bonus (though creating problems in other respects); and above all high world prices for sugar and cereals were keeping down the cost of export refunds. On the other hand, a substantial increase in common prices for 1981/2 seemed inevitable in view of the big cost increases which farmers had suffered in 1980. The need to take steps to avoid reaching the budgetary ceiling might thus appear less urgent; nevertheless, action could not be indefinitely deferred.

In view of the growing volume of exports of products eligible for export refunds, the world price level was playing an increasingly significant role in determining the cost of the CAP. Developments and prospects on world agricultural markets are discussed in the following chapter.

Bibliography

The subject-matter of this chapter and its bibliography, already extensive, could be kept within manageable proportions only by concentrating on developments at the Community or west European level; the following list omits many significant works on particular countries, which nevertheless are important elements in the overall context.

Commission reports are an essential source for this period, together with the press and the agencies European Report, Europe and Agra-Europe (the English and French editions of the latter being significantly different). Fennell (1979) and the information service *CAP Monitor* by Agra-Europe (London) provide comprehensive material on the workings of the CAP.

Major writings by economists have already been referred to in the penultimate section of the chapter. The series of OECD reports provide valuable analyses of both structural problems and methods of supply control.

Agra-Europe, London (continuous) *CAP Monitor.*

Agra-Europe, Paris (1980) *La PAC – Qui veut quoi?* (Suppl. au No. 1126 du 17 octobre 1980.

Atlantic Institute (1970) *A Future for European Agriculture.* (Panel report, rapporteur P. Uri). Paris.

Averyt, W. F. (1977) *Agropolitics in the European Community (Interest Groups and the CAP).* New York and London: Praeger.

Behrens, R. and de Haen, H. (1981) 'Aggregate factor input and productivity in agriculture'. *European Review of Agricultural Economics.*

Bergmann, D. (1975) *Matériaux et réflexions pour une réorientation de la politique agricole.* Paris: INRA-Economie.

Britton, D. K. (November, 1976) 'National policies within the CAP'. *Food Policy.* 405-12.

Commissariat Général du Plan (1980) *L'Europe: les vingt prochaines années.* Paris.

Commission of the European Communities

(a) Regular reports:

The Agricultural Situation in the Community (annual).

Bulletin of the European Communities (monthly).

Financial Reports on the EAGGF (Guarantee and Guidance Sections) (annual).

General Reports on the Activities of the European Communities (annual).

'Green Europe' Newsletter (monthly).

Reports on the application of the Council Directives on agricultural reform (annual).

(b) Special reports:

(1968) *Mémorandum sur la réforme de l'agriculture dans la Communauté Economique Européenne.* (The original 'Mansholt Plan') (COM(68)1000).

(1970) *Réforme de l'Agriculture.* (The proposals to the Council based on the Memorandum) (COM(70)500).

(1973) *Improvement of the Common Agricultural Policy*. In *Bulletin*. Suppl. 17/73.
(1975) *Stocktaking of the Common Agricultural Policy*. In *Bulletin*. Suppl. 2/75.
(1976) *Restoring balance on the milk market – Action programme 1977–80*. In *Bulletin*. Suppl. 10/76.
(1978) *Progressive establishment of balance on the market in wine – Action programme 1979–1985*. In *Bulletin*. Suppl. 7/78.
(1978) *Economic effects of the agri-monetary system*. (COM(78)29 final).
(1980) *Reflections on the Common Agricultural Policy*. In *'Green Europe' Newsletter*, No. 13.

Court of Auditors (1980) *Annual Report concerning the financial year 1979*. Official Journal No. C 342 of 31 December 1980.

Deutscher Bundestag (1980) *Jahresgutachten 1980/81 des Sachverständigenrates zur Begutachtung der gesamtwirtschaftlichen Entwicklung*. 9. Wahlperiode, Drucksache 9/17.

Economic and Social Committee (1975) *Progress Report on the Common Agricultural Policy*. Brussels.

Fabiani, G. (1979) *L'agricoltura in Italia tra sviluppo e crisi*. Bologna: Mulino.

Fennell, Rosemary (1979) *The Common Agricultural Policy of the European Community*. London: Granada.

Heidhues, Th., Josling, T., Ritson, Ch. and Tangermann, S. (1978) *Common Prices and Europe's Farm Policy*. London: Trade Policy Research Centre.

House of Lords, Select Committee on the European Communities (1980) *The Common Agricultural Policy*. London: HMSO (156).

Irving, R. W. and Fearn, H. A. (1975) *Green Money and the Common Agricultural Policy*. Ashford: Centre for European Agricultural Studies.

Keller-Noellet, M. J. (1979) 'La politique commune des structures: un tournant?' In Tracy and Hodac *eds. Prospects for Agriculture in the European Economic Community*. Bruges: College of Europe.

Koester, U. and Tangermann, S. (1977) 'Supplementing farm price policy by direct income payments'. *European Review of Agricultural Economics*. 4.1, 7-31.

Mahé, L. P. et Roudet, M. (1980) 'La politique agricole française et l'Europe verte: Impasse ou révision?'. *Economie rurale* **135** (1), 12-24.

Mansholt, S. (1979) *The Common Agricultural Policy – some new thinking*. London: The Soil Association.

Marsh, J. (1976/7) *European Agricultural Policy: A Federalist Solution*. Reprint from *New Europe*.

Marsh, J. (1980) *Memorandum* in House of Lords report.

Moulias, J. (1980) 'Réforme de la politique agricole commune'. *Economie rurale* **137** (3), 47-50.

OECD (1964) *Low Incomes in Agriculture*. Paris.

OECD (1965) *Agriculture and Economic Growth*. Paris.

OECD (1970) *Capital and Finance in Agriculture*. Paris.

OECD (1972) *Structural Reform in Agriculture*. Paris.
OECD (1978) *Part-time Farming*. Paris.
OECD (various years) *Review of Agricultural Policies*. Paris.
Priebe, H., Bergmann, D. and Horring, J. (1972) *Fields of Conflict in European Farm Policy*. London: Trade Policy Research Centre.
Rieder, P. (1980) *Milchproduktion unter Kontingentierung*. Arbeitsdokument des Institutes für Agrarwirtschaft, ETH Zürich.
Ries, A. (1978) *L'ABC de la politique agricole commune*. Paris: Nathan.
Tangermann, S. (1980) *Memorandum* in House of Lords report.
Thiede, G. (1975) *Europa's Grüne Zukunft*. Düsseldorf: Econ-Verlag.
Thiede, G. (1979) 'L'agriculture européenne et la révolution technique'. In Tracy and Hodac *eds*. *Prospects for Agriculture in the European Economic Community*. Bruges: College of Europe.
Van Riemsdijk, J. F. (1973) 'A system of direct compensation payments to farmers as a means of reconciling short-run to long-run interests'. *European Review of Agricultural Economics*. 1.2, 161-89.
Vedel (1969) *Rapport*. Paris: Seclas.
Wageningen Memorandum on the Reform of the European Community's Common Agricultural Policy (1973). Trade Policy Research Centre and Agricultural University of Wageningen.

Chapter 15

Western European agriculture in the world context

Western Europe plays a significant role on the world agricultural scene. On the one hand, its imports account for a large part of world agricultural trade: the export prospects of many countries in the world, both developed and less-developed, are significantly affected by the agricultural policies of western European countries, above all the policy of the Community as the major entity. On the other hand, the Community's exports of certain agricultural products have taken on increasing importance. The Community's attitude is crucial for GATT negotiations and for international commodity agreements; and with the prospect of continuing food shortages in the world, important issues have arisen as to the role of western European agricultural potential.

Patterns and trends of trade

Despite a century of protectionism in most western European countries, and technological progress that had greatly raised agricultural productivity, western Europe was still in the 1970s a large importer of agricultural products; but the volume and pattern of its imports inevitably reflected past developments in its own agriculture. Western Europe's agricultural imports – excluding trade within the region – accounted for nearly a quarter of world agricultural trade. A large part of these imports – probably about a half – consisted of tropical or semi-tropical items which cannot be produced in Europe or not in any great quantity: rice, tropical fruits, tobacco, coffee, tea, cocoa, oilseeds, fats and oils – also sugar, which is a special case, being a tropical product when produced from cane and a temperate product when made from beet. Such products accounted for the greater part of agricultural imports from Central and South America, Africa and Asia (see Table 15.1).

For 'temperate' foodstuffs, including sugar, the growth of agricultural production in western Europe increased the region's degree of self-sufficiency and reduced its volume of imports (see Tables 12.3, 14.2 and 15.2 for the EEC). By the late 1970s, imports of grains for direct human consumption consisted almost exclusively of hard wheat from North America, required for English-style bread, and of durum wheat for making *pasta:* in both cases the quantities imported from third countries had fallen. Imports of sugar by the Community were limited to the quantities guaranteed under the Lomé Convention, and were offset by EEC exports to the world market which by 1980 were more than twice as high. After 1974 the Community's imports of beef and cattle were practically limited to the amounts laid down under quotas granting low or zero rates of levy and other special arrangements under GATT, the Lomé Convention and an agreement with Yugoslavia: the former large trade with South America was much reduced, though imports of cattle and calves from eastern Europe, mainly into Italy, remained quite substantial. In 1980 the Community became a net exporter of beef. EEC imports of mutton and lamb from Australia and New Zealand were subjected to voluntary export restraints in the context of the market regulation adopted in 1980 (see chapter 14). Imports from third countries of poultry and eggs were already much reduced by the early 1970s, as intensive specialised

Table 15.1: Pattern of imports of agricultural products by OECD European countries* in 1978

	Intra OECD Europe	Rest of world	*of which:* Canada US	Australia New Zealand	Latin America	Africa
			billion US dollars			
Cereals (incl. rice)	4.9	4.4	3.5	–	0.1	0.1
Feedingstuffs for animals	2.2	3.1	1.1	–	1.3	0.2
Sugar	1.1	1.2	–	–	0.5	0.5
Fruit and vegetables	7.8	6.0	0.7	0.2	1.4	1.4
Wine, whisky beer, etc.	3.5	0.3	0.1	–	0.1	0.1
Tobacco	1.2	2.1	1.1	–	0.4	0.2
Coffee, tea, cocoa, spices	2.3	9.4	–	–	4.0	4.4
Oilseeds, etc.	0.2	4.7	3.4	–	0.8	0.3
Oils and fats	1.8	1.8	0.3	–	0.3	0.4
Live animals and meat	9.2	3.2	0.4	0.6	0.7	0.1
Dairy products and eggs	5.7	0.3	–	0.2	–	–
Other products	5.8	3.1	0.9	0.2	1.0	0.3
All agricultural products†	45.7	39.6	11.5	1.2	10.6	8.0

* The Community of Nine and the seven EFTA countries, plus Greece, Spain and Turkey.
† SITC 0, 1, 4, 22, 29.
Source: OECD, *Statistics of Foreign Trade,* Series B, 4/1978.

production of these items, using purchased feedingstuffs, had expanded to meet nearly all domestic demand. As for dairy products, by 1980 imports were virtually confined to the limited quantities of butter still provisionally accepted by the Community from New Zealand, small amounts of Cheddar cheese imported into the UK under special arrangements, and imports of Emmenthal-type cheese mainly from Switzerland and Austria. In agreements with Mediterranean countries, the Community granted tariff concessions on various fruits and vegetables; as has been seen in the previous chapter, support to EEC producers was increased as a counterpart to these concessions. Imports of wine from the Mahgreb countries were reduced.

This rising self-sufficiency reflected the ability of agriculture in north-western Europe to expand its output at rates generally faster than the growth of internal demand. (Southern Europe was experiencing different trends: here the growth of consumer income had led to a rate of growth in demand for livestock products which domestic agriculture, handicapped by soil and climate as well as by low technology and poor structures, had been unable to meet. This had made Italy into a major importer of food and feed from within western Europe and from other regions; Spain, Portugal and Greece were following similar paths.)

However, the growth of livestock production throughout western Europe was achieved only with greatly increased use of purchased feedingstuffs. It was

Table 15.2: Trends in imports of major agricultural products by the Community of Nine from the rest of the world

	1972/3 – 1973/4	1977/8 – 1978/9	Index
	thousand tons		%
Total cereals*	22 269	18 719	84
of which: Wheat*	6 100	4 674	77
Barley*	2 106	1 094	52
Maize*	12 446	12 020	97
Rice	421†	759	180†
Wine	7 586	6 023	79
	1973–4	1978–9	
Sugar	2 039†	1 637	80†
Soya: beans	7 720	11 280	146
meal	3 365	6 026	179
Beef and veal**	791††	460	58
Mutton and lamb**	245	167	68
Pigmeat**	288	288	100
Poultrymeat**	68	70	103
Eggs	33†	28	85†
Butter	156†	121	78†
Cheese	83†	78	94†

* Data on the origin of imports are misleading as cereals imported through Dutch or Belgian ports are often declared by the country of final destination as 'intra'-EEC trade. Figures here are derived from the difference between total EEC imports and registered intra-EEC *exports*.

† Data only available for 1973/4 or 1974.

** Including live animals and processed products in carcase weight equivalent.

†† This average consists of 1 080 000 tons in 1973 and 502 000 in 1974: in the latter year the Community introduced import restrictions.

Sources: Commission, *Agricultural Situation* reports, and Eurostat data.

calculated that the Community in 1977/8 was using imported feed to an extent equivalent to an expansion of 11% in its production base or to 11 million hectares of agricultural land (Thiede, 1980). There was a large increase in imports of soya (as beans, meal or cake) especially from the United States, though Brazil and other South American countries were developing their exports; indeed the bulk of the world supply was produced in the United States, and variations in the US crop could produce wide fluctuations in the world price (as in 1972–3 when world shortages and US export controls sent world prices soaring). Imports of maize remained substantial, coming mostly from the United States, though their growth was checked during the 1970s by the rapid development of cheaper 'cereal substitutes', especially cassava (also referred to as 'manioc' or 'tapioca').

This increased recourse to imported feedingstuffs became a cause of concern within the Community, partly because of the high degree of dependence on a few sources – a worry intensified by the US controls on soya exports in 1973, which stimulated measures to promote production of vegetable protein feeds – and partly because cheap 'cereal substitutes' were displacing Community-grown feed grain (as well as imported maize), while intensive livestock production on the basis of imported feed (though generally efficient and economic) was adding to imbalances on the Community market. In 1979 the Community asked

Thailand – the main supplier of cassava – to limit its exports in return for Community aid for diversification of its agriculture, and in December 1980 decided to seek 'deconsolidation' of the 6% tariff on cassava through negotiations with other suppliers. There was also increasing discussion (intensified by problems likely to arise from an olive oil surplus if Spain should join the Community) of a charge on vegetable oils and fats. (The Commission's 1980 *Agricultural Situation* report included a special chapter on 'Policy for animal feedingstuffs'.)

Agricultural trade among the member states of the Community, stimulated by Community preference, was particularly dynamic, with a rate of growth exceeding that of imports from the rest of the world (Table 15.3). Towards the end of the 1970s, however, part of the momentum seemed to have been lost; intra-EEC trade in cereals began to decline.

Table 15.3: Trends in trade in food, beverages and tobacco* by the Nine

	1958	1972	1979	Increases 1958-72	Increases 1972-9
		million ECU†		*indices*	
Trade between the Nine	2 280	10 464	28 584	459	273
Imports from rest of world	7 190	11 945	24 540	166	205
Exports to rest of world	1 858	4 825	13 192	260	273

* SITC 0 and 1. This does not include oilseeds, fats or oils, so that the growth in imports of soya, for example, is not reflected.

† See note to figure 13.1.

Source: Eurostat, *Monthly external trade bulletin.* Special number, 1958-1979 (Table 7 b).

Agricultural exports by the Community rose faster than imports from the rest of the world (Table 15.3): sugar, cereals, dairy products, beef and wine were major growth items in volume terms (Table 15.5). About a third of Community exports consisted of items which were not subject to common market regimes, and it has been estimated that about two-thirds consisted of processed products. (Exports included items such as coffee, tea and cocoa for which the raw material was not even produced in the Community.) Canned meats, dairy products, biscuits, confectionery, canned fruit, wine and whisky were important items (Table 15.4). The high-value processed products tended to go mainly to other industrialised countries; the North American market took substantial amounts of canned pork, wine, whisky and – in spite of US dairy protectionism – cheese. However, a rising share of exports were going to developing countries, particularly the African ACP countries and Mediterranean countries. These exports consisted mainly of soft wheat, barley, sugar, milk powder and condensed milk. The oil-rich countries of the Middle East were the fastest growth area for Community exports, though the absolute amounts were not so large. The USSR and other eastern European countries bought cereals, sugar, fats and oils, meat and dairy products, but trade relations with this area were irregular and clouded by political constraints. In particular, sales of butter to the USSR, heavily subsidised through the export refund mechanism, became a controversial issue, and in 1980 the

Community supported the American policy of restricting food exports to the USSR following the latter's intervention in Afghanistan.

Table 15.4: Pattern of exports of agricultural products by the Community of Nine in 1978

	Intra-EEC	Rest of world	EFTA*	*of which:* Canada US	E. Europe	Africa
			billion US dollars			
Cereals and cereal preparations	4.7	2.3	0.3	0.1	0.4	0.8
Sugar	0.8	1.2	0.2	0.1	–	0.4
Fruit and vegetables	4.6	1.1	0.6	0.1	0.1	0.2
Wine, whisky, beer, etc.	2.6	3.2	0.4	1.4	0.1	0.2
Live animals and meat	8.8	1.2	0.2	0.3	0.1	0.2
Dairy products and eggs	5.0	2.3	0.1	0.1	0.1	0.7
Other products	10.1	5.0	1.4	0.7	0.4	3.0
All agricultural products†	36.6	16.3	3.2	2.8	1.2	5.5

* Austria, Finland, Iceland, Norway, Portugal, Sweden, Switzerland.
† SITC 0, 1, 4, 22, 29.
Source: As Table 15.1.

Table 15.5: Trends in exports of major agricultural products by the Community of Nine to the rest of the world

	1972/3 – 1973/4	1977/8 – 1978/9	Index
	thousand tons		%
Total cereals	10 838	12 512	115
of which: Wheat	6 048	6 674	110
Barley	3 618	4 944	137
Wine	3 668	6 364	173
	1973–4	1978–9	
Sugar	1 884*	3 472	184*
Beef and veal†	243	319	131
Pigmeat†	291	198	68
Poultrymeat†	148	224	151
Butter and butteroil	118*	355	300*
Cheese	189*	242	128*
Skim milk powder	339*	546	161*
Whole milk powder	192*	360	188*

* Data only available for 1974.
† Including processed products and live animals in carcase weight equivalent.
Source: As Table 15.2.

For most of the products subject to common market regulations, and at most times, EEC prices were above world prices so that exports had to be assisted through export refunds – calculated in the case of processed foodstuffs on the basis of the content of agricultural raw material (such as cereals, milk powder or sugar). The resulting budgetary problems have been discussed in the previous chapter. The extent to which the development of EEC exports would continue to require this assistance depended both on price levels within the EEC and on prospective outlets and prices on the world market. For cereals, the Community was essentially a 'price-taker' on the world market; for sugar, dairy products and increasingly beef, its own export volume and pricing policy played a major role (as the section below on world commodity markets will show).

'Export vocation'

The growth in agricultural production in EEC countries, and the importance of agriculture in the economy and in the export trade of several EEC countries, made it seem inevitable that exports to the world market would continue to rise. The COGECA (*Comité Général de la Coopération Agricole*) frequently stressed the need for the Community to develop an export policy, for instance in a memorandum submitted to the Commission in September 1978, which declared:

> With the prospect of a reasonable and moderate expansion of European agricultural output – which corresponds to the basic needs of the people and farmers in Community countries – regular export of a substantial part of this production is a need which cannot be disputed. It should be added that some countries of the Community with a particular export vocation have an overriding economic need to promote the development of their exports of agricultural and food products in order to achieve equilibrium in the trade balance and balance of payments. [Nouyrit (1979) page 327]

The COGECA therefore requested 'total independence' for the Community in fixing export refunds, introducing 'permanent' refunds for cereals (and sugar), establishing an annual export programme supported by budgetary provisions, grouping firms engaged in exporting dairy products so that they should be able to make joint offers on certain markets, and with regard to the ACP countries a special long-term policy of export refunds together with credits to these countries to enable 'permanent and competitive commercial presence'.

Difficulties in this approach were underlined by Finn Gundelach, the Commissioner for Agriculture, speaking at the closing session of the 1979 College of Europe symposium where these issues had been discussed. While recognising that the Community had a 'vocation to export', and the importance of this for the balance of payments of a number of member states, he referred to the problems encountered in GATT due to the view of trading partners that the Community was exporting too much and picking up part of their markets:

> We cannot have it both ways. You can't both tell me 'be open in GATT negotiations, accept division of labour in agriculture' and then come and tell me 'export more', which can only be done with the use of the taxpayers' money, i.e. with export restitutions. [Gundelach (1979) page 423]

Gundelach stressed Europe's tradition of processing primary and semi-manufactured commodities into more sophisticated products, and Europe's consequent dependence upon trade:

> On that ability to process, to use our real raw material, the human factor, and to trade in a world which is reasonably open, rests our whole democratic social system.... We cannot have a free-trading philosophy with regard to industry, and when we come to agriculture suddenly become self-sufficient, and refuse to accept the concept of division of labour. [Ibid., page 424]

In view of the mounting concerns, however, the Commission pursued its work on the possibilities for developing an export policy for agriculture. The rest of this chapter seeks to describe the international context within which such a policy would have to operate.

World commodity markets and the role of western Europe

During the 1960s and 1970s, world markets for the major agricultural commodities were generally oversupplied. Those for cereals and sugar were however subject to wide fluctuations (see Figure 15.1), as was that for soyabeans for reasons already mentioned. The general problem of rising instability on world markets was discussed, with reference to each major commodity, in an OECD report of 1980; FAO reports also describe the situation on world commodity markets. Major developments on three markets where western Europe plays a particularly important role – cereals, sugar and dairy products – will be briefly set out below.

CEREALS

The underlying feature of world cereals markets during the 1960s and 1970s was an excess of productive capacity over *effective* demand – the food shortages of developing nations being only partly translated into commercial demand, yet being met to only a small extent by concessional sales under food aid programmes (which accounted for about 12% of world trade in cereals in the late 1970s). World market prices for cereals, up to 1972, were low relative to support prices in the Community and other European countries. They would have been lower still in the absence of stockholding by the major exporting countries. During the 1960s, stocks of wheat and coarse grain held by the United States and Canada, and to a lesser extent by Australia and Argentina, averaged over 100 million tons, rather more than the annual volume of world trade. In 1972, the United States was holding 27 million tons of wheat and 47 million tons of feed grains, Canada had respectively 16 million and 6 million tons.

A change of unprecedented scale occurred in 1972 when the USSR, whose irregular cereals imports had already on occasion upset the world market, experienced a particularly poor harvest and purchased over 21 million tons of grain, mostly in the space of a few months in the autumn of 1972. By conducting these purchases secretly with different American companies, the USSR was able

Figure 15.1: Prices of international significance for major agricultural commodities, in US dollars per 100 kg

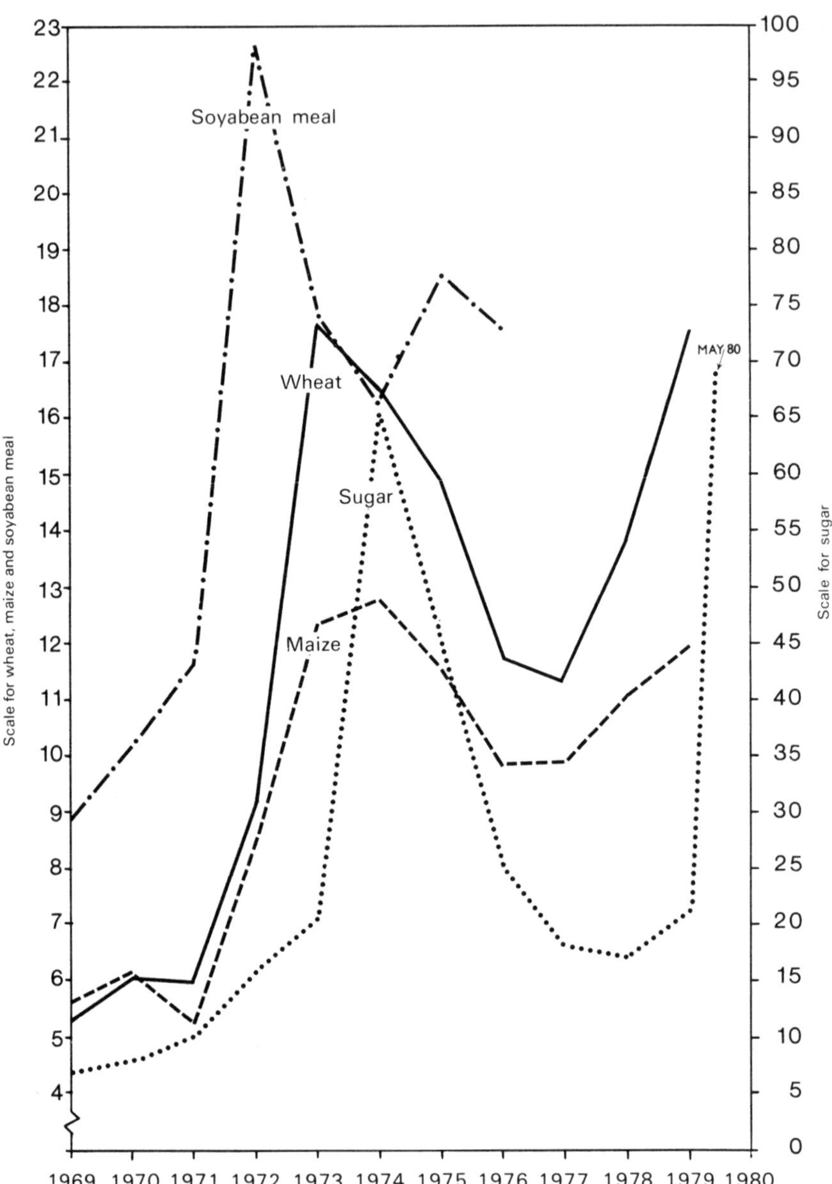

Wheat: US hard winter No. 2, f.o.b. Gulf.
Maize: US Yellow No. 2, f.o.b. Gulf.
Sugar: raw, f.o.b. and stored, Caribbean ports.
Soyabean meal: US, 44%, f.o.b. Baltimore (series not available after 1976).
Source: FAO (direct communication).

to obtain the bulk of these quantities at the existing low world price level. Once this became known, however, US and consequently world prices shot up to record heights. Contributory factors were a fall in production of rice and coarse grains in the Far East, so that China and India also had increased requirements. By 1974 world stocks of cereals were halved and the United States suspended its supply control programmes. To ensure greater stability in trade, the United States concluded a five-year bilateral agreement with the USSR according to which the latter would purchase annually at least 6 million tons, and was authorised to purchase up to 8 million tons without consulting the US government in advance: much higher amounts were in fact agreed each year.

World prices of grain fell back from early 1975 onwards, though they did not return to their previous level. Prices to importing countries were influenced by the rise in freight rates, which had jumped to a peak in 1973/4 – largely because of the increased demand for shipping resulting from the extra trade in grain – and which later, after a temporary fall, climbed steeply again under the influence mainly of rising oil prices.

The Republican administration in the United States (with Earl Butz as Secretary of Agriculture) pursued in the period 1973–6 a farm policy based on market orientation and unrestricted production. In 1977 however the Carter administration, with Bob Bergland as Secretary of Agriculture, was faced with increased stocks, falling prices and rising costs: it decided to reintroduce, in 1978, the 'set-aside' programme, to create farmer-owned grain reserves, and to raise domestic price supports.

1980 was a year of renewed uncertainty on world markets, when the United States – who had previously agreed with the USSR on the export of up to 25 million tons of wheat and maize in 1979/80 – decided after the Russian intervention in Afghanistan to limit supplies to the 8 million tons specified in the five-year agreement: this embargo was supported by Canada, Australia and the Community. The USSR, having suffered a poor harvest, nevertheless imported substantial quantities of grain from Argentina and other sources. China meanwhile had been emerging as an increasingly significant importer, signing long-term agreements with Canada, Argentina and, in October 1980, with the United States. Several developing countries, including some with oil revenue such as Mexico but also others such as Brazil and Egypt, were importing increasingly large quantities of grain.

The operation of the CAP market regime ensured relative price stability within the Community despite the fluctuations in the world market. In 1974/5, when the import price rose above the threshold price, import levies were suspended and for a while *export* levies were in operation, keeping down prices within the Community and discouraging exports to the world market. This fulfilled the Treaty aim of ensuring adequate supplies to consumers at reasonable prices. On the other hand, as was pointed out by American critics in particular, by such action the Community did not contribute to greater stability on the world market.

As has already been seen, exports of grain by the Community to the world market tended to rise. Part of these exports consisted of food aid: the Community programme consisted of 720 500 tons per annum from 1976 onwards; bilateral programmes by member states added another 566 500 tons in 1980. Requests from developing countries were greatly in excess of this amount: the constraint lay in the funds which the member states were prepared to provide under the Community budget (in 1980 the 720 500 tons were estimated to cost 135 million EUA at internal Community prices).

Many efforts were made to project world supply and demand for foodstuffs, particularly by the FAO and the US Department of Agriculture, with attention mainly to cereals. The unsatisfied food needs of hungry people in developing regions remained enormous, but only part of these needs were likely to make themselves felt in international trade. Reviewing various studies, OECD (1980, page 114) concluded that 'there are no problems in terms of physical resource availability to grow sufficient cereals for most commercial situations likely to prevail up to 1990': the United States, which provided over half the export supplies, 'has sufficient excess capacity at current price levels to be the residual supplier'. However, OECD stressed the risk of increasing instability. Weather changes in any major part of the world could lead to large variations in production: the capacity to absorb such variations might be diminished as the margin of security, so far provided mainly by idle acreage in the United States, was shrinking, and the United States was increasingly unwilling to bear the main burden of holding reserve stocks. International burden-sharing in respect of stocks was thus becoming an important issue: it played a major role in the unsuccessful negotiations for a new International Wheat Agreement (see later section).

SUGAR

The most unstable of all the major agricultural commodity markets was that for sugar. Until 1974, the greater part of world trade was conducted under special arrangements – UK imports under the Commonwealth Sugar Agreement, US imports under its Sugar Act, and exports by Cuba to other Communist countries. Fluctuations in supply were thrown on to the relatively small residual free market and thereby amplified. The free market was subject to recurrent booms and slumps: a substantial price increase in 1963, for example, gave a short-lived stimulus to production and was then followed by a prolonged period of low prices which discouraged investment in the countries producing cane sugar. In 1972 there was a drop in production and exports by Cuba. World demand, on the other hand, was rising steadily; moreover the USSR experienced a bad sugar-beet harvest in 1972 and in the following years became a substantial importer. The shortfall between world production and demand was probably not very great – it was estimated at about 1.2 million tons, only about 0.5% of world output – but it was sufficient for prices to rocket upwards in 1973/4, while stocks fell. The free world price reached levels over six times the revised prices negotiated early in 1974 under the Commonwealth Sugar Agreement.

Both US and EEC policy were permanently affected by this development.

The United States abandoned its policy of controlling imports through quotas allocated to each exporting country. In the Community, as has been observed in chapter 13, the United Kingdom was enabled to import sugar with Community-financed subsidies, while the high prices lasted. Moreover, the boom occurred when the Community was negotiating the replacement of the CSA by an agreement under what was to become the Lomé Convention, and determining its own sugar policy. This situation helped the ACP countries to get a settlement which broadly satisfied them. It also provided arguments for raising domestic production quotas and guaranteed prices, leading to a growth in the EEC sugar area, output and net exports: it was estimated (Harris, 1980) that the EEC share of the free world market rose from under 5% in 1972 and 1973 to about 20% in 1979 and 1980.

In 1975, however, prices plummeted, so that EEC exports again had to be financed through export refunds. This development led to complaints in GATT by Australia and Brazil: arbitration panels noted that Community production and exports of sugar had increased despite the Community's quota system, that the increased exports had been effected through the use of subsidies and that this system constituted a 'threat of prejudice' in terms of the relevant GATT provision (Article XVI:1). This issue remained under discussion in GATT at the end of 1980.

In 1980, however, mainly as a result of very poor harvests in several important producing countries, world prices again rose well above the Community level, the rise being fuelled by speculative influences. The International Sugar Agreement (see later section) was powerless to prevent this development. Once again the CAP mechanisms went into reverse, export levies being applied to stabilise prices within the EEC and to discourage the export of excessive quantities. This situation inevitably influenced once again the discussions then taking place on the Community's market organisation, the existing regime being due to expire on 30 June 1981. The Commission however recalled that since 1900 the world market had been characterised in eight years out of ten by surplus production and low prices: as this pattern had been valid also for each decade since 1950, there was in the Commission's view no reason to believe that it would change. As has been seen in the previous chapter, the Commission's proposals for the new regime – still under discussion at the end of 1980 – envisaged some reduction in the level of the 'maximum' quotas.

DAIRY PRODUCTS

While the world markets for cereals and sugar were characterised by fluctuations, with a cyclical pattern in the latter case, that for dairy products was beset with a structural surplus. The problems of overproduction facing the Community and other western European countries were common to the OECD area as a whole: in all the high-income countries, milk yields per cow were rising and deliveries to dairies were rising faster than utilisation. For some dairy products, markets were expanding; consumption of butter, however, was tending to fall throughout the OECD area, as a result of concerns over health risks associated with cholesterol, together with competition from margarine.

Thus during the 1980s all OECD countries had rising stocks and growing exports of dairy products (the US remained approximately in balance for butter with a very restrictive import regime, but had large surpluses of skim milk powder which were disposed of mainly under the PL 480 Food Aid Programme).

As has already been seen, the Community was more than self-sufficient in dairy products. Its imports of butter and cheese mainly took place under special arrangements.

For speciality products, cheeses in particular, remunerative outlets could be found in other high-income countries. Disposal of the large intervention stocks of butter and skim milk powder, however, posed difficult problems. Commercial outlets for butter were practically non-existent apart from eastern Europe, and, as has been already mentioned, the sale of butter to Russians at prices lower than those available to Community consumers became a controversial matter. The world market price for skim milk powder was generally depressed, though it rose sharply for a brief period around 1973 when the shortage of soya created demand for skim milk powder as an alternative protein livestock feed. The Community's exports, however, were a sufficiently significant element on the world market for its pricing policy to influence the world price level, and from 1975/6 to 1978/9, export refunds were kept at the same level: subsequently, with an improvement in world prices, it was possible to reduce them.

Food aid offered an outlet for dairy produce, principally for skim milk powder which in principle seemed ideally suited to meeting protein deficiencies in developing countries. There were however difficulties in practice: even skim milk powder could only be used effectively if regular supplies were available. The Community's food aid programme included from 1976 onwards 150 000 tons of skim milk powder and 45 000 tons of butter oil annually. In view of the need for suitable distribution structures in the receiving countries, it did not seem possible to increase these amounts. Most of this food aid was channelled through institutions and agencies with recognised distribution capacity.

*

* *

The foregoing discussion will have made clear that imbalances on world agricultural markets were the result of powerful technological and economic forces (including economic developments resulting from political factors, as in the case of the oil crisis). It was difficult for inter-governmental action to make much impact on this situation. Nevertheless, continuous efforts were made to promote trade liberalisation and to improve conditions on world markets, under the GATT and through international commodity agreements.

The General Agreement on Tariffs and Trade

Though in principle the GATT made no distinction between agricultural and other products, in practice agriculture always required special treatment. Article XI of the Agreement permitted import quotas on agricultural products if domestic production restrictions were in force; the United States in 1955 obtained a waiver going beyond this provision to legalise its use of import quotas for protecting domestic dairy producers (although its dairy programmes did *not* control supply). In early rounds of trade liberalisation, no progress could be made on agriculture. The pessimistic views of the Haberler Report in 1958 as regards the prospects for agricultural trade liberalisation, in the face of agricultural protectionism, have already been mentioned in chapter 11.

(A) NEGOTIATIONS ON THE FORMATION OF THE EEC, 1960–62

The formation of the EEC brought into application Article XXIV of the Agreement which, in sanctioning customs unions, stipulated that the common external duties must not on the whole be higher or more restrictive than the general incidence of duties collectively applied by individual member countries before forming the union. Negotiations under Article XXIV(6) between the EEC[1] and its trading partners took place in the context of a GATT Conference in 1960-62 and led to concessions in the form either of a 'binding' of duties under the common tariff (i.e. an undertaking not to increase them) or of reductions. The United States thus obtained duty-free entry (or nearly so) for soyabeans, soyabean meal, other oilseeds and cotton. A second stage of this Conference consisted of multilateral negotiations for new reciprocal tariff concessions (the 'Dillon Round'). Reductions were agreed in industrial tariffs, but there were no significant moves on agricultural products.

US hostility to the CAP intensified after the Community in January 1962 had agreed on the basic system of agricultural support. The American Secretary of Agriculture at the time, Orville Freeman, launched a series of attacks such as the following (from a speech to the National Council of Farmer Co-operatives at Miami Beach on 8 June 1962):

> Protective devices adopted or proposed by the Common Market centre around the use of a variable levy fee. To some people these levies appear as a gate on a dam which can be raised or lowered depending on the amount of water needed on the other side. To others, these levies appear to be more like a moving high-jump bar which rises to disqualify even the most proficient competitor.

The US Trade Expansion Act of 1962 increased the power of the President to reduce tariffs in exchange for concessions by other countries, and Section 252 of

[1] In fact the member states and not the Community itself were 'Contracting Parties' to the GATT, but they were represented by the Commission, acting on the basis of Council mandates and in consultation with a special committee of representatives of member states (known as the '113 Committee' by reference to the relevant article of the Treaty of Rome).

the Act directed the President to take appropriate steps to eliminate restrictions on US agricultural exports, including if necessary withholding concessions to the countries concerned.

(B) THE 'KENNEDY ROUND', 1963–7

The Resolution of the GATT Ministerial Conference in May 1963 which initiated the 'Kennedy Round' stipulated that this multilateral trade negotiation would include agricultural products. But the preparatory work, before the negotiations formally opened in May the following year, made evident a basic difference in outlook between the Community and the United States. The Community proposed a fundamentally new approach in the agricultural sector: negotiation on the basis of the *montant de soutien* ('support-level binding'). The level of support would have been calculated for each country (or the EEC) as the difference between a 'world reference price' and the price obtained by domestic producers: this difference would have been bound, though the contracting parties would have been free to choose their methods of support.

The Community approach also stressed the need to organise world markets. This had been French policy for some time: the French Minister of Finance, Wilfrid Baumgartner, declared at a GATT Ministerial meeting in November 1961:

> [The existing situation] is characterised essentially by the accumulation of surpluses at a time when there are enormous unsatisfied needs. Moreover, to sell their produce, producing countries engage in a ruinous price war which leads them to subsidise the consumption of foodstuffs in countries which are their main industrial rivals. It is obvious that we must adopt a new policy, and try to reconcile the aspirations of each country by looking for solutions based not on Free Trade – which is impracticable – but on the principle of market organisation. [Press release GATT/633]

The plan was further developed by the French Minister of Agriculture, Edgar Pisani. The Community proposal to GATT was an ambitious one: it envisaged a 'World Commodity Agreement' covering cereals, beef and veal, some dairy products, sugar and possibly oilseeds, to stabilise prices at fair and remunerative levels; moreover, producing countries would be required to take steps to avoid surplus production. Food aid would also play a role, though its limitations were recognised.

The Community felt that its *montant de soutien* scheme, together with the market organisation proposal, represented a radical departure from previous agricultural trade negotiations, and one which took account of the determining role of agricultural support policies. The United States however wanted modifications in the EEC's variable levy system and a lowering of the degree of protection this system provided. The US consequently rejected the Community approach, criticising it for failing to provide for reductions in existing trade barriers while eliminating price competition as a factor in future trade. (As the US in any case did not attain its objectives, it was subsequently questioned whether, by agreeing to negotiate on the basis of the Community proposal, the US might not have exerted more influence over the subsequent development of the CAP, bearing in mind that at this point the Community had not agreed on its internal level of cereal prices.)

The GATT negotiations then centered on three product groups: cereals, dairy products, and beef and veal. The results in agriculture, when the Kennedy Round was concluded in 1967, were modest. The Community made tariff concessions on a limited number of processed foodstuffs and on some products of interest to developing countries. The main outcome was an agreement on cereals, less ambitious than the scheme originally proposed by the Community, providing only for maximum and minimum prices together with a food aid programme of 4.5 million tons annually, of which the Community was to contribute 1 million tons. On this basis the International Grains Agreement came briefly into force on 1 July 1968 (see below).

(C) NEGOTIATIONS ON THE ENLARGEMENT OF THE COMMUNITY, 1973–4

While the United States kept up pressure on the CAP, the next important stage in the Community's relations with GATT was the opening of new negotiations in 1973 under Article XXIV(6) in view of the enlargement of the Community. Agreements were reached by July 1974 with thirteen countries, the concessions by the Community in the agricultural sector being limited to a number of bindings of the previous tariff of the Community of Six; the only concession concerning a product subject to levy related to imports of high-quality Cheddar cheese from Canada. Agreement could not be reached with the United States and Australia on cereals. An agreed declaration, however, stated willingness by the three parties to continue seeking solutions for trade in cereals through international negotiations.

(D) THE 'TOKYO ROUND', 1973–9

The multilateral trade negotiations (MTNs) under GATT, initiated with a declaration by the Trade Ministers of 102 countries meeting in Tokyo in September 1973, were the most extensive yet undertaken. The Tokyo Declaration envisaged, as regards agriculture, 'an approach to negotiations which, while in line with the general objectives of the negotiations, should take account of the special characteristics and problems in this sector'. This text represented a compromise between the US insistence that agriculture must be included and the Community position that it must be treated as a special case – a continuing divergence which became a major issue in the subsequent work and contributed to lengthy delays. While the United States remained anxious to get EEC import levies lowered and to restrict the EEC's use of export refunds, the Community made it clear – in the 'overall approach' to the negotiations laid down by the Council in June 1973 – that the CAP was one of the elements 'basic to its unity and fundamental objectives' which could not be called into question and could not constitute a matter for negotiation. The Community saw the objectives of the MTNs as being the expansion of trade within more stable world markets in accordance with existing policies, and the best way of achieving that objective was held to be the organisation of orderly world marketing arrangements by means of appropriate international agreements.

The Community's position was thus consistent with that which it had adopted in the Kennedy Round as regards the organisation of world agricultural markets, but with the CAP now firmly established it was no longer offering any such concession as the binding of the *montant de soutien.* The specific mandate for negotiation, drawn up by the Council in February 1975, set out fairly detailed proposals for international commodity agreements. For cereals, sugar and rice, the Community proposed a stock-holding policy based on co-ordination of national stock policies, together with minimum and maximum prices; for dairy products the Community proposed a system of minimum and maximum prices alone. In each case the price range was to be ensured by undertakings on the part of importing and exporting countries respectively. Food aid could play a role, but this was left undefined.

The MTNs did not start in earnest until after the US Congress finally passed the necessary Trade Act at the end of 1974; this was followed on the Community side by the establishment of the Council mandate in February 1975. Progress however remained slow, largely because of US–EEC opposition over agriculture, as regards not only the substance, as already outlined, but also the procedure. While the Community insisted that agriculture should be dealt with as a separate entity in the Agriculture Group (one of seven set up to conduct the negotiations), the United States wanted agricultural matters to be considered also in other groups, in particular that on 'Non-Tariff Measures' (in the US view, import levies constituted a 'non-tariff barrier'). The pace of the negotiations in general accelerated only after an agreement on procedures between the US Trade Representative, Robert Strauss, and the EEC Commissioner for External Relations, Wilhelm Haferkamp, in July 1977. The essential negotiations were concluded in April 1979.

The outcome of the Tokyo Round was important for world trade in general. Tariffs were cut by about a third on a world-wide basis. In the agricultural sector, however, tariff reductions were mainly limited to tropical products. In this sector, the Community – which had already, in 1971, introduced its 'generalised system of preferences' in favour of developing countries – implemented additional tariff and non-tariff concessions on 1 January 1977, in advance of the overall outcome of the MTNs. No basic agricultural products subject to common market regulations were affected, except very indirectly, by these concessions.

A number of limited concessions affecting agricultural products were made on a bilateral basis. The most significant of these concerned the United States, though they fell far short of the original US aims in the MTNs. These concessions (on the basis of the 'Strauss list') related to high-quality ('Hilton') beef, turkey cuts, table grapes, prunes, rice and tobacco; in the last two cases, adjustments were made to the relevant common market regimes to give compensation to EEC producers. US requests concerning citrus fruit, fruit juices and almonds were not accepted, as these were regarded as particularly sensitive products from the CAP point of view. On the other hand the Community obtained concessions from the US for cheese, spirits and beef and veal. Levy-free quotas for high-quality beef, as well as for beef for processing, were also granted by the Community to Argentina and Uruguay. Australia also obtained a levy-free quota for high-quality beef, and Australia and New Zealand (besides Canada which already had a concession) were to be enabled to export specified quantities of Cheddar cheese to the EEC. In return the Community obtained concessions benefiting its exports of processed foodstuffs to Australia and New Zealand and of cheese and alcoholic beverages to Canada.

These concessions met some specific requests which were important to the countries concerned. However, they did not represent a major inroad on agricultural protectionism. Commenting on the outcome, Warley observed:

> Trade in farm products among the developed countries was liberalised only marginally. No Minister of Agriculture had difficulty in assuring his farmer-constituents that he had surrendered little existing protection.... We rediscovered that internal national agricultural support and associated trade policies cannot be changed fundamentally by economic and political pressures exerted from without, but only from mounting disaffections from within. [Warley (1981)]

The Community in fact considered that a major result of the MTNs was implicit recognition by its trading partners, in particular by the US, of the principles and mechanisms of the CAP. The 'specificity' of the agricultural sector had been admitted; the United States had greatly toned down its initial demands and declared that it did not intend to 'undermine' the CAP.

There were however important elements in the MTN agreements besides tariff concessions. Such an element was the Code on Subsidies and Countervailing Duties, which gave more precise definitions to existing GATT rules under Article XVI. As regards 'primary products', signatories agreed:

> .... not to grant directly or indirectly any export subsidy on certain primary products which results in the signatory granting such subsidy having more than an equitable share of world export trade in such product, account being taken of the shares of the signatories in trade in the product concerned during a previous representative period, and any special factors which may have affected or may be affecting trade in such product. [Code on Subsidies and Countervailing Duties, Article 10]

'More than an equitable share of world export trade' was defined to mean any case where the effect of an export subsidy is to displace the exports of another country; with regard to new markets, traditional patterns of supply in the region concerned would be taken into account; and the 'previous representative period' would normally be the three previous years. The Community received assurances from the US that the wording was not intended to freeze market shares in world agricultural trade, but to prevent the build-up of pressures that might lead to a trade and subsidy war.

It did not prove possible to reach agreement, as the Community had hoped, on a new international grains agreement, although discussions on this continued for some time without success in the UNCTAD framework (see below). 'International Arrangements' were however established for dairy products and for 'bovine meat' (including cattle). These provided in each case for an International Council to keep the world market situation under review and, in case of serious market disequilibrium, to 'identify possible remedial solutions for consideration by governments'. For dairy products, three Protocols laid down minimum export prices (at rather low levels) for milk powder, butter and certain cheeses (in the case of milk powder, food aid was exempted from the price provisions).[1] Both these Arrangements made provision for consultations 'aimed at avoiding or resolving problems that arise'.

A number of participants in the MTNs advocated the creation of an overall forum in which information could be exchanged and consultations could take place to prevent problems arising in agricultural trade and to deal with any that did arise. A proposal for an International Agriculture Consultative Council – the so-called 'cathedral' scheme – was discussed without agreement being reached. The final agreement on the MTNs recommended that the Contracting Parties should 'further develop active co-operation in the agricultural sector within an appropriate consultative framework' and establish such a framework as soon as possible. In November 1980 the GATT annual conference decided that this role would be played by GATT's consultative group of senior officials from eighteen countries.

International commodity agreements

The disappointing results of international commodity agreements, both in the 1930s and in the years up to about 1960, have been pointed out in earlier chapters of this book. Subsequent developments were hardly more encouraging so far as temperate foodstuffs are concerned.

WHEAT

Successive International Wheat Agreements up to 1968 were essentially multilateral contracts under which exporters and importers agreed to buy or sell wheat from each other at prices within an agreed range: they did not include buffer stocks. As has been mentioned above, an International Grains Agreement came into force on 1 July 1968 as a result of the Kennedy Round, with a higher floor price for wheat than previously: it broke down within a year under the pressure of surpluses. Subsequent efforts to reach agreement on a new price range failed.

As has been seen, the conclusion of a new cereals agreement was one of the major items proposed by the Community under the Tokyo Round, but this was not achieved. The negotiations took place mainly in parallel to the MTNs under the International Wheat Council, and subsequently under UNCTAD: the interest

[1] There already existed a 'Gentleman's Agreement' under OECD providing for a minimum price for whole milk powder and a GATT Arrangement on floor prices for skimmed milk powder and milk fats.

of developing countries in the outcome became a major factor, and differences of view between the developed and developing nations were largely responsible for the breakdown of the talks in February 1979. The mechanisms envisaged involved a price band and reserve stocks: at certain price levels, specific action would be taken, involving acquisition or release of stocks. The price band favoured by the developed countries was regarded as too high by the developing countries, many of whom were becoming increasingly concerned at the cost of grain imports. As regards the stocks, while the developed countries considered that 25–30 million tons would be necessary and provisionally pledged about 18 million tons themselves, the developing countries did not accept that they should contribute a large share of the reserve and pledged only about 2 million tons – they also requested aid to help finance stock-holding.

There remained only certain administrative provisions of the International Grains Agreement, under the operation of the International Wheat Council, together with, more significantly, the food aid provisions. In early 1980 a new Food Aid Convention was signed under which commitments were raised from 4.2 to 7.6 million tons (of which the US pledged 4.7 million and the EEC countries 1.65 million).

SUGAR

The International Sugar Agreements operated on the basis of export quotas together with stocks held in producing countries in order to maintain prices within a stated range. They were unable to cope with wide market fluctuations: the 1968 Agreement was operative only until late 1971, when in a shortage situation all export quotas were withdrawn. In 1974, as has been seen, prices climbed to an all-time peak. In 1977 a new Agreement was reached, which came provisionally into effect (pending ratification, which in the case of the US took until April 1980) on 1 January 1978. Like its predecessors, it aimed to maintain prices within a stated range (11 to 21 US cents per pound) by regulating the flow of exports to the free market through export quotas together with the accumulation or release of stocks. An original feature of the 1977 ISA was that members agreed to establish a minimum level of 'special stocks' and keep these under the control of the International Sugar Organisation until needed to defend the ceiling price.

The Community did not adhere to the 1977 ISA. Its attitude had been defined in the Council mandate for the Tokyo Round, which emphasised buffer stock provisions but did not allow for export quotas. Under a revised mandate in the final stages of the negotiation, in September 1977, the Community delegation offered 'undertakings similar or equivalent' to those on export quotas entered into by other exporters, and to take part in all other aspects of the agreement, i.e. especially stock-holding, under the same conditions. This offer was rejected by the other participants. In 1980 the issue of accession to the ISA remained under discussion within the Community.

In 1978 and 1979, world prices remained below the lower limit of the ISA. Other countries considered that the non-participation of the Community and its subsidised exports (see section on sugar earlier in this chapter) were largely responsible: Community sugar exports became a substantial factor on the free market.

In 1980, as has been seen, world prices rose sharply, and although all the special stocks were put on the market and export quotas were suspended, the ISA was unable to prevent prices breaking through its upper price limit.

*

* *

New impetus was given in the context of the 'North-South dialogue' to negotiations for schemes for various commodities of interest to developing countries – the so-called 'Integrated Programme', under which buffer stocks and price stabilisation would be financed by a Common Fund. These schemes, together with previous Agreements for other tropical products, are beyond the scope of this book.

For the temperate foodstuffs, it had proved difficult to introduce effective commodity agreements, or to operate effectively those which did exist. The basic difficulties were brought out by Josling (1979), who stressed the need to be realistic with regard to commodity agreements. To be effective, they must persuade governments to do something that they would not otherwise do, such as buying or selling stocks, limiting exports, buying from certain sources, and observing certain trading prices. The only reason for governments to act against their direct interests in this way was that by doing so within the context of an agreement, others might be persuaded also to modify their behaviour. Josling (1979, page 277) commented, 'It follows that it is quite implausible for any government to imagine that it can participate in agreement and yet avoid any situation in which it is obliged to act against its own interest'.

The world's food

It would be beyond the scope of this work to analyse trends and prospects in the world food situation. Table 15.6 however indicates some significant data. During the 1960s and 1970s, the developing regions of the world achieved quite fast growth in food production, largely as a result of the 'Green Revolution' – the introduction of new varieties of wheat and rice together with increased use of fertilisers, pesticides, etc. and irrigation; India in particular was able to increase cereal production and build up stocks as a protection against harvest failure. Yet population growth was such that little increase in food supplies per head was achieved – there was even a reduction in Africa, where production increased less than in the other regions. In Asia and Africa, the supply of calories per head remained at levels which indicated undernutrition among a large part of the population; there was rarely a year without harvest failure and famine in some area. Lack of purchasing power prevented much of this potential demand from becoming effective in the market.

The oil crisis adversely affected the majority of developing countries (the countries which benefited from oil price increases mostly had small populations): the rise in their oil import bill far exceeded the amount of development aid they received. Moreover, the technology of the Green Revolution was largely dependent, directly or indirectly, on energy. The cost of fertilisers in particular was affected by the rise in energy costs.

Table 15.6: World food production and calorie supplies

	Food production in 1979		Calorie supplies in 1975–7
	total	per head	per head per day
	indices: 1969-71 = 100		*kcal*
Developing market economies	127	101	2219
Far East	127	101	2053
Near East	132	103	2657
Africa	116	90	2208
Latin America	135	106	2552
Developed market economies	121	113	3329
Western Europe	118	113	3378
North America	126	117	3519
Oceania	136	120	3418
USSR and E. Europe	120	112	3465
China	137	120	2439
World	125	106	2590

Source: FAO (1979) *Production Yearbook.*

Thus the ability of developing countries to feed their own rising populations remained in question, yet their ability to finance imports had been impaired. The solution to their food problems still had to lie primarily in developing their own agriculture; food aid could play only an accessory role. Nevertheless their import needs seemed bound to rise.

At the same time the USSR and other eastern European countries, whose own agriculture seemed unable to meet the demands of their people for better-quality diets, and also China with its vast population, were becoming substantial buyers on the world market.

In a basically hungry world, an abundance of food increasingly seemed a valuable asset. In the United States, mainly responsible for making up deficits in the world grain supply, there began to be references to 'food power'; the slogan 'a bushel of wheat for a barrel of oil' was heard (though officially disowned); following Russian intervention in Afghanistan, the US placed an embargo on supplies of grain to the USSR. The Community supported this action. On the other hand, in December 1980 the European Council (the Heads of State or Government) agreed to make available on favourable terms substantial quantities of various foodstuffs to Poland, following political unrest and economic difficulties in that country.

For all developing countries with food shortages, the rising cost and the fluctuations in the prices of food imports were becoming a matter of increasing concern. Food issues became significant in North-South relations with the World Food Conference called by the United Nations in 1974 in the light of a critical world-wide situation. In 1972, poor harvests in several sub-continents had caused world food output to fall for the first time in more than twenty years, world prices of cereals were rising steeply and stocks were being depleted, while the oil crisis was beginning to make its impact. In the following years, the issue of world food security was further discussed, particularly in the World Food Council set up following the 1974 World Food Conference by the UN General Assembly, and in the FAO. As has been seen above, the failure of the negotiations for a new International Wheat Agreement was primarily due to dissatisfaction of developing countries with the stock level and price range envisaged. The concept of a food financing facility for developing countries began to be discussed, and a scheme to this effect was put forward by the International Monetary Fund.

The world's food problem – and the possibilities open to western European agriculture to contribute to solving it – thus seemed increasingly bound up with progress towards a new world order. Most food-deficit developing countries without oil revenue could increase their food imports only at the expense of other items necessary for their development, unless they could increase their own export earnings, or obtain the food imports in the form of food aid, or receive some other form of finance. The Community, besides other types of development aid, was helping to stabilise the export earnings from primary commodities of the ACP countries through its 'Stabex' scheme. However, the economic situation of the developed countries after the oil crisis made them generally unable or unwilling to increase their aid efforts significantly. Growing difficulties were also being experienced with the competition of manufactured goods from countries with lower labour costs. A significant 'recycling' of the 'petro-dollars' earned by countries with large oil resources and small populations, which could break through these vicious circles, had not so far occurred.

Nevertheless, the means of developing a Community export policy for agricultural products was being increasingly discussed, as has been indicated earlier in this chapter. During 1980 the Commission was actively studying the subject, and though it had not yet made proposals, a special chapter on the Community's agricultural exports was included for the first time in the 1980 *Agricultural Situation Report.* It was evident that these issues would become increasingly important, although the outcome, and the ways in which western European agriculture might be affected, remained unclear.

Bibliography

Problems and prospects in world agricultural markets were analysed in successive OECD reports, together with Simantov's paper of 1979. FAO studies are also valuable, though mostly less directly relevant to the problems of the high-income countries; the 1979 issue of the annual *Commodity Review* assessed the outcome of the 'Tokyo Round' multilateral trade negotiations and also the likely impact of the second EEC enlargement.

Current trade issues in relation to agricultural policy, taking into account the results of the Tokyo Round, were discussed in several papers in two conferences: the College of Europe Symposium in 1979, directed and edited by the author, and the conference in memory of Theodor Heidhues in Göttingen in November 1980.

Buchholz, H. E. (1979) 'The Multilateral Trade Negotiations and EEC Agriculture'. In Tracy and Hodac *eds.*

Cohen, S. D. (1975) *The European Community and the General Agreement on Tariffs and Trade.* European Community Information Service, Washington DC.

Commission of the European Communities (1979) *Final report on the GATT multilateral trade negotiations.* COM(79)514 final.

Commission of the European Communities (1980) *Food aid programmes for 1980.* COM(80)57 final.

Corbet, H. (1981) 'Agricultural Priorities after the Tokyo Round Negotiations'. Paper to Theodor Heidhues Memorial Conference, Göttingen (publication due).

Curzon, G. (1965) *Multilateral Commercial Diplomacy.* London: Michael Joseph.

Desouches, F. (1979) 'L'agriculture française face à l'élargissement de la CEE vers le Sud'. In Tracy and Hodac *eds.* (1979).

Economic and Social Committee (10 March, 1977) *Opinion on the common agricultural policy in the international context.* Official Journal No. C 61.

FAO (annual) *State of Food and Agriculture.* Rome.

FAO (1979) *Commodity Review and Outlook: 1979-80.* Rome.

Fernon, B. (1970) *Issues in World Farm Trade: Chaos or Co-operation?* London: Trade Policy Research Centre.

Gale Johnson, D. (1973) *World Agriculture in Disarray.* London: Fontana.

GATT (annual) *Activities in . . .* Geneva.

GATT (1979) *The Tokyo Round of Multilateral Trade Negotiations.* Geneva.

Grogan, F. A. *ed.* (1972) *International Trade in Temperate Zone Products* (an Agricultural Adjustment Unit Symposium). Edinburgh: Oliver & Boyd.

Gundelach, F. (1979) 'Prospects for the Common Agricultural Policy in the World Context'. In Tracy and Hodac *eds.* (1979).

Harris, S. (1980) 'US and EEC policy attitudes compared towards the 1977 International Sugar Agreement'. *Journal of Agricultural Economics.* **XXXI** (3), 351–68.

Hathaway, D. E. (1981) 'Food in North-South Relations'. Paper to Theodor Heidhues Memorial Conference, Göttingen (publication due).

Heidhues, T. (1977) *World Food: Interdependence of Farm and Trade Policies.* London: Trade Policy Research Centre.

Hillman, J. S. (1978) *Non-tariff Agricultural Trade Barriers.* Lincoln and London: University of Nebraska Press.

Hillman, J. S. (1981) 'United States agricultural trade policy evolution and European Interaction'. Paper to Theodor Heidhues Memorial Conference, Göttingen (publication due).

Hillman, J. S. and Schmitz, A. *eds.* (1979) *International Trade and Agriculture: Theory and Policy.* Boulder: Westview.

Hubert, F. (1979) 'Résultats des négociations commerciales multilatérales au GATT'. In Tracy and Hodac *eds.* (1979).

Josling, T. (1979) 'The CAP and International Commodity Agreements'. In Tracy and Hodac *eds.* (1979).

Mackenzie, M. (1979) 'EEC's Food Aid Programme'. In Tracy and Hodac *eds.* (1979).

McQueen, M. (1979) 'Trade preferences for developing countries versus most-favoured nation tariff reductions: an appraisal of the EEC's schemes'. *Journal of Agricultural Economics.* **XXX** (3), 345–58.

Nagle, J. C. (1976) *Agricultural Trade Policies.* Saxon House.

Nouyrit, H. (1979) 'Perspectives de développement des principales productions agricoles françaises en relation avec le marché mondial'. In Tracy and Hodac *eds.* (1979).

OECD (annual) *Review of Agricultural Policies.* Paris.

OECD (1976) *Study of Trends in World Supply and Demand of Major Agricultural Products.* Paris.

OECD (1980) *The Instability of Agricultural Commodity Markets.* Paris.

OECD (1981) *The Outlook for Agricultural Policies and Markets.* Paris.

Pasca, R. (1979) 'Conflicts arising from the EEC enlargement: an Italian perspective'. In Tracy and Hodac *eds.* (1979).

Simantov, A. (1979) 'Outlook for World Agricultural Markets: A General View'. In Tracy and Hodac *eds.* (1979).

Smith, I. (1980) 'Europe's Sugar Dilemma'. *Journal of Agricultural Economics.* **XXXI** (2), 215–23.

Thiede, G. (1980) 'Berechnungen über die Versorgungslage der EG-Länder mit landwirtschaftlichen Erzeugnissen'. *Berichte über Landwirtschaft.* **58** (3), 356–77.

Tracy, M. and Hodac, I. *eds.* (1979) *Prospects for Agriculture in the European Economic Community.* College of Europe, Bruges.

Trade Policy Research Centre (1973) *British, European and American Interests in the International Negotiations on Agricultural Trade.* London.

Trager, J. (1973) *Amber Waves of Grain.* New York: Arthur Fields.

University of Minnesota (1978) *Speaking of Trade: Its Effect on Agriculture.* Agricultural Extension Service, Special Report No. 72.

Warley, T. K. (1978) 'What chance has agriculture in the Tokyo Round?' *The World Economy.* (2), 177–94.

Warley, T. K. (1981) 'Agricultural Trade Policy Issues in the 1980s'. Paper to Theodor Heidhues Memorial Conference, Göttingen (publication due).

Conclusion

At the end of this lengthy review, covering events over a wide field in time and space, it is necessary to take stock. The 'schematic summary' (see pages x-xi) illustrates the pattern of developments over the centuries in the various parts of western Europe, converging ultimately in the European Community and its common agricultural policy. This final chapter will try to assess the challenges and responses which have been examined in the book, to indicate the factors responsible for the various responses, and to outline some issues which seem likely to be significant in the years ahead.

Challenges and responses – an assessment

The nature of the first challenge, in the late nineteenth century, is now clear, though at the time it was not generally recognised. The opening-up of virgin lands in overseas countries, together with the spectacular improvement and cheapening in the means of transport, meant that the comparative advantage in products suited to extensive farming – grains above all – had shifted to the New World. The most advantageous response for European countries was to concentrate on products for which nearness to the market was an important advantage, in particular, dairy products, meat, fruit and vegetables (these, moreover, were products for which, as a result of rising incomes, the trend of demand was favourable). If agriculture, even when adapted in this way, could no longer provide adequate incomes for the farming population, then the necessary course was to facilitate a shift of manpower from farming to the growing industrial sector.

In fact Denmark and the Netherlands were almost the only countries to take positive action in order to adapt their agriculture. They reaped an immediate benefit: their exports of livestock products conquered the markets of Britain and Germany, their farmers prospered, and agricultural progress made an important contribution to national economic growth. Great Britain did nothing, and the result was a forced adaptation: arable farming went into decline and the agricultural population, already small, fell still further. Though the extra labour thus made available probably contributed to industrial expansion, the process took place only with severe hardship for a large part of the farming community. The need to economise compelled farmers to run down their capital equipment and lower their standards of cultivation. At a later stage, however, Britain's large farms were well placed to take advantage of mechanisation and other new technology.

France, Germany and Italy reacted defensively to the challenge of overseas competition, raising their tariffs on grains in particular and thereby maintaining or even expanding grain production. This protection was retained even after overseas competition had spent its force; in Germany the rates of duty were actually raised still higher by the Bülow tariff of 1902. It probably retarded the shift of manpower from agriculture to other sectors, and in the case of Germany set up an obstacle to industrial expansion by impeding the conclusion of commercial treaties. In all three countries, this action was taken mainly at the behest of the larger farmers and in their interests: support for grain prices gave little benefit to the peasantry, most of whom had little grain to sell. Though trade barriers were also erected for livestock products, the effects were much less significant than in the case of grains. The maintenance of grain prices through tariffs prevented consumers in these countries from enjoying the benefits of the cheaper produce made available by overseas countries. Preoccupation with tariff policy diverted attention from more constructive lines of action, and apart from the development of co-operatives, little was done to make agricultural production and marketing more efficient. Farm structures remained practically unchanged, except for some early efforts at consolidation in Germany. Some farm labour moved into industry in Germany; in France there was relatively little reduction in the farm labour force.

The second challenge, the crisis of the 1930s, forced all western European countries into protection, employing in addition to tariffs a variety of new and more effective devices. In most cases, their domestic markets were thereby insulated from outside pressures. But restrictive measures in one country harmed another, thereby prolonging the depression on the international market and perpetuating the need for protection. Behind these protective barriers, domestic agriculture was enabled to expand. This led in France to overproduction of wheat and wine, necessitating measures of intervention on the domestic market. In Nazi Germany, all aspects of agricultural production, marketing and trade were systematically organised, and increased self-sufficiency in food was aimed at. In Italy under Mussolini, encouragement was given to increased output of wheat in particular. In Britain, where markets were generally left open to imports from the Empire, subsidies granted at a high level encouraged a recovery in wheat cultivation and a great expansion of sugar-beet. In almost every case

(Nazi Germany excepted) these measures were taken piecemeal; no consistent agricultural policy was developed and the structural problems of agriculture continued to be neglected.

World markets were disrupted by the protectionist policies of the 1930s, and agricultural exporting countries were forced into various measures of intervention which aggravated the situation. International action was ineffective.

After the Second World War, the emphasis laid on price support, initially in order to raise output and more permanently as a means of maintaining farm incomes, further stimulated the growth of output beyond the amounts which domestic markets could absorb, causing steady growth in self-sufficiency. Despite this support, farm incomes failed to rise sufficiently, and disparities within agriculture, between farms of different sizes and between regions, remained considerable. In France, Germany, Italy and other continental western European countries, many farms were still too small or too fragmented, or both, to be viable.

With economic integration in western Europe, a new challenge arose. The inclusion of agriculture in the common market of the European Economic Community was essential – economically because otherwise distortions would have arisen, and politically because of the need to balance national interests. The various national agricultural policies therefore had to be unified. Agreement on common market organisations, painfully achieved in the early 1960s, and based on the principles of market unity, Community preference and Community financial responsibility, was a vital element in the development of the Community. But the common agricultural policy was faced with the underlying problems of maladjustment in the agricultural sector, and solutions required the consensus of all the member states. The accession in 1973 of the United Kingdom, together with Ireland and Denmark, added to the difficulty of developing common policies, in view of the different historical development and the often divergent interests of the UK.

Price policy remained the central feature of the CAP. The inflation provoked by the oil crisis of 1973 caused steep increases in the costs of farm inputs and led initially to substantial increases in common prices. Before long, problems of overproduction and rising budgetary costs necessitated a 'prudent' price policy. Still, output kept on rising, though farm incomes were squeezed. Numerous adjustments were made to the market regimes for the various commodities in order to contain expenditure; at the same time, additional elements or derogations – several of which aimed at redressing the balance of advantage as between 'northern' and 'southern' produce – added to the complexity and the cost of the regimes. Different rates of inflation and monetary instability made the system of common prices increasingly difficult to operate.

The need for policies to improve farm structures had been increasingly recognised by western European governments even before the formation of the Community. Still, under the CAP, a common structural policy took much longer to organise than a price policy. The 'Mansholt Plan' of 1968 was an ambitious attempt to adjust agriculture's productive capacity to available outlets while creating modern, viable farms. Conceived and implemented ten years earlier, to benefit from the period of dynamic economic growth which would have facilitated the transfer of people from the land, the plan might have succeeded. But when finally more limited 'common measures' were implemented, the recession provoked by the oil crisis was already beginning; the movement out of agriculture and the reduction in farm numbers slowed down. With insufficient transfers of land, farm modernisation aids appeared to be causing intensification of output and aggravating market imbalances. A major new development in the Community's structural policy occurred with the adoption, in 1978 and 1979, of a 'Mediterranean package'; but this, providing special aids to develop agriculture in backward regions, was not designed to deal with the basic problem of adjustment throughout the Community.

The growth in productivity in western European agriculture caused increased self-sufficiency, imports of competing temperate foodstuffs from the rest of the world being reduced – though imports of livestock feedingstuffs expanded rapidly – while exports of sugar, grains, dairy products and even beef increased. World markets, however, were unstable; prices were generally well below western European levels, and exports by the Community were usually possible only with costly subsidies. The International Wheat Agreement had broken down, and attempts to establish a new agreement foundered mainly on 'north–south' issues; the 1977 International Sugar Agreement was ineffective (partly because of the refusal of the Community, although an increasingly large exporter, to participate). Food aid in grains by the Community was limited by budgetary constraints; in dairy products, by limited absorption capacity in developing countries.

With price policy and the commodity regimes under severe strain, and with structural policy proving ineffective to achieve its main goal of promoting basic adjustments, the dilemma facing agricultural policy was acute. By the end of 1980, with the CAP accounting for over two-thirds of the Community budget, the overall ceiling imposed on the budget by the '1% VAT rule' – which could be altered only with the assent of all national parliaments – had brought to a head the issue of the cost of the CAP and provoked urgent discussion of 'reform'. The UK's demand for a fairer deal under the Community budget led to agreement to consider 'structural changes' in the Community, which inevitably would intensify pressures for changes in the CAP. Alternative policy measures had been suggested, including production quotas, degressive price scales related to individual farm output (the 'quantum' system), direct income aids in place of price support, and other schemes. The Commission's 'Reflections', presented in December 1980, envisaged a new principle for the CAP in that producers would be required to bear the costs arising from additional production beyond a certain point. Price support appeared to be reaching its limits.

Factors influencing the responses

An adaptation in agriculture of the scale needed over the past hundred years required positive action by governments and by farmers themselves. The example of Denmark showed what could be done: as a result of reforms in the late eighteenth and early nineteenth century, and as a result of the measures taken during the Great Depression to promote livestock farming and to create an efficient co-operative system, the conditions for a prosperous agriculture, independent of support, were laid down.

If we ask why in the majority of countries little or no constructive action was taken, a number of reasons can be given. Perhaps the first is that the need for adaptation was not always recognised. In the late nineteenth century, the increase in overseas competition was often not seen for what it was: a basic and irreversible transformation in the agricultural scene. Consequently, France and Germany resorted to protection partly in the hope that the crisis would pass; in Britain, inaction during at least the early stages of the depression was largely due to the belief that bad weather and other natural causes were the factors responsible. One reason why Denmark's reaction was so much more positive was that a sufficiently large body of opinion foresaw the increase in overseas competition in grains before it even began; the same was true for the Netherlands.

In times of crisis, moreover, the most obvious means of helping agriculture was to restrain the fall in agricultural prices – this was particularly so when other sectors were benefiting from tariff protection. Price supports of various kinds were also used in periods when increased food production was required, whether to meet a shortage, to safeguard food supplies in the event of war or to relieve the balance of payments. These price supports afterwards proved difficult to remove, even when they had served their immediate purpose. By maintaining or stimulating uneconomic production, they perpetuated the need for their own existence. In cases where increased production was a deliberate aim of policy, it afterwards proved difficult to keep the expansion within suitable limits. The aim of raising output was by no means incompatible with measures to raise agricultural efficiency, but in practice the emphasis laid on price policy as the means of stimulating output diverted attention from other types of action and even concealed the need for reform.

In general, governments found it easier to satisfy their farmers with short-term benefits than to embark on long-term programmes of structural reform. Farmers themselves, through their organisations, tended to concentrate their demands on immediate advantages, neglecting long-term action. The return to protection in France, Germany and Italy in the late nineteenth century was largely the work of organised pressure-groups, composed mainly of large farmers and landowners with a vested interest in grain prices. The influence of farm organisations was built up in all countries during and since the 1930s. The NFU in Britain, the FNSEA and the *Chambres d'Agriculture* in France, the *Deutscher Bauernverband,* the *Union des Paysans Suisses* and similar organisations in other countries became powerful bodies generally committed to maintaining and improving the standard of living of their members in what appeared to them the most direct way – higher prices and increased outlets for their produce. The farm organisations were generally unwilling to envisage far-reaching changes in the structure of agriculture (a notable exception was the *Centre National des Jeunes Agriculteurs* in France around 1960). At the level of the European

Community, the COPA saw its defence of farm interests primarily in terms of price support. Consumer representation was generally much less influential.

There was often resistance – not only from the farmers themselves – to a reduction in the number of farms and in the size of the farm population. Belief in the advantages of rural life has always been widespread: it may be expressed in sentimental terms – Jules Méline in France remains an outstanding example – or in a carefully elaborated theory, as with Adolf Wagner in Germany; it appeared in an extreme form in Nazi racial philosophy. The desire to preserve a large rural population had an important influence on policy in France, Switzerland and Austria; it was present to some extent in almost every country. Yet it cannot be said to have been a decisive factor, and it had little effect on policy when more material interests were at stake. In France and Germany, when practically all other agricultural products were protected, wool was left open to the full blast of overseas competition, in deference to the interests of manufacturers; the result was a decimation of the sheep population in western Europe. Yet the shepherd and his flock play an even more important role in our mythology than other members of the farming community.

Bearing in mind these various factors which played a role in shaping the responses of the past hundred years, it may be asked whether different courses could have been followed. The initial sections of this book have described the earlier developments which produced the situation of the late nineteenth century, when the first of the modern challenges had to be met. In view of the economic, social and political circumstances that had been created, the various national responses seem almost inevitable. Most countries of the Continent did not have a sufficiently well-developed agrarian structure nor a sufficiently well-educated farm population to carry out a positive adjustment; moreover, in most countries, political power was still in the hands of a landed aristocracy whose main interest lay in the price of grain. In each of these respects, Denmark and the Netherlands were exceptions; so too was Britain, though here the role of agriculture was so secondary to the interests of the urban population that the response was laissez-faire. Further, once these initial responses had been made, subsequent actions – in the 1930s and after 1945 – also seem largely inevitable.

The development of agricultural policy in western Europe might thus be held to support a theory of historical determinism. There was more opportunity for absolute rulers in the eighteenth century, if they had wished, to alter the course of events. The reforming action of Crown-Prince Frederik and his government in Denmark did have a lasting and decisive effect, in agriculture as in other areas. If monarchs in other countries had shown the same courage and foresight, much of later history might have been different. Subsequently, the scope for change was restricted by the resistance of vested interests in conjunction with the egoism of politicians seeking re-election – the adverse side of a democratic society. The actions even of prominent Ministers such as Jules Méline appear to have been largely a product of their times. Efforts to change direction when the context was unfavourable generally proved short-lived – as in the case of Napoleon III's attempt to impose Free Trade on France, or Caprivi's trade treaties which were largely countered by the agrarian interest.

Future issues

This book has been concerned with the past, but with the aim of gaining a better understanding of the present; and the analysis of developments up to the present time may give some idea of the issues which are likely to be significant in the coming years, although actual developments cannot be predicted. Some of these issues relate to western Europe itself; others concern its relationship to the world at large.

A basic issue is to what extent society, through its democratic institutions, is prepared to accept continued income transfers to agriculture: in the European Community, this question has arisen specifically in the context of the Community budget, where already the European Parliament is involved and probably before long the ten national parliaments will have crucial decisions to take. It is inconceivable that support for agriculture should come to an end; it has also become apparent that it has its limits. It would generally be recognised that there are grounds for intervention to stabilise agricultural markets, also that there is justification for assistance to the farm sector in view of its difficulty in adjusting to its declining relative role in the economy, bearing in mind the low mobility of farmers, especially when employment prospects in other sectors are poor. There appears also to be public willingness to maintain a healthy farm sector and prevent rural depopulation: aid to farmers in the role of 'guardians of nature' may be increasingly acceptable. At the same time, farmers will be expected to preserve the environment and to keep within acceptable limits the pollution associated with some features of modern farming.

It will become increasingly important to distinguish between the problems of different types of farm. The 'commercial' farm sector, consisting of large, well-equipped and well-managed farms, accounts for the greater part of agricultural output though the number of farms concerned is relatively small. Such farms have been adversely affected by the cost-price squeeze of recent years (including the rise in interest rates); they will continue to demand price increases and to oppose production restraints which would affect them directly. The dynamism of this sector will cause continued production growth; the farms concerned will continue to apply technological innovations and to take advantage of opportunities that arise such as the availability of relatively cheap imported livestock feed. They will become increasingly integrated with the 'agri-business' world – the food processing industry as well as the industries producing farm requisites.

Some farmers with holdings which are at present barely viable may be able to rise into the commercial category, though the high price of land and high interest rates on capital makes this difficult. Many holdings will still be unable to provide a living in line with income expectations on a full-time basis. In some regions this problem may be eased by a new rural-urban balance: continued decentralisation of industry provides on the one hand employment and income for farm families, on the other, relief from urban congestion and pollution, and opportunity for workers in secondary and tertiary sectors to live in more pleasant surroundings – possibly to practise spare-time farming. The duality of farming and other occupations may be diminished as families and even individuals increasingly combine different activities. In remote and backward regions, however, problems will remain, and special assistance will be required.

With such developments, the concept of an income gap in general terms between agriculture and other sectors becomes increasingly meaningless: the reduction of disparities within the agricultural sector appears more clearly as the main goal. It would be possible in principle to conceive policies more specifically designed than current price or structural policies to support developments along these lines – policies providing, above all, more selective assistance where it is needed, and taking account of a situation where many farm families are not dependent on income from farming alone. Regional policy too would obviously be an important element. However, in view of what has been said above about the difficulty of changing course in policy-making, it would be unwise to expect major changes.

The role of agriculture in relation to the energy crisis will no doubt receive more attention. With high energy costs, the possibility of producing crops for conversion into energy is already being studied, though it is doubtful whether any temperate-zone crop can produce a positive energy balance (sugar cane is so far the only crop in the world which has given clearly positive results). More promising developments relate to the use of parts of the 'biomass', which are at present wasted, for small-scale production of energy on the farm.

The pressure of supplies on available outlets will be maintained, as technological progress continues and agricultural productivity goes on rising. Better adjustment to market demands, including the production of high-quality foods, may provide some opportunities. But in western Europe, population growth is slow; and in north-western Europe (trends in southern Europe are different) there is little scope for increased *per capita* demand for foodstuffs in general; health concerns may cause reduced demand for specific foodstuffs (butter, sugar) and even – in a sedentary society – reduced *per capita* consumption overall.

Trade relations with the outside world will thus become increasingly significant. Western Europe is unlikely to increase its imports of temperate foodstuffs; attempts will probably be made to check the rise in imports of livestock feedingstuffs (which add significantly to production potential). The second – southern – enlargement of the Community, if the accession of Greece is followed by that of Portugal and Spain, will lead to re-alignments of trade within the enlarged Community as well as between it and the outside world, and some difficult adjustments will be involved.

Western Europe – the Community in particular – will seek to develop its agricultural exports. It will not be easy to find remunerative outlets for all the produce that is likely to be available. The future of world markets is uncertain: so far they have generally been oversupplied, and prices have usually been below those at which western Europe can profitably sell. Still, the USSR, other eastern

European countries, China, and some developing countries with oil revenue, have emerged as major importers. A crucial issue is how far the food needs of hungry people in the Third World will be translated into effective demand. If that were to happen to any great extent, world market prospects could look very different: the potential needs are great, and the capacity of the United States, so far the main residual supplier, is not unlimited.

The world's economic balance has been transformed by the vastly increased price of oil. Many of the countries which have benefited from oil revenue have small populations, while other developing countries with large populations and food deficits have been adversely affected; yet the ability and willingness of the West to help such countries with their problems has been impaired. The outlook for the world food problem, and for western Europe's potential contribution to solving that problem, thus seems to hang very largely upon the emergence of a new world order.

List of tables and figures

Chapter 3 France

Tables

Chapter 4 Germany

Tables

Chapter 5 Denmark

Tables

Part II The Crisis of the 1930s – The Second Wave of Protectionism

Chapter 6 General

Tables

Figures

*

* *

Symbols and conventions

In the tables throughout, certain conventions have been used as follows:

1978/9 : indicates a crop or marketing year.
'1978/9' : indicates the average of three years around the year shown.
1979-80 : indicates the average of the years shown.
. . : indicates data not available.
– : indicates nil or negligible.

One billion equals 1000 million.

Tons are metric tonnes unless otherwise specified.

As data in many tables have been rounded, minor discrepancies may appear between totals and their parts.

Note on units of account used in the European Community

(a) Common farm prices were originally expressed in Units of Account (UA). This unit was defined in terms of a quantity of gold such that its value was identical to the US dollar. This identity ended when in 1971 the convertibility of the US dollar was suspended. After currencies 'floated', the value of the UA (though its definition remained unchanged) was in practice calculated by reference to the currencies participating in the European 'joint-float' or 'snake' (i.e. observing a maximum spread of 2.25% at any given time). This meant that the value of the UA came to be determined by the 'strong' currencies (it was in practice equivalent to the European Monetary Unit of Account (EMUA) which was the basis of operation of the 'snake') and to be increasingly overvalued in relation to the average trend of member states' currencies.

The Community budget was originally expressed in UA. In Eurostat publications, price and value data were shown in 'Eur': this was the same as the UA.

(b) For budgetary purposes, a European Unit of Account (EUA) was progressively introduced. The general budget of the Communities, including FEOGA, was first expressed in EUA in 1978. The EUA was a 'basket' – in effect a weighted average – of all member states' currencies. Its value was therefore less than that of the UA. The relationship of each member state currency (and of the US dollar and other currencies) to the EUA was calculated daily on the basis of market exchange rates.

Eurostat and Commission statistical publications went over to the EUA for value data, and data for former years were recalculated using the definition of the EUA.

As the only remaining function of the UA was in relation to common farm prices, it was frequently referred to as the 'agricultural' unit of account (AUA).

(c) When the European Monetary System was introduced in March 1979, it was based on a new unit, the *European Currency Unit* or ECU, intended to become the sole unit of account in the Community. The definition of the ECU was the same as that of the EUA; the central rates of the member states' currencies were fixed in ECU at rates corresponding to their values in terms of EUA on the day of introduction of the ECU (for sterling, which continued to float, an 'imputed' central rate was adopted). Currency values in terms of the EUA/ECU continued to be calculated daily.

Common farm prices were expressed in ECU beginning with the 1979/80 marketing year. As the former UA was worth 1.208953 ECU, all common prices were adjusted by this factor, and 'green rates' (see chapter 14) were also correspondingly adjusted so as to leave prices in national currencies unchanged.

The Community budget, including FEOGA, continued to be expressed in EUA until the end of 1980: from 1 January 1981 the ECU was substituted in all Community acts.

In statistical publications, however, different practices arose. Some prices and value data for former years were expressed by Eurostat and by the Commission in ECU, usually as a simple relabelling of series previously given in EUA; but where common prices were concerned, data were raised throughout by the factor of 1.208953, while in Eurostat trade series the ECU was treated as equal to the UA or US dollar up to 1970. Definitions (where available) and conversion tables in each source have to be scrutinised to ascertain which practice has been adopted.

Author Index

n note; t table
Figures in bold type refer to Bibliographies

General Index

f figure; m map; n note; t table
* see also under individual countries